Animal Sciences

Pasha A. Lyvers Peffer
Michael L. Day
The Ohio State University

Kendall Hunt
publishing company

Kendall Hunt
publishing company

www.kendallhunt.com
Send all inquiries to:
4050 Westmark Drive
Dubuque, IA 52004-1840

Contents

Acknowledgements

We offer our thanks to Kimberly Cole, Ph.D., Mike Davis, Ph.D., and Joe Ottobre Ph.D. for their editorial contributions. In addition we thank our colleagues for providing supportive content.

Importance of Animals

"Each species is a masterpiece, a creation assembled with extreme care and genius."
—*Edward O. Wilson*

This text concerns the use of animals and introduces basic principles and practices that allow humans to successfully coexist with animals in captive and controlled environments. The importance of animals is depicted throughout history and modern society as there are rich displays of human and animal interactions in engravings, sculptures, paintings, and drawings, some of which date to forty thousand years ago. Humans and animals sustain a relationship of mutual benefit. Although modern uses of animals differs throughout the world, humans have relied on animals as a source of food, clothing, knowledge, energy, power, status, transportation, companionship, entertainment, service, and capital. Animals in turn rely on humans to provide food, shelter, companionship, and protection.

Animals as a Source of Food and Fiber

Food

The interrelationship between humans and animals is one that began with the reliance on animals as a source of food. Human's earliest ancestors ate plants: fruits, vegetables, tubers, and nuts. But at some point in time, humans began to incorporate meat into their diet; most likely as scavengers initially, and then as hunters. Diversification of the diet to include animal protein is suggested to

■ **Fig. 1.1** The proboscis monkey consumes a diet nearly exclusive of plants, whereas the chimpanzee consumes a more diversified diet of plant and animal products. Accordingly, the gut capacity of the proboscis is greater to account for the greater intakes of less digestible plant products. (*Left:* © Kjersti Joergensen, 2014. Used under license of Shutterstock, Inc. *Right:* © Patrick Rolands, 2014. Used under license of Shutterstock, Inc.)

underlie the development of the complex human brain. According to Joyce (2010):

> … eating nothing but raw vegetable matter, as most apes and most probably our earliest ancestors did, means you've got to have a really big gut. Think cow-like … then our ancestors discovered meat. Meat contains lots of calories in a small package. More meat allowed our guts to shrink. Digestion was not so labor-intensive. So where did the extra energy go? The brain.

Animals as a source of food continue to represent a primary use in modern society. It is generally accepted that humans are omnivorous, relying on both plant and animal products for optimal nutrition. The United States food guide for a healthy diet includes choices for low-fat, or lean, meat and poultry as well as milk, yogurt, and cheese. The inclusion of animal products (meat, including fish and poultry) in the diet is practiced by the majority of the United States population, with 99 percent of adults reporting in a national survey that they consume animal products.

As a percent of total calories consumed, the United States consumes 27 percent of its calories in the form of animal products, excluding animal fats. This percentage is less for world calorie consumption, where 17 percent of total calories consumed are in the form of animal products. Meat and milk represent the greatest contributions to animal product consumption both within the United States and on a global scale. Meat and milk are also the predominant sources of protein consumed in the United States, where animal products represent nearly 70 percent of the total protein consumed. When animal product consumption as a percent of calories or protein consumed is compared within the United States to global animal product consumption, the United States consumes twice the calories and twice the amount of protein from animal products. This pattern of animal product consumption is associated with the economic status. Developed, food-importing countries consume two-fold greater egg and milk products and 1.5-fold greater meat products when compared to the least developed countries. The influence of socioeconomic status on food consumption patterns has been observed in the United States, Europe, Asia, the Middle East, and the Mediterranean. With an increase in income, there is an increase in lean meat, fish, and other seafood consumption as well as dairy consumption (yogurt, cheese, and low-fat milks).

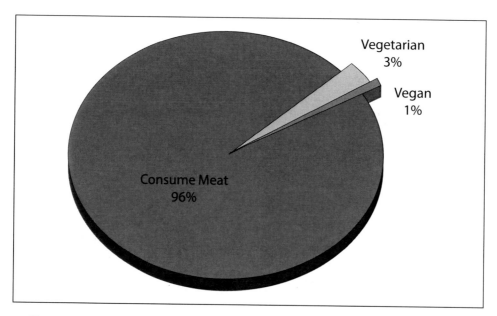

■ **Fig. 1.2**   Distribution of dietary habits in the United States. Data collected from a national telephone poll of 2030 adults aged eighteen or older. Meat consumption refers to consuming one or more of the products: beef, pork, fish, and/or poultry. Vegetarians reported never consuming meat as defined above. Vegans reported never consuming meat, dairy, or eggs. (Vegetarian Resource Group, 2012).

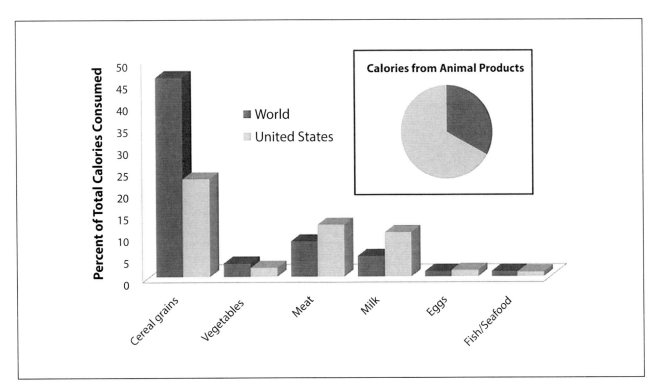

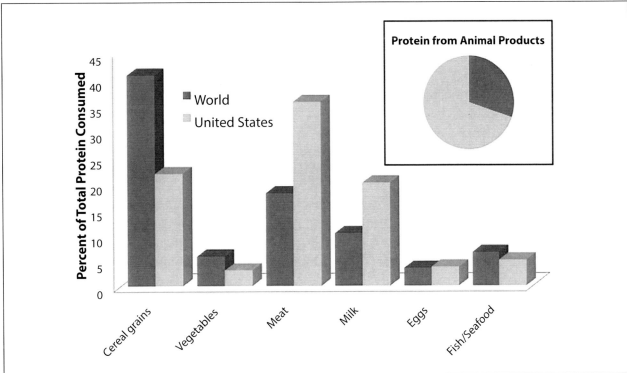

■ **Fig. 1.3** Cereal grains are the greatest contributor to world calorie supply; however, within the United States, animal products surpass cereal grains in percentage of total calories (top). For the United States, 64 percent of the total protein consumed is from animal products (bottom). When the United States animal product consumption is compared to the World, the United States consumes twice the amount of calories (top inset) and twice the amount of protein (bottom inset) from animal products. (Food and Agriculture Organization of the United Nations, 2009).

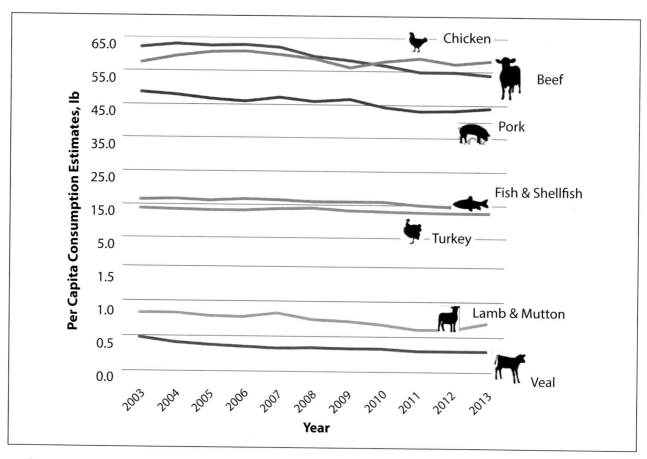

■ **Fig. 1.4**   Trends in per capita consumption estimates of meat. Per capita consumption estimates are determined from availability and disappearance data and represent boneless retail product weight. (United States Department of Agriculture-Economic Research Service, 2013).

Consumption of animal products is estimated from food production and disappearance data. During the last fifty years, total red meat consumption (beef, pork, lamb, and veal) has exceeded poultry consumption (chicken and turkey). Total red meat consumption peaked in 1971 and has gradually decreased; however, total meat consumed in the last five decades has increased as a result of increased poultry consumption. Today, chicken, beef, and pork are the top three meat products consumed in the United States. Choices of animal products consumed are influenced by income, age, health perceptions and awareness, ethnicity, religion, economics, convenience, and perceptions of risk and animal welfare. Low-cost supply and perceptions of healthfulness have attributed to a slight shift away from beef consumption in favor of chicken; whereas, veal consumption is limited by animal welfare concerns and affordability.

Changes in dairy product consumption have also occurred since the 1970s. Fluid milk consumption has decreased while cheese consumption has increased. The majority of fluid milk consumed is low-fat, whereas the majority of cheeses are American (cheddar, Colby, Monterey jack) and Italian (mozzarella, provolone, and parmesan) style. The decline in fluid milk consumption has occurred as Americans are less likely to consume milk with mid-day and evening meals compared to past generations. Despite increased cheese consumption, American dairy consumption habits are suboptimal relative to dietary guidelines. Globally, fluid milk consumption by adults is a cultural phenomenon. Whereas milk consumption persists beyond weaning in the United States and many European countries, many Asian, African, and Latin American cultures do not consume significant quantities of milk. Exceptions are nomadic, pastoral tribes of East

Africa that rely on milk as a dietary staple in lieu of regular meat consumption. While milk from cattle is primarily consumed in the United States, water buffalo, goat, sheep, camel, yak, mare, sow, reindeer, and llama milk have been consumed throughout the world. In developing countries, animals are raised for their multi-purpose use and need to reproduce under more adverse conditions, attributing to the use of water buffalo, camel, goat, and sheep in dairying. In India, 60 percent of milk consumed is collected from water buffaloes.

Fiber

Animal fibers as a source of textiles include the wool from sheep, mohair and cashmere from goats, angora from rabbits, and the fibers from llamas, alpacas, camels, and yaks. Many are not used in commercial fiber production but restricted to local use. Although mostly replaced by cotton and synthetic fibers, animal fibers represent a natural renewable resource for textiles. Animal fibers come in natural color variations or may be dyed depending on fiber type. Due to their texture, strength, warmth, durability, and availability, animal fibers demand a higher price compared to synthetic textiles. Globally, wool is the most used of the animal fibers, representing

5 percent of total textile fiber production. With long established diversity in sheep breeds and wool quality, wool use spans the market from carpets and upholstery to high-end apparel.

Animal Use beyond Food

Land Management and Transportation

Conservation agriculture, which promotes sustainable use of land, relies on animals for land management. The grazing habits of animals are used throughout the world for erosion control, range or pasture management including plant diversification, and noxious weed control. Grazing animals can subsist on land that cannot be cultivated and in turn provide products such as meat and milk that can be used.

Draft animals are considered renewable and affordable technology for small land holders throughout Asia and Africa. In regions of sub-Saharan Africa only 10 percent of land is cultivated through machinery whereas 25 percent of land relies on draft animals for cultivation. In the United States there is a renewed interest in draft animals as an opportunity to mitigate the environmental impact of traditional farming practices. However, within the United States the use of

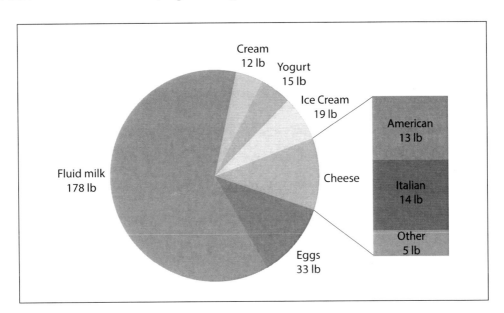

■ **Fig. 1.5** Dairy and egg consumption estimates. Estimates are determined from availability and disappearance data . Fluid milk is total beverage milk and includes whole (48 lb), low-fat (99 lb), and skim milk (27 lb). Yogurt includes frozen products, while ice-cream includes low-fat varieties. Swiss, blue, and Hispanic cheeses are included in the other category. (United States Department of Agriculture-Economic Research Service, 2010).

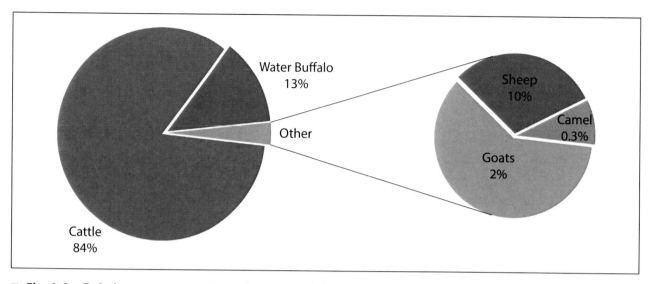

■ **Fig. 1.6** Relative percentage of world milk supply by major species. Depending on location, climate, culture, and socio-economic conditions, milk is derived from many different animal species. In addition to the species depicted, horse, yak, reindeer, and llama milk are also consumed. (Food and Agriculture Organization of the United Nations, 2014) .

■ **Fig. 1.7** Horses were the primary means of power in early logging industries. Within Alabama, horse and mule logging is still practiced as an alternative to mechanized timber harvesting. The use of animals for logging allows selective timber harvest on small land tracts. (© Donna Beeler, 2014. Used under license of Shutterstock, Inc.)

draft animals is a niche area unlike in Africa where animals contribute to the economic stability of farming systems and local communities.

Animals for draft can also be purposed for transportation. In the United States, transportation infrastructure supports the use of motorized vehicles in travel, and animal powered transportation is limited to select religious affiliations and tourist areas. Although animal use in transportation is limited

within the United States, animal powered transportation is vital to the transportation needs of other countries. In India during the late 1990s, it was estimated that animal drawn carts had an economic value of five billion dollars and the cost to replace with more modern means of public transportation was estimated at $6 billion. In Sub-Saharan Africa the number of donkey drawn carts was estimated at one thousand in the 1960s and has grown to seventy-five thousand in recent years and is predicted to continue to increase in the future.

By-products from the animal industries may one day reduce the dependence of the world on oil. In attempts to expand the petroleum base, research was conducted to refine chicken fat for use as a biodiesel. Animal fats are a lower cost biodiesel as they are not considered edible by humans, unlike vegetable oils.

Research
Research in the animal industries is integral toward the discoveries into how biological systems function for the advancement of all animals, including humans. Since the end of the nineteenth century animals have contributed to over 50 percent of total scientific discoveries and 2/3 of the Nobel Prizes awarded were for discoveries involving animals. Rats and mice are the predominant animals used in

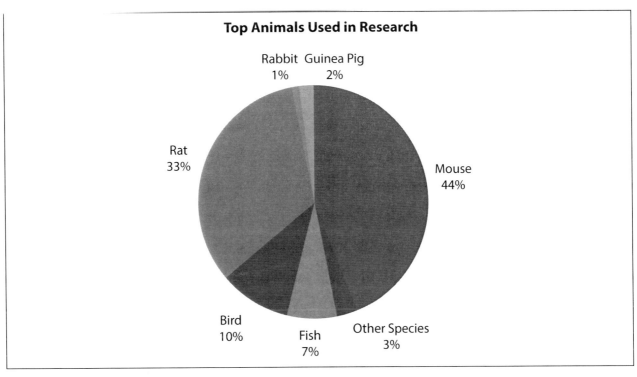

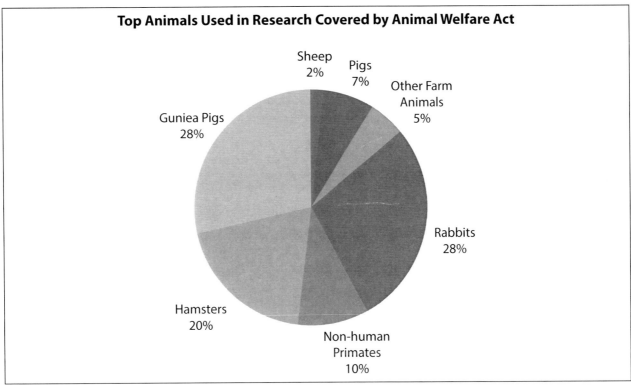

■ **Fig. 1.8** *Continues.*

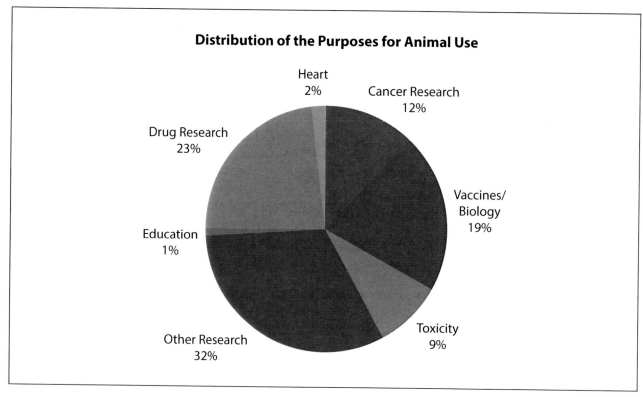

Distribution of the Purposes for Animal Use

Heart 2%

Cancer Research 12%

Drug Research 23%

Vaccines/ Biology 19%

Education 1%

Toxicity 9%

Other Research 32%

■ **Fig. 1.8**   Percentage of animal species commonly used in research (top), percentage of animal species covered by the Animal Welfare Act that are commonly used in research (middle), and the distribution of the purposes for the use of animals (bottom). Food animals are included in the category "Other species" when all animals are considered (top). Rats, mice, birds, and fish are excluded from protection through the Animal Welfare Act, however, oversight of their use and care is guaranteed through institutional animal care committees. (United States Department of Agriculture-Animal Plant Health Inspection Services, 2010).

fundamental research areas, especially biomedical research. Federal law requires that animal testing is conducted before most clinical trials involving humans are allowed. The widespread use of rats and mice in research is favored due to their availability, affordability, prolificacy, early maturity, and ease of care and maintenance. Per the guidelines on the care and use of laboratory animals, only fifteen square inches of space is required for housing four mice, each weighing less than twenty-five grams. In comparison, 1.7 square feet of space is required to house a single newborn pig for research.

When considering the use of animals as research models for humans, no single animal mimics the human condition and thus animals are selected based on their biological relevance. Accordingly, non-human primates, dogs, guinea pigs, hamsters, mink, rabbits, pigs, sheep, goats, cattle, horses, and

various species of birds, fish, and amphibians have proven beneficial in studies of genetics, embryology, developmental biology, physiology, immunology, and toxicology. The use of food animals in biomedical research is limited, but expanding, due to their contributions toward advancing knowledge in essential nutrient requirements, therapies for cancer, mechanisms of diabetes, and mechanisms of immunity. The anatomical and physiological similarities between pigs and humans have promoted the use of the former in determining the role of genetics, nutrition, and exercise in obesity and related complications, including heart disease. Sheep are providing insight into the role of maternal environment on fetal development and disease later in life and chickens may hold the key to understanding ovarian cancer in women. Furthermore, due to the widespread use of food animals by humans, research remains crucial to the continued improvements in efficiency of production, product quantity and quality, and healthfulness of food obtained from these animals.

To continue to foster the human-animal relationship, knowledge of the principles of behavior, nutrition, genetics, reproduction and production of the various food animal species as influenced by environment and interactions with humans must be disseminated to each generation. Understanding of the relevance of these biological disciplines to producing healthy, well managed animals in harmony with environmental and social concerns in order to fulfill their crucial role in providing affordable, high quality food for humans is essential.

Domestication

CHAPTER 2

"The wolf, disarmed of ferocity, is now pillowed in the lady's lap."

—Edward Jenner

The development of modern animals occurred over three stages: evolution, domestication, and production. The first stage, evolution, may be considered the longest stage of development as it considers the origins of animals and their continuing ability to change over time. Domestication is a newer event that occurred as a consequence of fixed evolutionary behaviors. The production stage is the most recent stage of development that reflects how humans have altered animal development through selective practices. This chapter will introduce the first two stages of development: evolution and domestication. You will learn of the processes that contribute to each stage and gain an appreciation for the events that have shaped modern human-animal interactions.

The Origin of Animals

Today's animals trace to the animals that first appeared during the Cambrian Period, 540 million years ago. In what is now known as the Cambrian explosion, in reference to the number of diverse animal fossils identified from this period to date, the complex body systems of animals originated. Mammals would first appear during the Triassic Period approximately 250 million years ago, followed by birds 150 million years ago during the Jurassic Period. The genus *Homo* did not appear until the Tertiary Period 65 million years ago. The Tertiary Period is viewed as the period of mammals and birds, both of which thrived during this geologic time. The Tertiary Period also marks the appearance of numerous land bridges, including those that linked North America to Asia. The appearance of land bridges allowed animals to cross into new regions that were previously separated by water, exposing animals to new environments that would shape their evolutionary paths.

> Even-toed artiodactyls (including pigs, cattle, sheep, and goats) share common ancestry with the odd-toed perissodactyls, which includes horses). Diversification of the orders occurred approximately 53 million years ago.

Principles of Evolution

Evolution is the process by which changes occur over successive generations. Both an animal's genetics and its environment are the catalyst for evolution. The theory of evolution is one that has withstood scrutiny due to the well-constructed and executed studies of many early biologists including William Charles Wells, Patrick Mathews, Alfred Wallace, and Charles Darwin. According to the basic principles of evolution, animals will evolve as they inherently possess the ability to vary, reproduce in excess, and are exposed to a continually changing environment.

Darwin is the most recognized of the scientists associated with evolutionary theory. Though many scientists of the time shared his ideas on the origin of animals, Darwin's greatest contribution to the theories of evolution was that of natural selection. Darwin recognized that there is a natural ability for organisms to vary, thus animals can be selected for whereby the morphology of animals is changed and new variations in animal form and function are

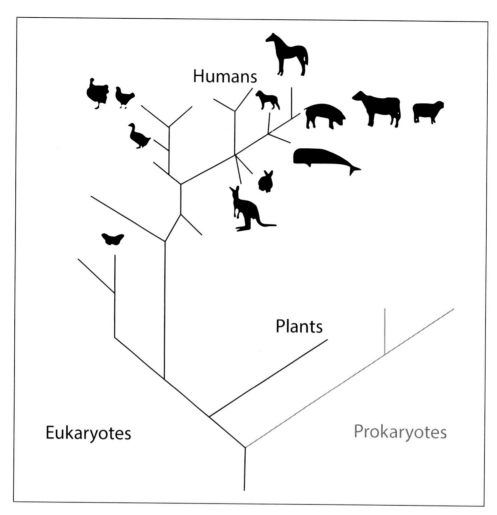

■ **Fig. 2.1** The phylogenetic tree of animals depicts the relationships of animals based on their genetics and reveals common ancestors among animal systems.

established. Selection may occur through irregular and unpredictable processes in the physical environment, as is the case of Darwin's natural selection, or may occur through controlled and deliberate steps as seen in artificial selection or selective breeding. Evolutionary theory is underscored by natural selection. Natural selection does not introduce variation into an organism, but works on existing variations that occur as a consequence of genetic changes. There is a natural tendency for animals to reproduce and overpopulate beyond environmental sustainability, which gives greater opportunity for variations to emerge, but mortality is high. However, when an animal emerges with modifications that offer an advantage and increases survival within the environment, there exists the potential to increase the occurrence of the modification within an animal population as long as the environment that favored the modification remains unchanged. Natural selection is a gradual process where small modifications that occur over time can ultimately lead to significant changes within an animal population.

Darwin's theory of evolution by natural selection is well accepted, but controversy often arises from Darwin's inference that all species arose from one or more common ancestors. Darwin's theory provides rational support for the observed instability of living organisms that allows animals to acquire different features within different environments.

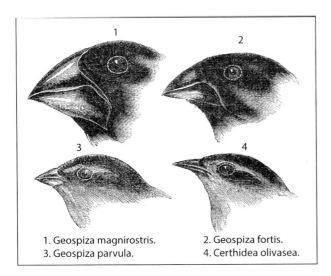

1. Geospiza magnirostris.
2. Geospiza fortis.
3. Geospiza parvula.
4. Certhidea olivasea.

■ **Fig. 2.2** Following his voyage to the Galapagos Islands, Darwin later reflected in his memoir on the possibility that a single species of Finch may have been modified according to its environment. The Finches of the Galapagos Islands are now commonly referenced as Darwin's Finches. (© Heritage Images/Corbis.)

Modifications that are maintained through natural selection are known as adaptations. The horse provides a well-studied and documented example. Progressive adaptations of the horse occurred over the last 65 million years to transform a small fox-like animal into the large, one hooved horse recognized today. The original ancestor of the modern horse was a browsing animal of fourteen inches in height that ran on four toes on the front feet and three toes on the hind feet. The body plan allowed the primitive horse to thrive in its marshy environment. But as the environment changed, the horse gradually acquired adaptations that allowed primitive horses to exploit their environment. As marshes gave away to grassy plains, adaptations in the limbs and feet supported long-distance speed. There was a reduction or loss of the first, second, fourth, and fifth toes; whereas the third toe flattened to form the hoof. The limbs lengthened and gained strength at the expense of flexibility as the joints became more restricted to accommodate the power and stride of each step needed to elude predators. Increased area, sharpness, and height of teeth occurred to grind grass material and allowed the transition from a forest browser to a grazer. Collectively, the adaptations that occurred over millions of years allowed for the survival of the horse and kept the animal from extinction.

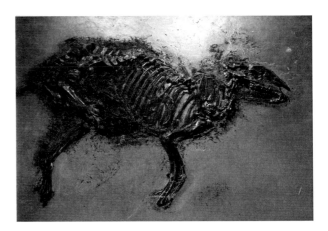

■ **Fig. 2.3** The earliest fossil remains of horses trace to the Eocene Epoch. Both Eohippus and Orohippus, evolved from Eohippus, coexisted during the later parts of the Epoch and remains have been identified in Oregon and Washington. (© Jonathon Blair/Corbis.)

Evolution during Domestication

While evolution is the process by which genetic diversity is introduced, domestication is the process by which striking variation arises and contributes to contrasting appearances within a species. Similar to evolution, an animal's environment and genetics contribute to its domestication. Similar to evolution, domestication is reliant on the ability of animals to vary due to genetic modifications, and the fact that animals can be selected to increase or decrease variations within a population. But during domestication it is the animal's social, and not physical, environment that is the primary force acting on existing variations. Surprisingly, all domesticated animals, regardless of species, share similarities in the variations that occur with domestication. Thus animals that show no recent relationship in evolution, evolved in similar directions following domestication. For example, floppy ears, curled tails, white head spotting, and unique coat color patterns appear with domestication in horses, cattle, sheep, pigs, rabbits, and dogs but are absent in wild counterparts. Collectively, these physical traits shared across animals are known as the markers of domestication.

Similar markers of domestication occur across species due to the selection of a single trait. Tameness, or docility toward humans, was the original target of selection for co-existence of animals

■ **Fig. 2.4**   Markers of domestication include white head spotting, floppy ears, and spotted or piebald coat colors. (*Left-1:* © withGod, 2014. Used under license of Shutterstock, Inc. *Left-2:* © Claudia Paulussen, 2014. Used under license of Shutterstock, Inc. *Right-1:* © tashh1601, 2014. Used under license of Shutterstock, Inc. *Right-2:* © oksana2010, 2014. Used under license of Shutterstock, Inc.)

within human environments. Historically, selection for tameness would have begun with natural selection. Through natural selection, animals predisposed to docility would show less stress when encountering humans. The selection for tameness is considered a crucial factor that altered not only the animal's appearance as stated above, but also developmental, reproductive, and growth traits of the animal. Artificial selection would play a greater role later in the domestication process. While under natural selection, unique features that appear as a consequence of genetic changes may not be continued within an environment, under artificial selection these genetic modifications may be chosen despite offering no survival advantage to the animal. Through artificial selection greater variation can be seen within a domesticated animal population than what would ever be achieved through natural selection. The variation seen in dogs is one of the greatest examples.

Unraveling the relationship between genetics and domestication began in the 1950s. In what is now known as the fox-farm experiments, a Russian scientist, Dmitry Belyaev, secretly undertook studies of the non-domesticated silver fox. Through selection for tameness, a model of domestication was established. Within four decades of selection, silver foxes selected for the sole trait of tameness began to show flop ears, spotted coats, and curled tails—markers of domestication.

Domestication Moves beyond Taming

Most animals suitable for domestication are currently domesticated. The exception is fish, which are debated as currently undergoing the domestication process for some species. Of the 148 large hooved animals, only fourteen have been success-

■ **Fig. 2.5**   Dogs are one of the greatest examples of the variation that can be achieved through artificial selection. In the absence of human input, genetic modifications that alter the physical appearance of dogs may never have persisted in nature. (© Erik Lam, 2014. Used under license of Shutterstock, Inc.)

fully domesticated. Domestication is less common in birds, with only 0.1 percent of species domesticated. According to Darwin, successful domestication required breeding of animals in captivity, was goal oriented, increased reproductive success of the animal, brought about atrophy or reduction in organ systems, enabled greater demonstration of adaptability, and was a process facilitated by subjugation to humans. Although selection for tameness was instrumental in early domestication events, domestication was more than taming. For the tame animal, acceptance of humans is temporary. Tame animals have not experienced successive changes over generations. The offspring of the tame animal will show wild tendencies, and the absence of domesticated markers.

Evidence suggests that early humans attempted to domesticate various animals, including gazelles, but attempts were not always successful. The relatively low number of domesticated animals occurred

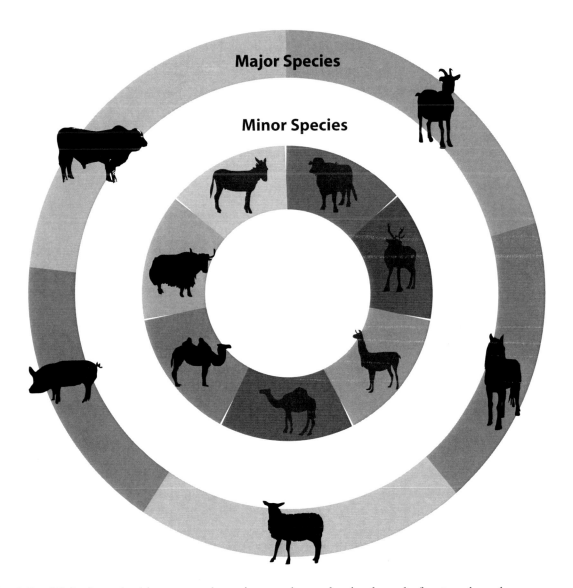

■ **Fig. 2.6** Of the large herbivorous and omnivorous hooved animals, only fourteen have been successfully domesticated. Successfully domesticated animals include the major five: goat, horse, sheep, pig, and cattle that have world-wide spread and influence. Minor species include: donkey, water buffalo, reindeer, llama and alpaca, Arabian camel, Bactrian camel, yak, and guar and mithan (not pictured). Minor species are considered of importance by limited numbers of people within limited or confined areas.

Social behaviors that support or oppose domestication success	
Support	**Oppose**
Large social groups with an hierarchal (leader and follower) structure (flocks, herds, packs)	Territorial structure with males in separate groups
Promiscuous matings with male dominance	Monogamous matings
Signal reproductive readiness through postures	Signals reproductive readiness through color markings
Precocial young	Altricial young
Short flight distance and low reactivity to humans	Long flight distance and high reactivity toward humans
Herbivorous or omnivorous	Carnivorous or specialized feeders
Low stress response to confinement	High stress response to confinement

due to the inherent traits of the animals themselves. While tameness was the main target for selection, many other traits were important to domestication success. Successful traits for domestication included: 1) large social group with a hierarchal structure that pre-disposed the animal to subjugation by humans. For social animals there must be a leader to reduce aggression within the social grouping. For domesticated animals, humans became the leader. 2) Promiscuous mating with posturing or other external signals of mating readiness that allowed successful human control of reproduction. 3) Adaptability to a wide range of dietary or environmental conditions that allowed animals to accept captive constraints with minimal stress. 4) Low reactivity toward humans. Animals with a greater flight distance or wariness toward human would be difficult to handle, which is counter-productive to the need for domesticated animals. 5) Precocial young that are well developed and require less maternal care when compared to their altricial counterparts. Collectively, these traits explain why large, solitary, carnivorous animals have not been successfully domesticated. The presence or absence of these traits does not always define domestication success, and animals may show some traits and not others yet are successfully domesticated. For example, there are many domesticated animals that give birth to altricial young, including: dog, cat, guinea pig, mouse, and pigeon, to name a few, despite this characteristic being consid-

ered unfavorable toward the domestication process.

It is not necessary for a species to demonstrate all behaviors for successful domestication. Consider that both horses and zebras are equids, yet the zebra is resistant to domestication attempts. Failure to domesticate zebras can be attributed to slight differences in their behavior, including high reactivity to humans.

Origins of Domestication and the Why, How, When, and Where

The domestication process is relatively new in human history, occurring as recently as seventeen thousand years ago by some estimates. The process cultivated the current relationship held between humans and other animals and has had a profound impact on the evolution of modern society. Indeed, the rise of human civilization is attributed to the domestication of animals, which freed humans from the continuity of securing food through migrating and hunting.

Domestication as defined by Price is:

A process by which a population of animals becomes adapted to man and a captive environment by some combination of genetic changes occurring over generations and environmentally induced developmental events recurring during each generation.

An animal's flight distance is the distance from an individual that causes the animal to flee and is commonly referred to as the animal's flight zone. When a human, or another potential threat, moves into an animal's flight zone, the animal will move away. The size of an animal's flight zone is greater for wild versus domesticated animals and is influenced by the animal's environment. Animals kept in more confined spaces will generally have smaller flight zones. For example, cattle that are raised on open range may have flight zones of up to 165 feet, while the flight zone of feedlot cattle ranges from seven to twenty-six feet. Other factors influencing the size of an animal's flight zone include the size of the animal's enclosure, quality and frequency of animal handling, and the way in which the animal is approached. Flight zones diminish in animals handled gently and often. On the other hand, if an animal is approached aggressively, their flight zone will increase. Domestication reduced the flight zone, making animals less likely to panic when approached by humans, but domestication did not eliminate the flight zone. Knowledge of an animal's flight zone shapes how animals are handled. An individual should work just on the edge

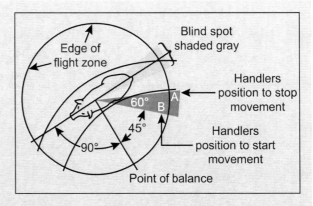

■ **Fig. 2.7**   Flight zone indicating the relative position of the handler to stimulate movement of the animal.

of the animal's flight zone. Standing just inside the flight zone will cause the animal to move away, while stepping outside of the flight zone will stop the animal's movement. To move an animal backwards the handler should stand in front of the animal's point of balance within the flight zone. To move the animal forward the handler should stand behind the point of balance within the flight zone.

Other definitions more loosely define domestication as a condition whereby humans control the breeding, feeding, and care of animals. However, these more contemporary definitions fail to capture the fact that not all animals in captivity are domesticated or capable of being domesticated. It is important to keep in mind that domestication required interplay between an animal's genetics and their environment, and although humans are given credit for the domestication process, it likely began with a period of unintentional human involvement. One of the first animals to have undergone domestication was the wolf. Estimates place domestication of the wolf between 17-15 thousand years before present. Some suggest that wolves drove their own domestication, which began initially through scavenging from human refuse. Over time, wolves adapted to this new environment would be less likely to flee from humans within the same environment. Through natural se-

lection, wolves less aggressive and more capable of begging for food would have adapted to feeding directly from human hands. It would be a few thousand years later before major animal domestication events triggered by deliberate human interactions would occur.

Why

The stimulus for human intervention in domestication is disputed, but the process coincided with the development of agrarian societies. These agrarian societies were necessitated through human population expansion. As the human population increased the ability to migrate decreased, and communities living near water resources developed. The number of animals available for hunting was overexploited within these established areas and a more stable source of food was required. Early domestication provided a reliable and accessible source of food. Further domestication events have occurred over the

■ **Fig. 2.8**   The origins of domestication remain uncertain. It is generally accepted that the wolf was the first species domesticated. It is unknown if domestication of the wolf was driven by man, or by the wolf. Some suggest that wolves drove their own domestication. Scavenging from human refuse, wolves less likely to flee thrived in this environment and were more likely to undergo the changes associated with the domestication process. (© Ron Hilton, 2014. Used under license of Shutterstock, Inc.)

millennia and involve the use of animals for transportation, economic status, social and religious ceremony, and research. The most recent domesticated animals are small animals with multi-purpose use. Additional domestication attempts of animals for food is subdued by the economic value of current domesticated animals. Efforts to establish new domesticated livestock have not been successful. Though animals such as deer are farmed, they remain undomesticated.

> It is generally accepted that no mass domestication events of animals for food occurred within the tropics where abundant food is garnered by hunting throughout the seasons.

How

Domestication occurred in phases. Animal keeping was the first phase, whereby a relationship between humans and animals was forced through human control of the animal's environment. The second phase was animal breeding. In the first phase, humans likely would have begun control of the animals within their wild environment by enclosing the animals and restricting their habitat. Alternatively, animal keeping may have involved the continual capture of young animals, use during the animal's productive life, and slaughter of the mature, wild animals. As human settlements were located near water resources, animals would have been attracted to these same water resources and therefore accessible to capture and/or confinement. Animal keeping would involve management and control of animal populations well before markers of domestication would appear. At some point there was a move toward the retention of the mature animal and their use in generation of offspring. Animals that survived within the captive environment would serve as stock for this second phase of domestication. Animal breeding represented intentional human selection of animals for farming.

When and Where

Archeological evidence traces the earliest origins of domestication to Mesopotamia, or the Near East; the cradle of civilization occupied by modern day Iran and Iraq. Reductions (cattle and pigs) or increases (chicken) in skeletal size, reductions in brain size, and changes in the size and shape of horns are all indicators of domestication events.

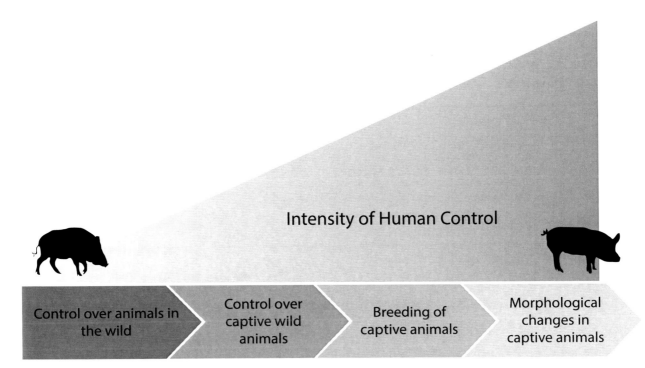

■ **Fig. 2.9** The transition from wild to domestic occurred in phases. With each phase, human control over animals intensified. As the relationship between humans and animals intensified, domesticated animals emerged.

However, these common archaeological markers of domestication may occur well after early stages of humans control over animals. Profiling of male and female ratios provides a more accurate estimate of domestication processes that predates morphological changes in animal populations. For example, animal populations kept by humans will have greater numbers of older females and few older males as females are retained for breeding, while males are killed for food at a young age.

Feral animals, animals that were once domesticated but have returned to a wild state, will regain some of the original features of the wild type that were lost with domestication. Recovery of wild traits lost in domestication of the pig includes elongation of the snout and increased hair coat covering. However, the reduction in brain size that occurred with domestication remains in the feral animal. For both pigs and mink, a reduction in brain size is seen in both domesticated and feral animals relative to their wild counterparts.

In addition to archaeological evidence, DNA is also used to reconstruct the time and location of domestication. Ancient DNA from the ancestors of domesticated animals is recovered, analyzed, and compared to that of domesticated animals. Often, the DNA of mitochondria are studied due to their abundance in preserved tissue and more rapid rate of change to reveal when and where domesticated animals began to diverge from their wild counterparts. Unlike archaeological evidence that is limited to the study of morphological changes occurring with the skeleton, DNA evidence is able to identify the wild populations from which domesticated animals trace, and the genetic changes occurring under human-animal interactions. The timeline of domestication pieced from DNA evidence often predates archaeological findings and brings attention to the fact that domestication was not instantaneous. Domestication was a gradual event as animals and humans co-evolved, and a precise time of domestication is not achievable.

A combination of archaeological and genetic evidence has revealed that multiple domestication events were common for many animals. Cattle, donkeys, and chickens were domesticated at least two distinct times, with the domestication of chickens

Archaeological Time Line of Domestication

Years Before Present	Species
15,000	
10,000	
9,000	
6,000	
5,000	
4,000	
3,000	
2,000	
1,000	
<500	
Current	

Domestication is a gradual event and therefore, a precise timeline can not be achieved.

separated over 3000 years between locations. Populations of sheep and goats underwent three domestication events. The domestication of sheep and goats closely coincided in time. Both animals were ideal targets of domestication due to behavioral and physical traits. Whereas sheep were domesticated within low lying regions, goats were more suitable for domestication in higher elevations. Domesticated

pigs trace to multiple domestication centers. At least four independent domestication events are reported across the Near East, Europe, Northern India, and South East Asia. The first of the domestication events occurred simultaneously within Eastern Turkey and China. Later domestication of wild pigs occurred in Europe. Prior to available genetic evidence, it was thought that pigs were domesticated in only the Near East and China, with domesticated pigs dispersed from these centers through human migration. Dispersal of domesticated animals through human migration and trade accounts for the global introduction of many domesticated animals. However, it is now known that while early human migration introduced domesticated Near East pigs to Europe, the original domesticated imports became extinct and were replaced with domesticated wild stock of European origin. Whereas the spread of domesticated sheep and goats from domestication centers occurred rather quickly in time, rates of dispersal occurred slowly for domesticated pigs owing to their limited mobility and adaption to pastoralism, which likely accounts for the increase in domestication centers.

Domestication events of cattle and horses remain debated. The large size of cattle rendered the animal less suitable for dispersal through human migration, at least until the smaller domesticated phenotype began to appear. Evidence supports the use of cattle through various human settlements but indicates that use for many was through hunting rather than human management. The most recent genetic evidence contends that all domesticated cattle of today are descendants from a small herd domesticated in the Near East with dispersal limited for the first 2000 years. The use of horses by human settlements dates nearly 7000 years before present, however, it is likely that horse use was also through hunting as supported by age and gender distribution in recovered artifacts. It was recently determined that less than 100 wild breeding mares could account for the genetic diversity seen in todays domesticated horses, but it is likely that these wild mares originate from different geographical locations.

Behavior

"Behavior is too important to be left to psychologists."

—*Donald Griffin (1915–2003)*

This chapter introduces the basic principles of animal behavior and the approaches to defining and studying animal behavior. The study of animal behavior is important for establishing acceptable animal practices that promote animal wellbeing. You will gain an understanding of the behaviors of animals as a consequence of genetics and environment, how behaviors develop, and how animals learn. Commonly observed behaviors and factors that contribute to animal development are presented as a distinction is drawn among how behaviors are classified.

Behavior: A Historical Perspective

Even before the process of domestication, the behavior of animals was of interest to humans. A basic understanding of predator-prey relationships was essential to early hunters who needed to understand the behavior of the animals hunted in order to guarantee food for human survival. These earliest observations of animal behavior are depicted in human artifacts. Human's observations of animal behavior and the knowledge gained were of further importance in successful animal domestication events.

The study of animal behavior as a science traces to Aristotle around 300 B.C. In his works, Aristotle systematically recorded his observations and ideas of animal behavior. By the seventeenth century, the field of behavior was concerned with the study of animal behavior in nature and the application of behavioral knowledge toward advancing domestica-

tion and breeding practices of animals. During this time, naturalist John Ray theorized that complex behaviors are innate and develop without learning. This era was also the time of philosophers, including René Descartes, who began to consider the relationships between humans and non-humans animals. Descartes proposed a mechanistic view of animal behavior, a view that non-human animals were incapable of consciousness or self-awareness, and only humans were capable of thought and language. Descartes' view on behavior prevailed until the nineteenth century when Charles Darwin wrote "The Expression of the Emotions in Man and Animals." This seminal work depicted how animals and humans express and relate their emotions to others and cast doubts on the view imposed by Descartes.

Despite early conjectures, the study of animal behavior was labeled as a pseudoscience by some and would not receive the credit that other fields of animal study did. In the late twentieth century, Hafez completed the first text on the behaviors of domesticated animals, and in 1967 Fox proposed the teaching of behavior in agriculture and veterinary practice. By the mid-1990s, the behavior of agricultural species began to receive full attention, coincident with the development of large, intensive systems of concentrated livestock production. In 1997, the USDA opened the Livestock Behavior Research Unit at Purdue University with the mission:

> … to determine behavioral and physiological indicators of stress and/or well-being in food-producing animals and to develop management systems that maximize well-being in farm animals …

■ **Fig. 3.1** The scratching by the hind limb placed over the forelimb is an innate behavior that many birds and mammals display. This form of complex behavior occurs without learning. Many innate behaviors are considered instinctive. (© Cheryl ann Quigley, 2009. Used under license of Shutterstock, Inc.)

This center aims to understand the biological basis of animal behavior through promoting collaborations among ethologists, physiologists, immunologists, and others. Today, the study of animal behavior is recognized as a significant discipline toward improving human-animal interactions.

Factors Affecting Animal Behavior

Animal behavior is defined as the expression of the activities of an animal and includes all things that an animal may do: moving, grooming, reproducing, feeding, sleeping, and communicating; and involves individual and group interactions. The expression of behaviors is a consequence of genetic traits interacting with environmental factors. During the process of domestication, animals were inadvertently selected for traits that allowed for their control and management. There are several behavioral traits found in most domesticated animals that promote their success in a captive environment. These behaviors are underscored by the animal's genetics and have become fixed to some extent during the evolution of the species. Such behaviors include: gregariousness, social organization, promiscuous matings, precocial young, adaptability, limited agility, and docile temperament. For example, agricultural animals are generally docile, with attenuation of aggressive behavior considered a consequence of the do-

mestication process. In addition, they are gregarious and are most content in organized groups that are established through social dominance. These animals traditionally do not rely on pair-bonding for mating as some wild species do, including beavers, golden eagles, and swans. If this were the case, it would be necessary to have one bull for every cow or one boar for every sow, which would not allow for efficient production systems. The offspring of agricultural animals are precocial, considered well developed at birth without a requirement for extensive maternal care. Hoofed animals, including foals, calves, pigs, lambs, and kids, are born with their eyes open and are ambulatory shortly after birth. Precociousness is needed by the offspring of wild animals to elude predation, but maintains importance in agricultural animals for production efficiency.

Fixed behaviors are considered instinctive, or innate, behaviors and are those that an animal will exhibit without having any opportunity to learn. As these behavioral characteristics may be inherited, these behaviors can be altered by selection. In wild populations natural selection is the primary stimulus altering inheritance of behaviors. Natural selection also played a role in inherited behaviors during domestication. Animals that adapted to confinement remained healthy and reproduced, while animals that did not adapt did not reproduce, or were at least less efficient. Consequently, a higher proportion of animals in each successive generation were more adapted to a human controlled environment. Today, artificial selection plays a greater role in inherited behaviors for captive animals.

Laying hens in production systems may display what is characterized as non-aggressive feather pecking directed toward the plumage of other birds, often referred to as gentle feather pecking. Gentle feather pecking is considered a typical exploratory behavior of birds. However, the behavior can adopt a more aggressive form that results in deterioration of the plumage and ultimately cannibalism and death of the targeted bird. When birds that display a reduction in this behavior are selected for breeding, the incidence of feather pecking will decrease within a flock over subsequent generations, implicating the role of genetics in generational changes of this behavior.

■ **Fig. 3.2** Nest building behaviors are common in many species prior to hatch or birth and are considered instinctive. (© Four Oaks, 2009. Used under license of Shutterstock, Inc.)

While the genetics of an animal are established at conception, the environment will influence the animal's behavior its entire life. Environmental effects on behavior begin prior to birth. Studies in rodents and monkeys show that psychological stress on the mother during gestation, particularly during periods of rapid fetal development of the brain and nervous tissue, alters behavior in the offspring. Postnatal animals develop in synchrony with their environment. Facility design, space allocation, availability of food and water, social environment, and human interaction each contribute environmental influences on behavior. Behaviors that ultimately form in response to specific experiences are considered learned responses. Learned responses may range from associative learning, where an animal associates one stimulus to another, to the more controversial higher-order problem solving abilities.

Birds are highly susceptible to their post-hatch environment. Birds reared with human contact will associate with humans as their social order. When the bird approaches mating age, the bird may attempt to mate with humans. The act of mating is considered instinctive, the choice of a mate is learned and a consequence of the environment in which the bird was raised. For this reason, birds reared in captivity for wild release are reared without human contact.

Introduction to the Study of Behavior

The psychology of animals has been studied since the early 1900s; however, the original application was toward humans. The recognition of the need of an understanding of animal behavior led to the establishment of four major approaches to the study:

1. Comparative psychology—the study of the mechanisms controlling behaviors of non-human animals including: learning, sensation, perception, and genetics. This approach has been accepted and used for many years by psychologists, physiologists, and cognitive scientists.
2. Sociobiology—the study of the biological basis of social behavior. An emphasis is placed on the role of genetics in controlling expressed behaviors. Sociobiology considers the evolutionary advantage of inheriting behaviors.
3. Behavioral ecology—the study of the relationship between a behavior and its environment is behavioral ecology.
4. Ethology—the study of animals' behavior in their natural surroundings with a focus on instinctive or innate behaviors. Ethology is the scientific study of animal behavior in the animal's natural habitat.

Scientists studying ethology develop and apply ethograms, which are catalogs of the range of behaviors an animal exhibits in its environment. Ethograms aim to account for environmental circumstances that impact an animal's behavior that may be reflected by their domestication and/or captivity. Konrad Lorenz (1903-1989) is accredited as the founder of ethology, but the beginnings of the

■ **Fig. 3.3** The relationship between behaviors and the environment is a concern of animal scientists. Domesticated pigs raised in indoor confinement operations often display bar-biting behavior. The basis of this behavior is not fully understood, but it is suggested to be a manifestation of the rooting reflex exhibited by wild and feral pigs. (© Shawn Hine, 2009. Used under license of Shutterstock, Inc.)

science trace to Karl von Frisch and Nikolaas Tingergen as well. Their collective observations sought to understand processes of learning that contribute to behavior patterns in animals. The field of ethology was once limited only to the study of wild animals, but it has expanded to include domesticated species in their usual surroundings and is referred to as applied ethology. Applied ethology focuses on the behavior of animals reared by humans and aims to understand the expression of behaviors as indicators of the animal's welfare.

Physiological Aspects of Behavior

Animals are able to perceive their environment through sensory receptors of the nervous system which receives visual, auditory, tactile, and olfactory information. The body can initiate a response to such stimuli through: 1) the spinal cord, 2) the hypothalamus, and 3) the cerebral cortex. Reflex reactions are initiated at the spinal cord and are involuntary responses to stimuli. The hypothalamus is responsible for hunger and thirst sensations and emotions. The cerebral cortex plays a key role in memory and learning and has been associated with adverse behavioral changes in dogs.

■ **Fig. 3.4** Horses will display the flehmen response to receive olfactory signals indicative of estrus in the female. (© Marcel Mooij, 2009. Used under license of Shutterstock, Inc.)

Auditory

The hearing range of livestock and poultry is within the frequency range of humans, while dogs have the ability to perceive sounds at a much greater frequency. Livestock, and many other animals, have relatively large ears that can be directed independently. The ability of animals to move their ears allows them to detect and locate the source of a sound. The fact that they can move their ears independently also allows them to simultaneously hear sounds from opposite directions.

Olfactory

Many animals have a more acute sense of smell than humans as a consequence of an accessory olfactory organ known as the vomeronasal organ or Jacobson's organ. This organ is considered responsible for pheromone perception in animals and has been documented to influence exploratory behavior in rodents. Located between the mouth and the nasal cavities, it enables the animal to distinguish among odors that

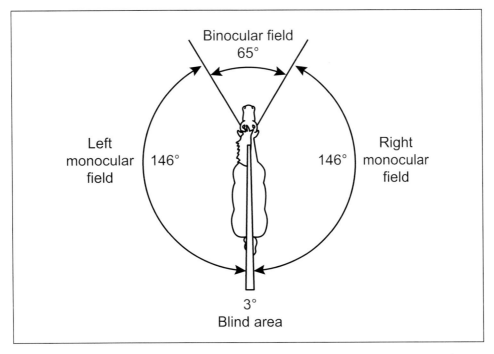

■ **Fig. 3.5** The monocular vision of grazing animals provides a nearly 360° field of vision.

humans are unable to detect. When a bull, ram, or stallion smells the urine of a female to determine if she is in estrus, he is inspiring the scent of the urine into the vomeronasal organ for identification. As the male is inhaling, he will curl his upper lip. This action of lip-curling is called the flehmen response.

Tactile

Animals are acutely aware of their bodies and have a well-developed tactile sense. *Bos indicus* cattle are able to easily detect insects on their skin and dislodge them by selectively shaking these areas due to a subcutaneous muscle layer. Horses are highly sensitive to the tactile sensation provided by the rider in the saddle. Touch is an important aspect of communication. Animals will groom, lick, and scratch each other; and horses will often stand beside one another, head to tail, mutually swatting flies. Grooming in baboons is a behavioral pattern suggested to reduce tensions within the group. Observations suggest that pigs prefer to be close and physically against each other for tactile comfort.

Taste

Most animals readily distinguish between the four tastes: bitter, sweet, sour, and salty. Also, they can distinguish among the intensities of these flavors and their combinations. Browsers, such as goats, are able to consume plants that are avoided by other animals due to their insensitivity to bitter tastes. Exceptions are poultry, which have a rather poor sense of taste and instead rely primarily on sight. Feed intake can be increased in poultry by coloring the feed red or blue. In comparison, flavor and odor have a greater influence on feed intake by cattle and horses.

Visual

A major distinction between different species is the size of their visual field. Eye placement on the side of the head provides grazing animals with an almost 360° field of vision. Grazing animals also have monocular vision. The wide field of monocular vision that grazing animals have allows them to watch a maximum area around them for potential predators. In contrast, predator species, including humans and birds of prey, have a larger field of binocular vision due to the frontal placement of their eyes. Binocular vision provides the animal with better depth perception, which allows predators to focus in on their prey. With monocular vision, blind spots are established directly in front and behind the animal. In the horse, these blind spots are equivalent to a loss of visual field four feet in front of the animal and ten feet directly behind the animal. Vision is communicated through contrasts and boundaries. Animals do not detect fine lines but discriminate against form. The fine structures of facial expressions are not detected by most animals, an exception being some primates, and a bull will readily mount an artificial dummy that lacks detailed structure as long as the scent of the female is present.

Communication

Auditory, olfactory, tactile, and visual inputs are important aspects of animal communication, along with vocalization. Animals can communicate in a variety of ways and understanding how they communicate is important for proper animal management, as well as for the safety of both the handler and the animal. Most communication between animals is within their own species, but some communication does occur across species. The vocalizations of animals have distinct meanings. For example, four to seven distinct vocalizations have been identified in the horse.

A snort generally means danger or fear, while a whinny or nicker may signify distress. Chickens also have a feeding call that serves to attract other chickens to the feed. Mothers of all species will often call to their young and vice versa. When vocally communicating with an animal low, soft tones of longer duration are suggested to have a comforting or calming effect. Short, high-pitched tones are generally excitatory. Animals also rely on visual cues and body language to communicate. Postural changes in addition to eye contact and movement of the ears or tail confer meaning.

Fields of Behavioral Study

Studies of animal behavior may be conducted on wild, feral, captive, or domesticated animals. Through the study of behavior the significance and function of a behavior can be examined. Studies should provide for objective or straight forward assessment of an animal's behavior that can be analyzed without the confounding and often anthropomorphic interpretations of the observer. The study

■ **Fig. 3.6** Horses are expressive and convey varying degrees of excitement, playfulness, submission, or aggression using movement. (© Winthrop Brookhouse, 2009. Used under license of Shutterstock, Inc.)

of behaviors asks not only what a behavior *is,* but what a behavior *does*. A fundamental question in the fields of behavioral study is: What are acceptable behaviors for a given animal under the conditions in which it is maintained? Because behaviors are a consequence of genetic traits interacting with the environment, the environment in which the animal is kept must be considered. By understanding behaviors observed during routine animal practices, decisions can be made that improve human and animal interactions and the productivity and well-being of the animal.

Ingestive Behavior

The study of animal's ingestive behavior is important for proper management of production systems, companion animals, and captive wildlife. In the majority of livestock species, most of the interest is in maximizing consumption in order to improve rate of gain and increase feed efficiency. In horses and companion animals the focus is optimizing performance and maximizing health and longevity. Health and longevity is also a focus for captive wildlife. Deviations in ingestive behaviors are an early sign of underlying health issues.

Integral to ingestive behaviors is knowledge of water and feed intake patterns. Horses, cattle, and sheep usually drink one to four times a day. The amount of water consumed, and how frequently the animal drinks, depends on the temperature, the type of feed, and feed intake. Silage and pasture forage are generally high in water content, while hay and many types of grain are much lower in moisture content. The greater the water content of the feedstuff, the less potable water needs to be consumed. Grazing patterns also vary among species and breeds, and may be influenced by temperature, wind, and quality of forage. Animals tend to graze most near sunrise and dusk and will generally avoid forage in areas of recent urination or defecation, particularly that of their own species. Horses and sheep graze closely to the ground. While horses tend to cover large areas while grazing, sheep and cattle travel less and lie down intermittently in order to rest and ruminate. Area grazed by cattle is further influenced by water location as cattle will decrease grazing distance to remain near water resources. In poultry and swine production systems, market animals are often fed ad libitum. For this reason, it is important that there is adequate feeder space per animal to minimize competition and aggression. Since they feed ad libitum, swine and poultry may drink or eat at any time, though specific meal patterns often develop. If made to adhere to meal patterns, pigs should be fed on a predictable schedule. When fed intermittently and at random, pigs show stress and an increase in aggressive behaviors.

Biological Rhythms

Biological rhythms are the patterns of activity that occur in cycles throughout an animal's life. Understanding biological rhythms can promote a better understanding of animal activity, sexual cycles, and physiological responses. Abnormal sleep and activity levels in animals also may be detected, although this often proves difficult with domestic animals due to modern confinement practices. The most studied and most often referenced cycle is the circadian rhythm, the twenty-four hour cycle of living systems that defines patterns of behavior and physical changes. The primary factor contributing to circadian rhythms is photoperiod, which establishes the diurnal and nocturnal patterns of animals. Diurnal species are most active during the day, whereas nocturnal species display the greatest activity at night. Disruption of the circadian rhythms of an animal can negatively impact animal health. For example, forcing a nocturnal animal to be active during the day can lead to disorientation, reduced appetite, and reduced growth. Biological rhythms are reliant on the endocrine system and influenced by a variety of additional factors, including pharmaceuticals and nutrition.

Resting Behavior

Resting behaviors range from lying positions to deep sleep, otherwise known as REM (rapid eye movement) sleep. Research has shown that all species of domesticated livestock, companion animals, and birds experience deep sleep; however, most of the sleep experienced by domestic animals appears to be a light sleep, the exception being swine and dogs. Deep sleep is characterized by muscle relaxation, lowered hearing and respiratory rates, and dreaming. Evidence of dreaming is similar to that of humans and includes: rapid eye movements under closed lids, facial and/or limb movements, and brain wave patterns. Animals progress from non-REM to REM sleep, and because deep sleep is more difficult to arouse from and predisposes the animal to danger, REM sleep occurs multiple times within a sleep cycle. Horses generally sleep only a few hours during each twenty-four hour period. Most of this sleep occurs while the horse is standing; however, to reach deep sleep, horses must be lying down in what is referred to as lateral recumbency. Deep sleep occupies approximately 10 percent of the horse's total sleep duration; this is in contrast to humans where deep sleep may occupy 25 percent of total sleep. In ruminant animals, deep sleep is controversial. For dairy cattle, resting may occupy twelve to fourteen hours of the animal's day, and dairy cattle have been observed to sacrifice time committed to eating in favor of resting time. However, the majority of resting occurs as drowsing, with only three to four hours committed to sleep. Within the period of sleep, deep sleep only occurs for two to eight minutes at a time. Limited duration of sleep is attributed to the movements and contractions of the rumen and the reticulum. These movements and contractions require the sternum to be held in place. Although animals can ruminate while in non-REM sleep, rumination ceases during deep sleep, which only occurs when the animal is lying on its side during lateral recumbency and relaxation of the sternum. Calves and lambs that do not yet have a functional rumen are often observed lying on their sides in deep sleep. The young of many animals spend greater amounts of time in sleep. Pigs and dogs are similar to humans in their sleep and rest characteristics. These species generally sleep more hours daily than do cattle, sheep, or horses. They also sleep more deeply and for longer periods of time than do the other species. Birds

■ **Fig. 3.7** Resting behaviors of cattle from counter-clockwise: long, narrow, and short. Cows in the long position rest with their head extended, the hind legs tucked near the body, and the fore legs may or may not be extended. Narrow cows differ from long in that the resting is positioned on the sternum. The short resting period is observed when cows enter into sleep and is distinguished by the tucking of the head into the body. The purpose of the various resting behaviors is not fully known. (*Top:* © george green, 2009. Used under license of Shutterstock, Inc. *Bottom left:* © sharon kingston, 2009. Used under license of Shutterstock, Inc. *Bottom right:* © Eric Limon, 2009. Used under license of Shutterstock, Inc.)

sleep while perched on a roost or standing. This requires the bird's muscles to be toned and balanced, which would in turn require a higher level of consciousness than for an animal that is lying down. Their ability to sleep in this position is due in part to the fact that when the bird's hock is flexed, the tendons in their legs pull their toes downward and around the perch.

Maternal Behavior

The study of maternal behavior includes mutual recognition by the dam and offspring. Many of the initial behaviors between a mother and offspring serve to establish a bond and encompass all physiological aspects of behavior: auditory, olfactory, visual, tactile, and taste. Additional behaviors reflect learning. For example, monkeys have been observed to exaggerate behaviors to facilitate learning by their offspring. The study of maternal behavior also includes an understanding of aberrant behaviors such as rejection of the offspring, cannibalism, stealing of offspring, and refusal to nurse. The production and continuation of livestock and companion species depends on the female's ability to care for her offspring. The ability of the female to care for her young may be affected by human intervention, previous experiences, heredity, and physiology. The temporary separation of a newborn calf or lamb from the mother immediately after birth due to a storm, herd or flock migration, or the birth occurring in a crowded pen are all circumstances that can interfere or inhibit maternal bonding and contribute to offspring rejection or mis-mothering.

Sexual and Reproductive Behavior

The study of sexual and reproductive behavior in animals includes the physiological basis of sexual behavior, which is dependent on both the endocrine and central nervous systems. This field of study examines the social and sexual experiences, as well as the environmental influences that promote typical mating behavior and how to deal with atypical mating behavior or unwanted sexual behaviors exhibited by neutered animals. The sexual behavior of domestic animals is influenced by many factors, including genetic selection, confinement rearing, management practices, and interaction with other species—particularly humans. In animals on pasture, courtship and mutual grooming can often be observed during the breeding season. Cows in estrus will spend more time than usual walking and less time resting or eating. A cow in estrus will allow both females and males to mount her within twelve hours prior to ovulation, and she may attempt to mount other cows as well. In the presence of a ram, a ewe in estrus will roam around the ram and will rub his neck and body, while vigorously shaking her tail. Male courtship behavior usually includes nosing the area near the vulva, flehmen, or nudging the female, particularly in the flank areas or from the rear. Males will be vocal, may stomp or paw the ground, and may even challenge another male. In avian species it is the male that generally initiates courtship and the mating process. The male will move among the females and will approach the one that appears to be the most receptive.

Herd Behavior and Social Structure

Domesticated animals tend to display a strong group instinct and isolation from their herd or flock can induce stress. The members of a herd or flock are arranged into a social hierarchy often referred to as pecking order. The term pecking order arises from the social dominance demonstrated by birds and first characterized in the 1920s. When unfamiliar birds are placed together in confinement, they will begin to peck one another to begin to establish a social hierarchy among themselves. The most dominant bird will be free to peck all the other birds, while all the other birds will not peck the most dominant bird. Similarly, the second most dominant bird will be free to peck all the birds less dominant than itself, but will not peck the most dominant bird. This will continue down the line to the bottom bird, which is the bird that can be pecked by all the other birds, but will not return the pecking behavior. Similar to observations noted in birds, when unfamiliar mammals are introduced fighting is common until the dominance order is established. These social hierarchies are usually linear, as that described for birds, but can become more complicated in larger groups. For example, circular social hierarchies may exist where individual A dominates B and B dominates C, however, C dominates A. Over time, pecking orders within a group may change as some animal's age or are moved in or out of a group. Social hierarchies underscore the order in which cows enter a milking parlor. The more dominant cows enter fist, but as the cow ages her place in the hierarchy declines and the order of parlor entrance is delayed.

Not all group behaviors are aggressive. Social play is an important part of group behavior, especially in horses. Play is important for the social development of horses and helps them to develop skills they will need later in life such as those required for obtaining feed.

Development of Behavior

The study of the development of normal behavior is an important area of study for those in the fields of veterinary medicine or animal sciences. From this study of behavioral development has arisen the concept of sensitive and critical periods. Distinctions between sensitive and critical periods are often obscure. Both terms represent times in an organism's life span that it is more sensitive to environmental influences and during which certain types of learning occur more readily and are more easily accomplished. Critical periods are suggested to occur more rapidly, end more abruptly, and be defined by a *point of no return,* whereby learning beyond this point does not occur. In contrast, sensitive periods are suggested to occur gradually and are represented by a point of maximal response where learning is accomplished more readily. It is generally accepted that it is during these periods when experience, or lack of it, will influence an animal's behavior later in life. By studying the development of behavior, scientists have gained a better understanding of factors that shape an animal's behavior such as socialization with other animals, as well as with people.

Learning and Training

There are four major categories of learning: imprinting, classical conditioning, habituation learning, and operant or instrumental conditioning. Imprinting is a relatively irreversible learning process that occurs quite rapidly within a few hours or days following hatch or birth in birds and mammals, respectively. This process usually includes the offspring learning who its mother is and to what species it belongs and is a relatively irreversible process. Recognition of the process of imprinting has shaped animal practices and contributed to the successful reintroduction of captive populations of rare or endangered species into their native environment. Whooping cranes, California condors, and Griffon vultures have been

successfully reintroduced following rearing of post-hatch chicks with puppets, which prevented imprinting of humans on the chicks during the critical period of learning. Classical conditioning, also referred to as Pavlovian conditioning, is the type of learning first demonstrated by the Russian physiologist Ivan Pavlov (1849-1936). Pavlov showed that a reflex-like response can be stimulated by a neural response. In his classical studies, dogs were conditioned to salivate at the ringing of a bell due to the learned association of the bell with food. Habituation learning occurs when an animal learns to ignore a stimulus that occurs rather often, such as wearing a halter or the sound and movement of traffic. Operant or instrumental learning is often thought of as learning by trial and error. This type of learning is in progress when an animal learns to act in a certain way to obtain a desired response or reward. The use of an automated feeder requires this type of learning. Learning by trial and error is often accompanied by reward or punishment. If a reward is to be used to increase the occurrence of that particular behavior, or if punishment is going to be used to decrease the occurrence of a behavior, then the reward or punishment should occur simultaneously, or directly after the behavior is demonstrated. If too much time elapses between the behavior and the reward or punishment, the animal will be confused and not connect the reward or punishment with that behavior. Reward or punishment in response to a particular behavior generally must be repeated several times for the animal to learn. It also should be remembered that occa-

■ **Fig. 3.8** Learning by reward, a form of operant learning, is commonly used in training dogs. (© Anke van Wyk, 2009. Used under license of Shutterstock, Inc.)

sional reinforcement or retraining may be necessary. Animals may forget, making repeated training an important basis for memory. Finally, the amount of emotional arousal may have an influence on the animal's ability to learn and retain what has been learned. For example, severely punishing an animal for a behavior it cannot help may cause fear and instead of learning to avoid the behavior, they may become afraid of the handler.

Classifying Behaviors

An understanding of animal behavior advances handling and management practices that are an important component of animal welfare. Domesticated animals are reared in an environment that differs from their wild ancestral environment. This requires animals to adapt. Successful adaptations can be assessed through the behaviors of the animals to minimize management errors and improve animal success. Toward an understanding of behavior, behaviors are often classified as normal or abnormal.

Abnormal behaviors are defined as behaviors that deviate from commonly practiced behaviors of animals. By definition, abnormal means a deviation from the normal. The terms normal and abnormal are subjective and differ according to the population under study. For example,

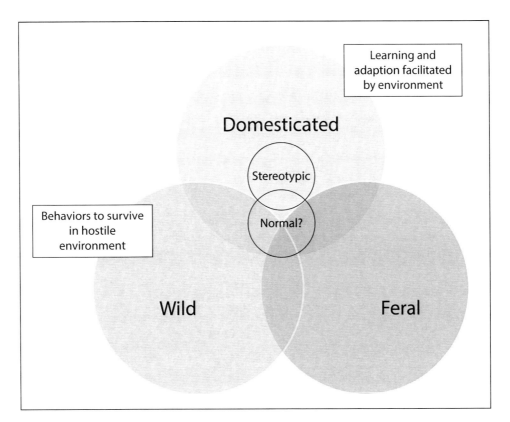

■ **Fig. 3.9** Behaviors classified as normal may reflect the range of behaviors commonly observed across wild, feral, and domesticated animal populations. However, it is not always possible to study all states for a given animal. Furthermore, behaviors are influenced by the animal's environment. Behaviors of the wild animal that developed in response to a hostile environment may be lost in the domesticated animal whose behaviors occurred through learning and adaptation to a human-made environment. Conversely, the development of stereotypies that occurred with the domesticated environment is not observed in wild animals. These stereotypies may reflect normal adaptations to survive changing environments.

domesticated animals will exhibit behaviors not commonly observed in wild ancestors. Consider the bark of a dog that is not observed among the wolf. When defining abnormal behaviors, one must consider that processes including domestication have influenced the range of behaviors expressed by animals. For a more accurate indication of behaviors, one can study the behaviors displayed across wild, feral, and domesticated animal populations. However, for many animals not all states are available for study resulting in an incomplete or narrow range of behaviors defined. It is not possible to completely preserve behaviors across wild and captive animals; instead efforts should be directed toward minimizing behaviors that are detrimental. Thus, behaviors considered abnormal may be tolerable unless they subtract from the survival, physical well-being, or social health of the animal. In general, deviant behaviors in captive animals are coping mechanisms and often manifest as stereotypies.

Stereotypies are considered to be repetitive behaviors with no apparent end goal or function. As stereotypies are observed in domesticated and captive animals but absent in wild animals, environment has remained a focus of the causative agent for their development. Stereotypies are thought to occur when animals are forced to adapt to an environment lacking in physical or social resources. The stereotypy reflects a manifestation of a behavior that would otherwise occur in a more enriched environment and supports the animal's adaptation to their environment. Stereotypies are more common within sub-optimal environments, and within these environments animals that display stereotypies have improved performance over animals that do not. An

Common behavioral syndromes in domesticated and captive animals

Animal	Syndrome	Occurrence	Physical outcome
Dairy cattle	Stall kicking		Leg injuries from repetitive trauma
Sheep	Wool chewing		Bald spots in fleece
Pigs	Tail and/or ear biting	More common following mixing of animals or introducing new animals	Cannibalism and death
Horses	Head shyness Stall kicking Weaving Cribbing	More common in horses reared in isolation. Cribbing is associated with endorphin release	Leg injuries may occur with stall kicking and weaving
Chickens	Severe feather pecking Pacing	Lack of appropriate stimuli	Cannibalism and death
Llamas	Berserk Male Syndrome	Excessive human handling following birth	Spitting and violence toward humans
Dogs	Tail chasing Self-mutilation General destructive behaviors	Lack of appropriate stimuli	
Cats	Clawing Wool sucking	Lack of appropriate stimuli	
Zoo animals	Pacing	Lack of appropriate stimuli	

■ **Fig. 3.10** Introduction of novel objects into the environment of captive animals encourages exploratory behaviors and reduces incidences of abnormal behaviors. (© Eric Isselee, 2009, Used under license of Shutterstock, Inc.)

elimination of stereotypies should not be the goal of an animal producer, but an understanding and correction of the factors initiating the stereotypic behaviors should be the aim.

As many stereotypic behaviors are suggested to manifest when exploration behavior is suppressed, alleviation of these behaviors occurs through enrichment practices. Enrichment practices aim to provide stimuli to animals. Studies show that the introduction of dog toys reduces incidences of unwanted behaviors in individually housed macaques. In lowland gorillas, enrichment increases foraging, social play, and solitary play while decreasing sedentary behavior. Knowledge of enrichment practices has contributed to improved design of captive environments of zoological species and domesticated food animals. However, it should be recognized that once a behavior develops, the behavior may continue as habit despite improvements within the animal's environment that should otherwise reduce the behavior.

CHAPTER 4

Nutrition

"Life is a chemical process"
—Antoine Lavoisier (1743–1794)

This chapter introduces the study of nutrition, including the classes of nutrients and the importance of each to the maintenance, growth, and health of an animal. The role of digestive physiology in directing nutrient requirements is also discussed. You will gain basic knowledge to differentiate between essential and non-essential nutrients, define the energy contribution of specific nutrients, develop an understanding of how nutrients are organized within the body for tissue formation and function, and differentiate between the digestive strategies employed by animals that allow them to occupy different niches within the food chain.

NUTRIENTS

"Let your foods be your medicines and your medicines be your food"
—Hippocrates (460–377 B.C)

Nutrients: A Historical Perspective

Associations between diet and the body trace to biblical times. In what may be considered the first nutritional study, the *Book of Daniel* (2500–2400 years ago) records the capture and treatment of Daniel by the King of Babylon. Daniel was to serve in the king's court and was therefore offered the finest foods and wine of the court. Daniel refused, asking to be fed only vegetables and water. The guard who was responsible for Daniel feared for his own life, but agreed to Daniels requests for ten days. After the "study" Daniel was permitted to eat as he chose as men consuming his diet appeared more fit and healthy than those receiving the king's rations. Nutrition remained a topic of interest during the ages of the Greek philosophers including Hippocrates. It was Hippocrates that made the first references to the medicinal properties of food and that the proper amount of nourishment was key to sustained health, sentiments that are still echoed today.

The origins of nutrition as a science began during the Chemical Revolution of the late 1700s and the observations of the French chemist Antoine Lavosier. Lavosier's interests in metabolism led to his defining of life as a chemical process. Lavosier's studies were cut short by his execution at the guillotine; however, he paved the way for the beginning of the nutritional sciences. To follow, in 1816, were the studies of Francois Magendie. The simplistic studies involved feeding only a single food that was previously deemed nutritious to dogs. Magendie concluded that alone, none of the foods tested provided all of the requirements of the dog, as with each test the dogs died. The implication, which remains as the cornerstone of nutrition today, was that diversity was the key to proper nutrition. At the time of Magendie's experiments it was unknown what essential components were contained within food. This information would remain unknown until the first noteworthy studies to identify food components were conducted in Weende, Germany in the 1860s. To support economic development, the German government began to shift the labor force from agriculture to industry. As a result, agriculture production needed to be more efficient. In order to facilitate the shift in labor, a research station was established in Weende to study livestock nutrition.

With a greater understanding of nutrition, food production increased and interest grew in this field of science. One major development that resulted from the formation of the Weende research station was the system of classifying the components of feed known as proximate analysis. This system classified feed components into five categories; water, crude protein, ether extract (crude fat), carbohydrates (crude fiber and nitrogen free extract), and ash (minerals). Researchers then focused on the important role of each of these nutrient categories in sustaining the functions of a healthy body.

Interestingly, vitamins were not an initial nutrient category. A breakthrough in understanding the importance of vitamins in the diet would not occur until the 1880s, when alarming numbers of Dutch soldiers developed symptoms pertaining to the weakening of the legs, shortness of breath, and in extreme cases, edema and heart failure. These were indicators of a disease classified as beriberi. Christian Eijkman, an army physician, was sent to investigate. Eijkman, suspecting the condition was bacterial, began to use chickens as test subjects and injected half with blood from beriberi patients and left the other half as a control. However, both groups showed symptoms of the disease. Eijkman repeated the trial and to his surprise, none of the birds in the second trial showed any symptoms indicating beriberi. Confounded, Eijkman began to examine differing factors between the two trials. After questioning the animal caretaker, Eijkman realized that the chickens that contracted beriberi in the first trial were fed a diet that consisted of white rice, whereas the unaffected chickens in the second trial were fed brown rice. The difference between the two grains was the removal of the hull. This led to the conclusion that the brown, hull intact, rice contained an element essential to health which the white rice, striped of the hull, did not. This essential element was discovered to be thiamine, and would later be classified as vitamin B_1.

Twenty years later, the 1900s marked the vitamin revolution, a period in time that coincided with the identification of several vitamins essential to animal well-being. Stephen Babcock, an agricultural chemist at the University of Wisconsin Agricultural Experiment Station, is credited by some with the initial studies that led to the discovery of vitamin A. Babcock fed heifer calves diets containing either wheat, bran, corn, or a mixture of all three grains.

The offspring of the corn-fed heifers were the most viable, whereas the offspring of heifers fed wheat were stillborn or died shortly after birth. It was concluded that either corn contained a component essential to the offspring's survival or wheat contained a toxic component. It was later determined that corn contained a vital component first identified as factor A and later renamed vitamin A. The studies performed by Eijkman and Babcock were some of the earliest controlled experiments in nutritional science.

Overview of Nutrition

Nutrition is defined as the use of foods or feed by living organisms. The discipline represents how the body uses nutrients consumed to sustain life and production by supporting cellular needs for work, growth, reproduction, maintenance, and repair of tissues. As feed costs may exceed 70 percent of the costs of production, nutrition plays a major role in animal industries.

> The term food refers to an item that supplies nutrients, usually in its natural state. Feed is used when referencing the food that is supplied to animal systems.

■ **Fig. 4.1.1** Stephen Babcock was an agricultural chemist whose primary interest resided with improving the quality of marketed milk. His research into single-grain feeding in cattle contributed to the identification of Vitamin A and the practice of grain feeding in the cattle industry. (Courtesy University of Wisconsin-Madison Archives.)

Nutrient Classification

A nutrient is any chemical element or compound in the diet that is required to support cellular needs and thereby support work, growth, reproduction, maintenance, and repair of tissues. There are six major classes of nutrients: water, carbohydrates, lipids, proteins, vitamins, and minerals. Within these classes, nutrients may be recognized as essential or nonessential. Essential, or indispensable, nutrients are required in the animal's diet as the body is incapable of producing adequate amounts of these nutrients to support the metabolic processes required for life. In contrast, nonessential, or dispensable, nutrients are not required in the diet. Essential nutrients are defined in each of the nutrient classes, with the exception being carbohydrates. The nutrients defined as essential may differ across species and differ according to physiological status or age. Collectively, more than forty nutrients are recognized as essential across animal species.

Water

Water is the most overlooked nutrient despite the fact it is essential to all living things. Water constitutes 50–75 percent of an adult animal's body mass, up to 90 percent of a newborn animal's mass, 90 percent of blood in mammals, and 60–65 percent of the mass of a cell. Interestingly, the percent of Earth covered by water is 70 percent and falls neatly within these ranges as well. The vitality of water was recognized by Max Rubner (1854–1932) who observed that a loss of as little as one-tenth of body water can result in death.

Water serves two primary functions for all terrestrial animals: 1) a component in cellular metabolism, as biochemical reactions that occur in an animal's body require water, and 2) a medium for thermoregulation, the process of maintaining a constant body temperature independent of environmental factors.

Water is involved in many of the chemical reactions that take place in the body and is considered the universal solvent. Thus, it is important that water be present for many biological functions, since most compounds readily ionize or dissolve in water. An organism may obtain water through consumption or generate metabolic water within the body as a result of metabolism. Metabolic water is a substrate in hydrolysis, and a product in oxidation. Water is also a transport medium and a diluent and is required for the transportation of semisolid digesta in the gastrointestinal tract, as well as transportation of solutes in blood, tissues, cells, and exogenous secretions, such as urine and sweat.

Sources of water available to animals include potable (drinkable) water and feeds. Contribution of water to an animal through feeds is dependent on the type of feed consumed and is influenced by plant maturity. For example, most grains are approximately 10 percent water, while pasture grasses may exceed 80 percent water content. However, the water content of forages varies with plant maturity,

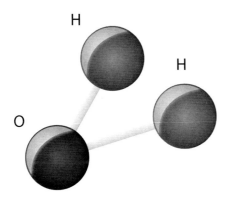

■ **Fig. 4.1.2** Water is the most vital of the nutrients, yet it may be the least recognized or remembered of the nutrient classes. Waters asymmetrical structure results in negatively charged oxygen and positively charged hydrogen, which underscores its solvent properties.

■ **Fig. 4.1.3** For most animals metabolic water only contributes 5-10 percent of the animals daily water balance. In some desert animals, such as the kangaroo rat, metabolic water can partially sustain the animal for weeks in the absence of drinking water. (© Dmitriy Karelin, 2009. Used under license of Shutterstock, Inc.)

■ **Fig. 4.1.4** Some animals are unable to sweat efficiently to cool the body. When excess heat needs to be dissipated in elephants, blood vessels dilate and increased blood flow through the capillaries of the ears occurs. Elephants may then flap their ears to effectively transfer heat into the environment. This form of heat dissipation is less efficient than the evaporative cooling accomplished through sweating or panting. (© Four Oaks, 2009. Used under license of Shutterstock, Inc.)

with more mature forages containing less water. There is an inverse relationship between water content of feeds and free water intake by an animal; whereby, as the water in the diet increases, the need for potable water decreases.

While water intake is intermittent, water lost from the body is continual. Water is lost from the animal through urination, defecation, and vaporization (sweat and respiration). Urine is a primary route of water loss. Losses of water through urine occur as a consequence of the constancy of metabolism and may be increased to offset increased intakes and maintain a steady state of body water. Water lost through vaporization is important to thermoregulation. For animals that need to maintain a constant body temperature, the animal must either generate heat if the environment is cooler than its body or have a way to dissipate heat if the environment is warmer than its body. Some animals dissipate heat through sweating or panting. The process of sweating is triggered by elevated body temperatures and activates sweat glands located beneath the skin.

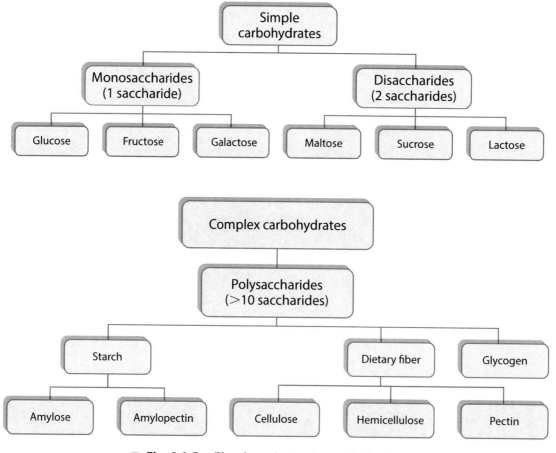

■ **Fig. 4.1.5**   Simple and complex carbohydrates.

These activated glands produce water that is evaporated away from the body. The process of evaporation requires energy as water changes state during evaporation. This energy is received in the form of excess body heat. As sweat leaves the body and evaporates in the surrounding air it cools the skin. Animals with a limited number of sweat glands accomplish similar temperature regulation by panting. In this process, water is evaporated not from the skin, but from the oral cavity to maintain a standard body temperature.

Carbohydrates

Carbohydrates serve primarily as a source of energy for cellular processes, providing 4 kcal/g. They are the primary constituents of plant tissues and comprise approximately 70 percent or more of the dry matter of forages. Carbohydrates are classified as simple or complex. Monosaccharides and disaccharides are simple carbohydrates commonly referred to as sugars, whereas complex carbohydrates include the polysaccharides starch and fiber.

Monosaccharides include glucose, galactose, and fructose. Glucose is the most important carbohydrate in biology and is used as a primary source of energy by most cell types. Glucose also is a component of the disaccharides maltose, sucrose, and lactose. Maltose is derived from the glycosidic linkage of two glucose units, sucrose from the glycosidic linkage of glucose and fructose, and lactose from the glycosidic linkage of glucose and galactose. Lactose is commonly referenced as milk sugar and is an important dietary carbohydrate for young animals. Starch polysaccharides include amylose and amylopectin, which are formed solely from glucose molecules; however, amylose is a straight chain of glucose molecules, and amylopectin is branched. The branched organization of amylopectin results in a compact storage form of glucose for plants. The cereal grains (corn, barley, wheat, oats) are rich sources of starch, which is a readily digestible carbohydrate.

Fiber is another form of complex carbohydrates and includes cellulose, hemicellulose, pectin, and lignin. These molecules are a main component of plants' cell walls and provide structural support to bolster the plant. Cellulose consists of only glucose units, whereas hemicelluloses, pectin, and lignin contain additional saccharides. Dietary fibers are considered nondigestible carbohydrates as they are resistant to digestive enzymes of animal origin. However, microorganisms present in the forestomach of ruminants and hindgut of nonruminants release enzymes that can digest dietary fiber. This process of microbial digestion is known as fermentation and results in the production of the volatile fatty acids (VFA's) butyrate, acetate, and propionate. Butyrate and acetate are major energy sources, whereas propionate may be used as a precursor in the production of endogenous glucose.

The body is capable of producing glucose from several precursor molecules; thus, carbohydrates are not regarded as dietary essentials. The production of glucose by the animal body is balanced by glucose use to meet cellular energy needs and storage as glycogen. The animal body is only capable of storing limited supplies of glucose as glycogen. These supplies are readily depleted in-between meals. The ability of the body to produce glucose helps ensure constancy in the supply of glucose to tissues that rely on this nutrient to function in the absence of intake.

> The processes by which the body uses the nutrients to support the needs of the living system is termed metabolism. Metabolism encompasses two collective processes: catabolism and anabolism. The body breaks down nutrients in the process of catabolism and builds new compounds in the process of anabolism. The two processes are interrelated and allow an animal to grow and/or sustain life processes from dietary nutrients that are much different from the animals own nutrient composition.

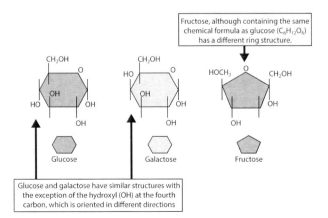

Fructose, although containing the same chemical formula as glucose ($C_6H_{12}O_6$) has a different ring structure.

Glucose   Galactose   Fructose

Glucose and galactose have similar structures with the exception of the hydroxyl (OH) at the fourth carbon, which is oriented in different directions

■ **Fig. 4.1.6**   Common monosaccharides.

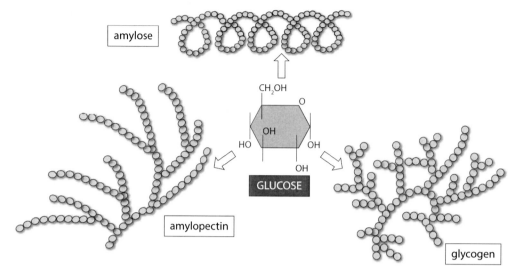

■ **Fig. 4.1.7** Starch occurs in linear (amylose) and branched (amylopectin) forms. The branched form of starch is structurally similar to glycogen stores in animal tissues; however, glycogen contains more glucose molecules due to greater branching of the glucose units.

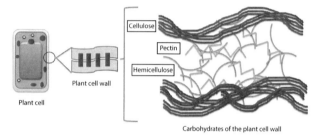

■ **Fig. 4.1.8** Fiber components of the plant cell wall.

Lipids

Lipids are a diverse group of organic compounds that are relatively insoluble in water, but are soluble in organic solvents. Thus, lipids are considered hydrophobic. Fats and oils constitute the largest percent of lipids in most feedstuffs. Lipids, as fatty acids, are the most concentrated source of energy providing 9 kcal/g. Along with providing energy, lipids are the source of essential fatty acids, integral components of cell membranes, and carriers of the fat-soluble vitamins, A, D, E, and K.

Approximately 95 percent of dietary lipid from animal products, cereal grains, and oils is supplied in the form of triglycerides. Triglycerides are compounds that have three fatty acids attached to a glycerol backbone. Complete hydrolysis, or breakdown, of triglycerides yields three fatty acids to one glycerol molecule. Hydrolysis is necessary for the uptake of dietary lipid by the animal body, but released

fatty acids are restored as triglycerides in specialized tissue known as adipose. Adipose tissue serves as an energy reserve. Unlike glucose storage, the storage of lipids is a near limitless storage of energy. In-between meals or following a prolonged fast when glucose availability is limited, fatty acids and glycerol are released from adipose triglyceride stores. Fatty acids provide a concentrated source of energy, whereas glycerol is used as a precursor to glucose production to help support the bodies need for a constant glucose supply.

The energy of fatty acids is a consequence of their structure. Fatty acids consist of hydrocarbon chains ranging from two to twenty-four or more carbon atoms in length with a terminal carboxylic acid group (-COOH). Most of the fatty acids found in animal tissues are straight chained (not branched) and contain an even number of carbons. When the carbons are connected only by single bonds and no double bonds exist, the fatty acid is saturated. The term saturated indicates that the fatty acids have the maximum number of hydrogen bonds, a property that causes saturated fats to be solid at room temperature, indicating an increased melting point. In contrast, unsaturated fatty acids have one (monounsaturated) or more (polyunsaturated) double bonds. In unsaturated fatty acids, the chain of carbons does not contain the maximum number of hydrogens possible, resulting in a molecule that is less structured. Consequently, unsaturated fatty acids are liq-

Volatile fatty acids

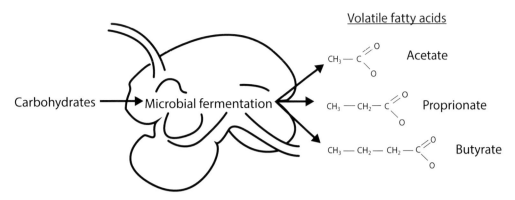

■ **Fig. 4.1.9** Volatile fatty acids are byproducts from the microbial digestion of carbohydrates that can occur within the rumen, cecum, and/or colon.

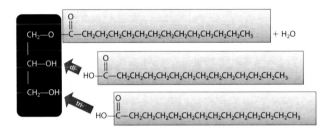

■ **Fig. 4.1.10** Acylglycerols are formed between fatty acids and a glycerol backbone. The addition of one fatty acid to the glycerol backbone is known as a monoacylglycerol, a second fatty acid yields a diacylglycerol, and a third fatty acid yields a triacylglycerol.

uid at room temperature. Unsaturated fats can occur as *cis* or *trans* isomers and refers to the orientation of hydrogens about the double bond. *Cis* indicates a structure in which the hydrogens adjacent to the double bond are oriented in the same direction (plane) and is the predominant isomer found in nature. *Trans* is a structure in which the hydrogens adjacent to the double bond are oriented in opposite direction (plane). Many dietary *trans* fats are produced commercially through hydrogenation, a chemical process in which hydrogens are added to unsaturated fats. An exception to this is the naturally occurring *trans* structure of conjugated linoleic acids (CLA), which are produced by the microorganisms of the rumen. There are many reported

Lipid Class	Functions
Fatty acids	Metabolic fuel (energy), cell membrane anchor
Acylglycerols	Fatty acid storage and transport as a triacylglycerol
Phospholipids	Cell membrane structure
Sphingolipids	Cell membrane structure
Ketone bodies	Glucose sparing metabolic fuel
Isoprenes	Coenzyme and vitamin backbone
Sterols	Cell membrane structure and vitamin and steroid hormone precursor. Found as cholesterol in the animal body.

Lipids are a diverse group of compounds that are related by their lack of solubility in water. Lipids can be grouped into classes according to structural similarities.

health benefits of CLA as they are considered to be a cancer fighting agent and have a role in reducing body fat. Humans and other nonruminants are incapable of producing CLA and must obtain these fatty acids by consuming foods resulting from ruminants, such as milk, butter, beef, and lamb.

> Nutritional regulations and labeling guidelines of *trans* fats in human foods excludes the labeling of naturally occurring *trans* fats such as CLA.

Two of the most important unsaturated fatty acids for animals are α-linolenic acid (omega-3) and linoleic acid (omega-6). These fatty acids are recognized as dietary essentials for many animals and are precursors for the synthesis of longer chain polyunsaturated fatty acids important to animal's health and well-being. It should be noted that cats have an additional requirement for arachidonic acid. Although most animals can convert linoleic acid to arachidonic acid within the liver, cats evolved as efficient predators acquiring arachidonic acid from the muscle tissues of their prey. Over time, cats lost the key enzyme necessary for the conversion of linoleic acid to arachidonic acid, thus necessitating this additional fatty requirement in their diet.

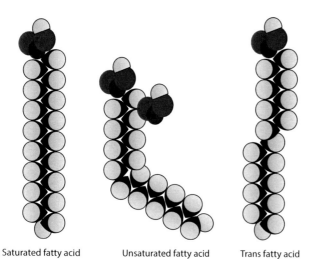

Saturated fatty acid   Unsaturated fatty acid   Trans fatty acid

■ **Fig. 4.1.11** Saturated, unsaturated, and trans fatty acids. In the absence of double bonds, saturated fatty acids show a linear confirmation. With the addition of a cis-double bond the fatty acid adopts a kinked confirmation. The addition of trans-double bond results in an unsaturated fatty acid with a more linear confirmation.

> Dietary essential fatty acids are recognized for nonruminant animals. In the case of ruminant animals, there is no defined dietary essential fatty acid requirement. Ruminant animals are suggested to receive adequate supplies of unsaturated fatty acids in their diets. Coupled with the ruminant's ability to conserve these fatty acids within their body, a dietary requirement has not been established.

Phospholipids are a third type of lipid that share similarities to the general structure of triglycerides. Both classifications of lipids contain a glycerol backbone. However, whereas triglycerides contain three fatty acids attached to the glycerol backbone, phospholipids contain only two fatty acids with a phosphate group replacing the third. Often, the phosphate group is bound to an additional organic compound. The addition of the phosphate group confers hydrophilic properties to the phospholipids. The hydrophilic head along with the two hydrophobic fatty acid tails support the most well-known function of phospholipids—a lipid bilayer that surrounds the cell. The lipid bilayer forms as the fatty acids tails orient to exclude water, exposing the hydrophilic head to the interior and exterior aqueous environments of the cell. Thus, phospholipids are the primary constituents of the cell membrane and help facilitate movement of molecules into and out of the cell.

Proteins

Proteins perform a variety of functions in the body, the most notable being tissue accretion and growth. Proteins are also present as structural components in cell membranes, muscle tissue, hair, skin, and hooves. Proteins are the product of gene transcription and are the molecules of enzyme-catalyzed reactions, muscle contraction, metabolic regulation, and immune function. All cells synthesize proteins, and without proteins life could not exist. Although proteins are not a primary source of energy for the tissues of most animals, they provide 4 kcal/g when catabolized. The relative contribution of protein as a source of energy is greater for carnivores, which receive little carbohydrate within their diet.

Proteins are composed of chains of amino acids. The basic structure of an amino acid is a central car-

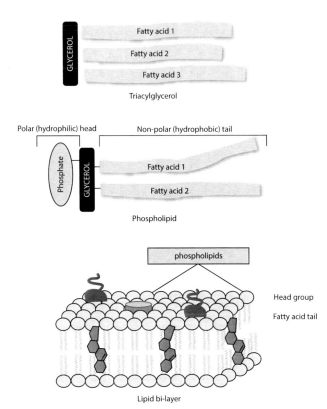

■ **Fig. 4.1.12** Phospholipids share similarities to the triacylglycerol structure. Both rely on fatty acids bound to a glycerol backbone. Whereas triacylglycerols contain three fatty acids, phospholipids contain two. The third position of the glycerol backbone is occupied by a phosphate. The phosphate is commonly associated with another organic compound. Phospholipids orient with the hydrophilic head group facing outward and the hydrophobic fatty acids facing toward the interior, producing a lipid bilayer essential to the formation of cell membranes.

by the linear sequence of amino acids. Secondary structure is the result of local folding that occurs due to interactions between closely located amino acids. Tertiary structure involves more distant interactions within a polypeptide chain, and quaternary structure results from interactions between different peptide chains. Biological function occurs with tertiary and quaternary structures, whereas the primary structure reflects the amino acid content and thus, the nutritional value of a protein.

> Dietary essential amino acids are recognized for nonruminant animals. In the case of ruminant animals, microbial fermentation supplies the necessary amino acids as long as the basic elements are available. These elements include carbon, hydrogen, nitrogen, and sulfur. Urea is a form of non-protein nitrogen that can be fed to animals to support microbial amino acid synthesis.

There are twenty amino acids commonly found in proteins, and all must be present for protein synthesis to occur. Although species differences exist, ten amino acids are generally recognized as dietary essential for many nonruminant animals based on the work of William C. Rose in the 1930s. The essential amino acids, in turn, are used for the synthesis of non-dietary essential amino acids. When the amount of an amino acid supplied by the diet is not

bon bound to a hydrogen, a carboxylic acid group, an amine group (-NH$_2$), and a side chain (-R group) that differs for each of the twenty amino acids commonly found in proteins. Amino acids are linked by peptide bonds. The linkage of two or more amino acids results in formation of a peptide. Polypeptides are made up of more than one peptide and are extended peptide chains. The term protein is reserved for large molecules of one or more polypeptides that show biological function. Proteins display different levels of complexity defined by interactions within and between polypeptide chains. Primary structure is determined

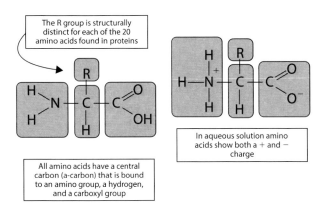

■ **Fig. 4.1.13** Amino acid structure. Each of the amino acids present in a protein will differ according to the elements of the R group.

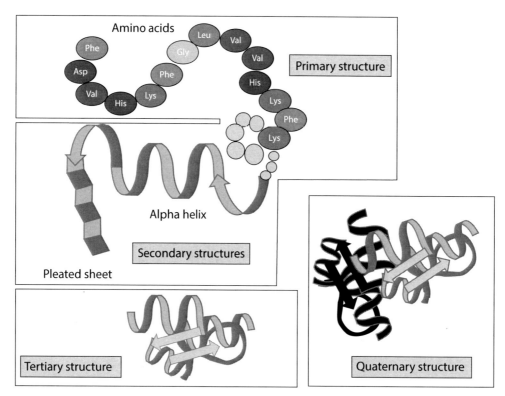

■ Fig. 4.1.14 Individual amino acids form peptide bonds for the primary structure of a protein. Further interactions of amino acids within a linear chain as a result of hydrogen bonds are responsible for the secondary structure of a protein. The tertiary structure occurs as a result of more complex interactions within the amino acid chain, including the formation of disulfide bonds. When more than one amino acid chain interacts, a quaternary structure is formed. Protein structure is the ultimate determinant of protein function.

Essential amino acids	Nonessential amino acids
Arginine	Alanine
Histidine	Aspartic acid
Isoleucine	Citrulline
Leucine	Cysteine
Lysine	Glutamic acid
Methionine	Glycine
Phenylalanine	Hydroxyproline
Threonine	Proline
Tryptophan	Serine
Valine	Tyrosine

The dietary essential amino acids reflect the ten amino acids defined for the diet of the rat. Additional amino acids, such as glycine and proline, that are non-essential in the rat are considerd dietary essentials in poultry.

sufficient to meet the animal's requirement for protein synthesis, the amino acid is said to be limiting. The first-limiting amino acid is the amino acid that is present in a diet in the least amount in relation to the animal's need for that particular amino acid. The lack of availability of that amino acid in the diet restricts the performance of the animal as it limits the use of all amino acids in synthesizing a protein. For example, the limiting amino acid of corn is lysine, with tryptophan a close second. On the other hand, soybean meal is deficient in methionine, but is rich in lysine and tryptophan. By combining corn and soybean meal in animal diets, a more adequate balance of amino acids is achieved.

The roles of individual amino acids beyond tissue accretion and growth are diverse. Arginine (Arg) plays an important role in cell division, the healing of wounds, removing ammonia from the body, immune function, and the release of hormones. Arginine is essential to the dog and cat at all stages of development. In other animals, including pigs and humans, arginine is considered conditionally essential and is required in the neonatal diet, but not the adult diet as it is produced by the urea cycle. Histidine (His) can be decarboxylated to form histamine in mammals. Histamine controls the constriction of certain blood vessels and also regulates the secretion of hydrochloric acid to the stomach. Lysine (Lys) is involved in calcium absorption, enzymes structure, and antibody response in chickens. Methionine (Met) supplies sulfur, which is essential for normal metabolism and growth. It is also a methyl donor, aiding in chemical and metabolic reactions in the body. Methionine can meet the total need for sulfur-amino acids in the absence of cysteine, another sulfur containing amino acid. Phenylalanine (Phe) is unique, as it can meet an animal's complete need for both phenylalanine and tyrosine. Tyrosine is a product formed from the degradation of phenylalanine and can satisfy up to 50 percent of the need for both of the amino acids, though it cannot serve as the sole source. In an inherited disorder, phenylketonuria, the enzyme responsible for phenylalanine degradation is non-functional. Left untreated, the condition results in neurological disorders. In humans, the condition is managed through restricting phenylalanine in the diet and supplementing tyrosine. Tryptophan (Trp) is a precursor for the vitamin niacin in some animals, the neurohormone melatonin, and the neurotransmitter serotonin, thus it is required for electrical signals between a neuron and another cell. Cataracts may develop under conditions of tryptophan deficiency. Isoleucine, leucine, and valine are the branched-chain amino acids. The branched-chain amino acids are targets of body building supplements in the form of whey proteins isolated from the dairy industry. Branched-chain amino acid degradation also is associated with a rare, inherited disorder called maple syrup urine disease. With this disease, a nonfunctional enzyme in the common pathway of the degradation of the branched-chain amino acids causes the buildup of certain metabolites in the urine, resulting in the characteristic odor from which the disease derives its name. Glycine (Gly) is the simplest amino acid in the body. Beyond supporting the synthesis of proteins, glycine is an inhibitory neurotransmitter in the central nervous system and is important in the synthesis of nucleic acids. Lastly, proline (Pro) is important in determining the three dimensional structure of proteins due to its hydrophobic nature. Proline also confers rigidity to a protein. As a major building block of collagen, hydroxylation of proline increases the conformational stability of collagen and protects against weak tissues.

> The quasi-amino acid taurine is essential for cats. Taurine is a sulfur-containing compound derived from cysteine. It is not a component of proteins, but exists as a free compound that is necessary for normal cardiac muscle function, reproduction, formation of bile salts for digestion, and vision. Lack of dietary taurine in feline diets is associated with irreversible blindness.

Amino acids also support the aforementioned need of animals to maintain a constant supply of glucose during fasting or may be directed toward lipid synthesis when in dietary excess.

Vitamins

Vitamins are a group of organic substances that are required in relatively small amounts and are essential for life. Each vitamin has its own specific function, and the lack of a certain vitamin in the diet can cause specific deficiency symptoms that may ultimately lead to the death of the animal. Vitamins can be divided into fat soluble and water soluble based on their solubility properties when first isolated.

The fat soluble vitamins include vitamins A, D, E, and K. Vitamin A is required for night vision, epithelial cells, and bone growth and remodeling. It is

■ **Fig. 4.1.15** Vitamin A deficiency and resulting hyperkeratosis. (Image courtesy of Tony Buffington and colleagues, The Ohio State University.)

■ **Fig. 4.1.16** Vitamin D deficiency in growing lambs can lead to the development of rickets. (Image courtesy of Tony Buffington and colleagues, The Ohio State University.)

important for reproduction and also appears to protect against cancer. Deficiency symptoms include night blindness, poor growth, reproductive failure, reduced egg production, inadequate bone remodeling, swollen and/or stiff joints, rough coat, and hyperkeratosis. Although carnivores require preformed vitamin A in their diet, other animals can use the precursor β-carotene to fulfill their vitamin A requirement. β-carotene is a member of the carotenoid family of compounds that are responsible for the colorations of yellow to red seen in fruits and vegetables. Intakes of carotenoids also contribute to the colorations of amphibians, reptiles, and birds. Vitamin D promotes calcium and phosphorus absorption, which is important for bone mineralization. Exposure to direct sunlight promotes conversion of cholesterol to vitamin D; therefore animals exposed to direct sunlight may not require dietary sources of vitamin D. However, the ability to convert cholesterol to vitamin D is highly variable and influenced by length of sunlight exposure, hair coat covering, and skin color. Deficiency symptoms include rickets in growing animals, osteomalacia and osteoporosis in adults, reduced egg production, and soft egg shells. Vitamin E is an antioxidant that is important in cell membrane maintenance and required for proper muscle structure and normal reproductive function. Vitamin E is located in the cell membrane to provide direct protection against the actions of radical molecules that would otherwise damage the cell by removing electrons from the lipid bilayer. Deficiency symptoms include nutritional muscular dystrophy characeterized by white or chalky streak-

ing of skeletal and cardiac muscle (white muscle disease) and reproductive failure. Vitamin K is required for normal blood coagulation. It is specifically required for the synthesis of prothrombin, a coagulating agent, in the liver. Deficiency symptoms include reduced blood coagulation in response to diminished concentrations of prothrombin, and association with spontaneous hemorrhages. Dicoumarol (a derivative of coumarin) is a natural antagonist to vitamin K that was discovered in damaged sweet clover hay by researchers at the University of Wisconsin. The anticoagulating properties of dicoumarol are useful in the medical profession in the prevention of blood clots. Dicoumarol also is effective as a component of rat poison. Exposed animals hemorrhage or bleed through the capillaries, which is the cause of death.

The remaining vitamins are classified as water soluble. The water soluble vitamins include vitamin C and the B vitamins. Vitamin C (ascorbic acid) plays a role in collagen formation and is involved in Vitamin E conservation. Many animals can synthesize vitamin C, but it is required in the diet for primates, humans, guinea pigs, bats, and some birds. Deficiency symptoms include scurvy, anemia, hemorrhaging, spongy gums, weight loss, swollen joints, and structural defects in bone, teeth, cartilage, connective tissues, and muscles. The overall effects of vitamin C deficiency are one of weak tissues, as vitamin C is necessary for hydroxylation of proline.

> Dietary essential B vitamins are recognized for nonruminant animals. In the case of ruminant animals, microbial synthesis by rumen bacteria fulfill the animals need. Elements of sulfur (for thiamin and biotin) and cobalt (for vitamin B_{12}) are required or a vitamin B deficiency can occur.

The B vitamins collectively support the use of carbohydrates, lipids, and proteins by the body by serving as essential coenzymes for various reactions. Thiamin (B_1) was the first vitamin identified and is necessary for the loss of CO_2 that occurs during energy metabolism. A role for thiamin in neurological function is also reported. Deficiency symptoms include cardiovascular disturbances, convulsions, weakness, reduced growth rates, emaciation and anorexia, and polyneuritis in chicks and opisthotonos (stargazing) in other animals. Riboflavin (B$_2$) functions in the coenzyme flavin adenine dinucleotide (FAD) and flavin mononucleotide (FMN). Deficiency symptoms include dermatitis and hair loss, reduced growth rates, diarrhea, ocular lesions, and curled toe paralysis in birds. Niacin (B$_3$) is a constituent of the coenzymes nicotinamide adenine dinucleotide (NAD) and nicotinamide adenine dinucleotide phosphate (NADP). Deficiency symptoms include inflammation and ulceration of the mouth, tongue, and digestive tract, as well as anorexia, irritability, diarrhea, dementia (loss of cognitive function), and dermatitis. Both riboflavin and niacin coenzymes serve as electron acceptors and donors in numerous enzyme systems that support energy transfer from the breakdown of the energy yielding nutrients. Pantothenic acid (B$_5$) is a component of coenzyme A (CoA) which is necessary in the metabolism of fatty acids. Deficiency symptoms include dermatitis, graying and/or loss of hair, poor growth and reproductive function, digestive disorders, spastic gait, and goose-stepping or posterior incoordination and paralysis. Pyridoxine (B$_6$) is a coenzyme involved in non-dietary essential amino acid synthesis and nitrogen metabolism. It also is involved in red blood cell formation and antibody production. Deficiency symptoms include convulsions in swine, hyperirritability, and reduced growth rates. Biotin (B$_7$) is an acceptor and donor of CO$_2$ necessary for lipid synthesis. Deficiency symptoms include reduced growth, hair loss, dermatitis, and bone deformities in chicks. Folate (B$_9$) is necessary for the transfer of single carbon units for amino

acid metabolism and purine formation. Deficiency symptoms include poor growth, pernicious anemia, and a decrease in white blood cell numbers (leucopenia). Cyanocobalamin (Vitamin B$_{12}$) is important in regenerating folate. In its absence, purine formation does not occur for DNA synthesis. The most notable effects are on red blood cell maturation. Deficiency symptoms are similar to folate deficiency and include pernicious anemia. Vitamin B$_{12}$ is also involved in energy and protein metabolism, and neurological disturbances; poor reproductive performance and reduced growth rates are also observed in a deficient state.

Pernicious anemia, also known as megaloblastic anemia, is characterized by larger than normal red blood cells. The condition occurs as a consequence of the cells preparing for cellular division; however, without adequate DNA synthesis the cells are unable to divide and remain in an immature state. Because animals store vitamin B$_{12}$ for extended periods of time, the condition is often associated with a lack of the factor necessary for vitamin B$_{12}$ absorption, not a lack of dietary supply of the vitamin.

Minerals

Minerals are inorganic, solid, crystalline chemical elements that are required by all animals. Minerals are classified as macrominerals or microminerals, a distinction that is based on the relative amounts of each that are required in the diet for normal life processes.

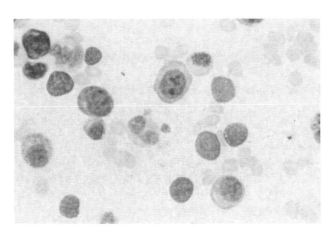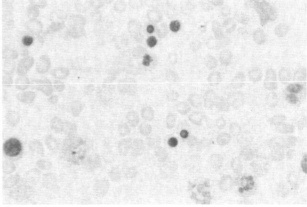

■ **Fig. 4.1.17** Megaloblastic anemia (left) characterized by large, dysfunctional red blood cells following vitamin B$_{12}$ deficiency and recovery (right) following B$_{12}$ supplementation. (Image Courtesy of Tony Buffington and colleagues, The Ohio State University)

The Periodic Table of the Elements

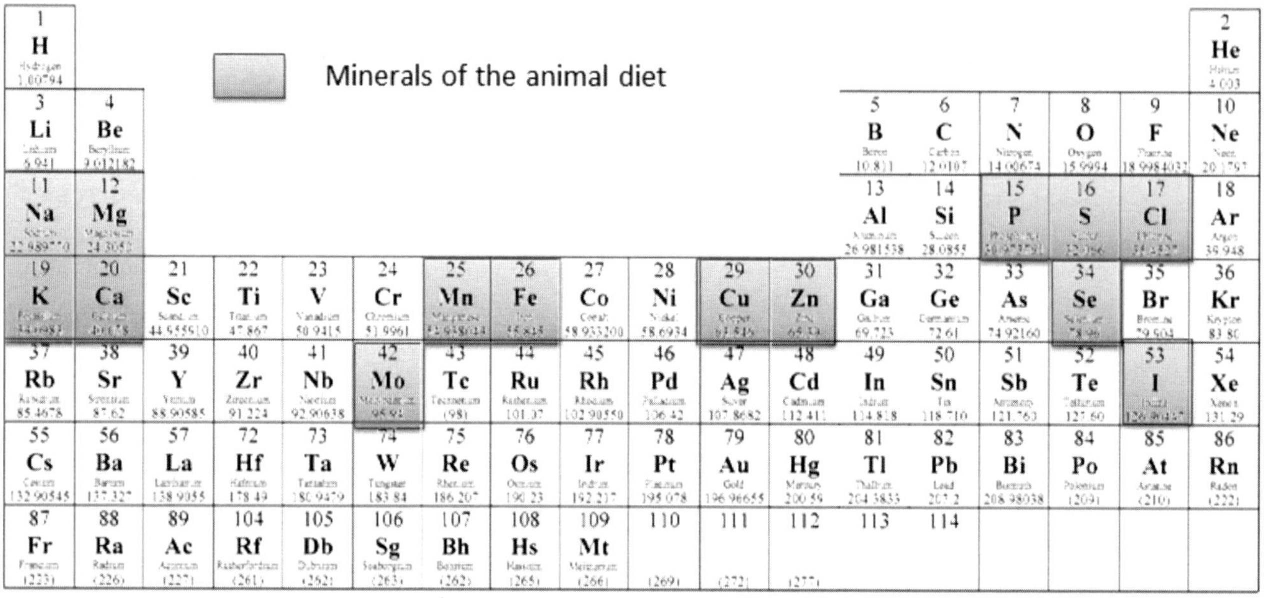

Fig. 4.1.18 Minerals required in animal diets are reflected in the first five rows of the periodic table of elements.

Macromolecules are required in greater amounts in the diet and include calcium, phosphorus, magnesium, potassium, sodium, chloride, and sulfur. Calcium (Ca) is essential for bone and teeth formation, and over 99 percent of calcium in the body is found within the skeletal system. As an intracellular ion, calcium is involved in transmission of nerve impulses and muscle contractions. Calcium is also required for milk production and egg shell formation. Deficiency symptoms include rickets in young animals, osteomalacia in adults, reduced egg production, thin-shelled eggs, impaired blood clotting, and milk fever in dairy cattle. Deficiency symptoms are similar to those observed during a vitamin D deficiency, as the main effect of vitamin D is calcium regulation. Phosphorus (P), commonly found in the body as the phosphate ion, is involved in formation of bones and teeth, is a component of the phospholipids of cell membranes, is found in nucleic acids, and influences the energy status of a cell. Deficiency symptoms include rickets in young animals, osteomalacia in adults, reduced growth rates, weight loss, and reduced egg production. Magnesium (Mg) is involved in bone formation and is important in the use of phosphate, thus it is involved in many aspects of metabolism. Deficiency symptoms include vasodilatation, hyperirritability, convulsions, loss of equilibrium, trembling, and ultimately death. In ruminant animals grazing spring pasture, the diagnoses of magnesium deficiency is associated with the condition of grass tetany. Potassium (K) is a major cation of intracellular fluid and is involved in osmotic pressure (water movement), acid-base balance, and transmission of nerve impulses and muscle contractions. Deficiency symptoms include lethargy with increased incidence of coma and death, diarrhea, decreased food and water intake, reduced growth and egg production in poultry, cardiac muscle abnormalities, and distended abdomen. Sodium (Na) is the principle cation of extracellular fluid. Similar to potassium, sodium is involved in osmotic pressure, acid-base balance, cell permeability, and nerve impulses and muscle contractions. With a reduction in sodium, water moves into cells causing the cells to expand. Deficiency symptoms include reduced growth rates, reduced feed efficiency, reduced milk production, and overall wasting. Chlorine (Cl) acquires an electron to become chloride, the major anion of the intra- and extracellular fluids. Chloride is essential for oxygen transport to tissues and carbon dioxide transport from tissues. Chloride is involved with digestive function as it combines with hydrogen to form hydrochloric acid in the regulation of stomach pH. Reduced growth rates are a common deficiency symptom. Sulfur (S) is a component of biotin and thiamine and is present in sulfur-

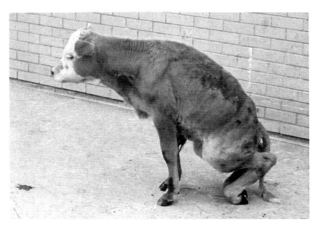

■ **Fig. 4.1.19** Posterior paralysis characteristic of copper deficiency. (Image courtesy of Tony Buffington and colleagues, The Ohio State University.)

■ **Fig. 4.1.20** Skin lesions associated with zinc deficiency. (Image courtesy of Tony Buffington and colleagues, The Ohio State University.)

containing amino acids, including methionine and cysteine. As a component of vitamins and amino-acids, most animals receive adequate sulfur through other dietary nutrients. Exceptions are ruminant animals that will require sulfur for microbial synthesis of the vitamins and minerals. A primary deficiency symptom is reduced growth rates due to the requirement of sulfur containing amino acids for protein synthesis. Reduced wool growth and poor feathering also occur due to the abundance of the amino-acid cysteine in these structures.

The remaining minerals are required in lesser amounts in the diet and are considered microminerals or trace minerals. Chromium (Cr) is involved in lipid, protein, carbohydrate, and nucleic acid metabolism. Deficiency symptoms are rare. Cobalt (Co) is a component of vitamin B_{12} and is needed for rumen bacterial growth. Deficiency symptoms include anemia, reduced appetite, reduced growth rates and body weight, and eventually death with prolonged deficiency. Copper (Cu) is required for bone and connective tissue metabolism, maintenance of myelin of nervous tissue, pigmentation of hair, and hemoglobin synthesis. Deficiency symptoms include loss of hair pigmentation, lack of wool, lameness, swelling of joints, fragile bones, anemia, and incoordination in lambs, goats and pigs, as well as cardiovascular lesions and hemorrhages in swine, chicks, and cattle. Fluoride, the anion of fluorine (Fl), protects teeth from decay and is suggested to slow the development of osteoporosis in mature animals. Tooth decay may occur due to deficiency; however, fluoride excess is of concern in infants and livestock. Toxicity is associated with reduced serum calcium and magnesium and

is accompanied by cardiac abnormalities and nervous distress. Iron (Fe) primarily exists in the body as the heme portion of red blood cell hemoglobin and myoglobin in muscle cells. It is also a co-enzyme involved in energy transfer. Deficiency symptoms include anemia due to less than normal amounts of hemoglobin and fewer and/or smaller red blood cells resulting in shortness of breath, reduced growth rates, rough hair coats, and reduced appetite. Increased susceptibility to stress and infectious agents have been noted in swine. Iodine (I) is needed for the formation of the thyroid hormones thyroxine (T_4) and tri-iodothryonine (T_3), both of which regulate metabolism. Deficiency symptoms include goiter, abortions and stillbirths, hairless pigs or wool-less lambs at birth, brittle hair, dry skin, reduced metabolism, irregular or suppressed estrous cycles in females, and reduced semen quality in males. Manganese (Mn) is a cofactor for several enzymes and plays a role in bone formation, growth and reproduction, amino acid and cholesterol metabolism, and fatty acid synthesis. Deficiency symptoms include reduced growth rates, testicular degeneration in males, impaired ovulation in females, and perosis (slipped tendon) in birds. Molybdenum (Mo) is involved in purine metabolism and also stimulates microbial activity in the rumen. Deficiency symptoms are uncommon. Selenium (Se) is an antioxidant that maintains the integrity of cellular membranes and is required for immune system function. The role of selenium as an antioxidant is intimately associated with the role of vitamin E, thus it is often difficult to distinguish between the mineral or vitamin deficiency. Deficiency symptoms include mortality in poultry, liver necrosis in pigs, white muscle disease in lambs and calves, reduced growth rates, reduced fertility, and paralysis. If the cardiac muscle is affected, treatment is ineffective. Selenium toxicity may occur as well. Toxicity is associated with reduced egg production and hatchability in poultry. Chicks that do hatch commonly have deformities. General hair loss occurs in pigs, loss of hair from the tail occurs in cattle, and loss of hair from the mane and tail occurs in horses. Complete loss of hooves may also occur. The toxicity causes a decline in appetite that may be severe enough to cause starvation and death. Selenium toxicity can occur in animals that graze plants that have accumulated selenium from selenium rich soils. Zinc (Zn) plays a role in lipid, protein, and carbohydrate metabolism and is necessary for bone and feather development, as well as immune system function. Deficiency symptoms include reduced growth rates, anorexia, poor hair or feather development and rough, thickened, or scaly skin, impaired bone formation, and delayed wound healing.

DIGESTION

"The physiologist who succeeds in penetrating deeper and deeper into the digestive canal becomes convinced that it consists of a number of chemical laboratories equipped with various mechanical devices."

—Ivan Pavlov (1849–1936)

Digestion: A Historical Perspective

In 1750, the French scientist Reaumur discovered that substances present in the digestive tract of birds were capable of liquefying meat. Reaumur had trained a bird to swallow sponges that were regurgitated shortly after ingesting. It was the liquid recovered from these sponges that Reaumur learned could dissolve meat. The substances contained within the liquid would later be isolated and defined. Following the studies of Reaumur, an Italian scientist by the name of Spallanzani conducted digestive studies using himself as the research subject. Spallanzani would ingest small linen bags containing food. He would later recover these bags and learned that digestion occurred without mastication, as the food would disappear from the bags despite being swallowed intact. The discovery of hydrochloric acid (Prout, 1824) and pepsin (Schwann, 1835) as important digestive compounds followed. At the same time William Beaumont (1833) carried out chemical and physiological studies of digestion on a patient under his care. The patient, a trapper injured following discharge of his gun, was left with a fistula (permanent opening) into his stomach. For the pay of $150 per year, the patient agreed to allow the experimentations of Beaumont to be performed. These studies were instrumental in revealing the nature of chemical and mechanical digestive processes.

Overview of Digestion

Most feedstuffs are consumed in forms that the contained nutrients are unavailable to the body for absorption, and the feed must be broken down to smaller molecules to be absorbed. The reduction in feed size and release of nutrients for absorption is the goal of digestion and involves mechanical and chemical processes. The digestive systems of animals has evolved with their diet and the type of digestive system that an animal has will ultimately determine the types of feed the animal can consume and utilize most efficiently. The variation in types of digestive systems allows animals to occupy different places within the food chain, which reduces competition for food sources. Animals are often classified according to their diet and their digestive system. Classifications according to the primary diets of animals results in the three broad dietary strategies of carnivorous, omnivorous, and herbivorous. Carnivores are flesh-eating animals, whose diet is composed primarily of non-plant material including meat, fish, or insects. Cats, birds of prey, snakes, and some fish such as sharks are all carnivores. Omnivores are animals whose diet consists of both plant and animal material. Dogs, domesticated poultry, pigs, most bears, and humans are all omnivorous species. Herbivores are animals that consume primarily plant material and have evolved specialized digestive anatomy or behaviors to maximize nutrients from highly fibrous plant matter. Herbivores include horses, cattle, rabbits, guinea pigs, llamas, alpacas, sheep, goats, elephants, and pandas.

 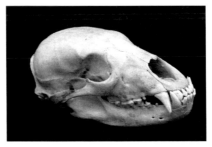 

■ **Fig. 4.2.1**   Animals may be classified as (left to right) carnivores, omnivores, or herbivores according to their diet. Dentition has evolved for each classification to maximize utilization of flesh and/or plant material. (© Michael J Thompson, 2009. Used under license of Shutterstock, Inc.)

The word carnivore is ambiguous and may reference taxonomy or dietary strategy, resulting in debate over the classification of dogs feeding behavior. As domesticates of the wolf, dogs are of the order Carnivora. This order is quite diverse and is not indicative of dietary strategy. Indeed, species within this order may demonstrate carnivory, omnivory, or in the case of the panda, herbivory.

Animal digestive systems are also classified anatomically and include nonruminants and ruminants. Nonruminant animals are considered simple stomached animals that have one glandular stomach, and include pigs, cats, dogs, and humans. The avian digestive system is a modified version of the nonruminant system. Herbivores including the horse, rabbit, and elephant also are nonruminant animals that are more commonly referred to as hindgut fermenters. Ruminants are animals that have a four compartmented stomach consisting of three forestomachs preceding the glandular stomach, which is analogous to that of a nonruminant. Ruminant animals include cattle, sheep, goats, and giraffes. Camels, llamas, and alpacas are other known ruminants; however, the stomachs of these animals have three compartments. Due to the differences in anatomy when compared to classic ruminants, the term pseudoruminant is sometimes applied. The ruminant system supports pregastric fermentation that is aided by rumination, the act of regurgitating, re-masticating, re-salivating, and re-swallowing the digesta. Other animals practice pregastric fermentation in the absence of rumination, which occurs in peccaries, kangaroos, and hippopotamus. For these animals, the stomach is enlarged (hippopotamus) or lengthened (kangaroo) and pouched to support fermentation.

Prehension, Mastication, and Swallowing

Digestion begins with prehension and mastication. The mouth, including the tongue, lips, and teeth are used to obtain, hold, grind, and mix food with saliva. While the horse, llama, and alpaca use their lips to obtain food; prehension also involves the tongue of cattle and sheep, and the snout of pigs. Dentition of mammals includes the front (incisors), canine, and cheek (molars) teeth. Mammals develop a set of deciduous teeth (baby or milk teeth) that are replaced with permanent teeth as the animal ages. The types of teeth that an animal possesses are specialized for prehension and mastication according to the diet. Carnivorous animals tear food, but grinding is negligible as the canines are adapted to tear flesh and the molars are pointed to crush bone. In contrast, herbivore dentition permits extensive mastication of the food. Both lateral and vertical jaw movement work to efficiently break down plant fibers. Whereas incisors are present in the upper and lower jaw of the horse, incisors of the upper jaw is absent in ruminant animals and instead a modified, dense, keratinized dental pad is found. In omnivorous species the incisors are used to bite off pieces of food and in rodents, the incisors grow indefinitely during the animal's life. This permits wearing of the teeth associated with gnawing on hard materials. Consider the incisors of beavers that are used in cutting wood. In male pigs, the canine teeth grow indefinitely during the animal's life and give rise to the tusks observed in mature boars. The molars of pigs have a flat surface that supports their grinding function.

The process of mastication can be controlled; however, the presence of food in the mouth is stimulatory. Salivary secretions, consisting of water, mucus, and enzymes, aid in the digestive process. Classical studies of Ivan Pavlov (1849-1936) revealed that the sight, smell, and presence of food in the mouth is not the only trigger for salivary secretions, but that the mere thought of food through a conditioned response can increase saliva as well. In humans 0.2 to 0.5 gallons of saliva is secreted daily, whereas in adult cattle over fifty gallons of saliva is secreted each day. Saliva lubricates ingested feed for swallowing and in ruminants, maintains fluid consistency and pH of the rumen and detoxifies plant toxins. In pigs, saliva also contributes the starch digesting enzyme amylase. Salivary amylase, however, is negligible in cattle and canines.

Mastication, following the secretion of saliva, aids in the formation of a bolus (rounded mass) of food. The tongue pushes this bolus to the back of the mouth and into the pharynx. The pharynx is the junction that opens to the esophagus and the trachea. When the animal swallows, a cartilaginous flap called the epiglottis covers the opening of the trachea to prevent food from entering the trachea and directs it instead to the esophagus. The esophagus transports food from the pharynx and down into the stomach by peristalsis (an involuntary, rhythmic response of smooth muscle). From this stage, the digestive processes will vary depending on the system.

Nonruminant (Simple Stomach)

Once swallowed, food enters the stomach. The shape and size of the stomach varies with species and is anatomically divided according to the type of glands and secretions produced. The cardiac and pyloric regions produce primarily mucus, which protects the stomach from acidic secretions of the fundic region. The fundic region secretes water, mucus, hydrochloric acid (HCl), and pepsinogen (the precursor to pepsin). These gastric secretions are stimulated by the sight, smell, and taste of food. The HCl secreted creates an acidic environment of a pH between 2 and 4. This acidic environment decreases the survival of foreign microorganisms that may have been ingested. In animals that consume carrion, an even lower pH is maintained as protection against pathogenic microorganisms consumed with decaying matter. The acidity of the stomach is also

necessary to denature proteins and activate pepsinogen to its active form pepsin to digest proteins to peptides. It is one of the few enzymes that functions in a highly acidic environment. Lipase is another enzyme found in the stomach. Similar to pepsin, its optimal activity is found in an acidic environment. Lipase begins the digestion of lipids by hydrolyzing fatty acids from the glycerol backbone. Chemical digestion by gastric secretions is accompanied by mechanical digestion accomplished by the churning actions of the smooth muscles in the stomach wall.

> Protein denaturation refers to the process of disrupting the quaternary, tertiary, and even secondary structures of ingested proteins. Because the structure of a protein determines its function, denaturation results in proteins losing their function. The primary structure remains. The digesting enzymes are necessary for release of amino acids from the primary sequence.

Chyme, the digestive material produced by the stomach, enters the small intestine through the pyloric sphincter. The pyloric sphincter regulates the passage of material into the intestine, controlling particle size and the rate of delivery. The small intestine can be subdivided into three smaller sections. The first is the duodenum, also known as the duodenal loop due to its anatomical positioning with respect to the pancreas. Bile produced by the liver and stored in the gallbladder is secreted into this section of the small intestine. Bile does not contain any digestive enzymes, but instead contains bile salts, which have detergent properties and are essential for the formation of emulsified particles of fatty acids, phospholipids, and cholesterol; which facilitates their absorption. Digestive secretions from the pancreas also enter the duodenum. The acidity of the chyme is adjusted by alkaline secretions of the pancreas that raise the pH in the duodenum to a more neutral 6.4. The increase in duodenal pH is necessary for the digestive enzymes of the pancreas to function. Enzymes including pancreatic lipase, which further digest lipids; chymotrypsin and trypsin, which hydrolyze proteins into shorter chain peptides; peptidases, which further hydrolyze peptides into amino acids; and pancreatic amylase, which breaks down starch, are of pancreatic origin. Further carbohydrate hydrolysis occurs by the actions of the disaccharidases: sucrase, maltase, and lactase, all of which are produced and secreted by the intestinal mucosa.

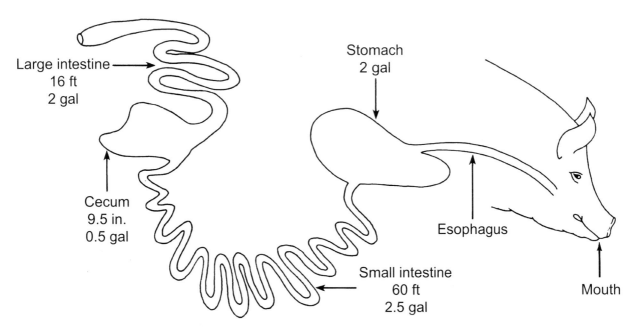

■ **Fig. 4.2.2** Digestive physiology of the pig, simple stomached nonruminants. (From *Animal Feeding and Nutrition*, 11th Edition, by Marshall H. Jurgens and Kristjan Bredgendahl. Copyright © 2012 by Kendall Hunt Publishing Company. Reprinted by permission.)

> Not all animals possess a gallbladder. Continual eaters, including horses and deer, lack the storage structure. For these animals, bile is released directly from the liver into the duodenum.

Due to the secretions of the liver and pancreas, a considerable amount of digestion occurs within the duodenum. Disruption of feed and release of individual nutrients is a prerequisite for absorption that occurs in the jejunum and upper ileum. The digestion of polysaccharides to monosaccharaides, lipids to fatty acids, and protein to amino acids is necessary for absorption to occur. The efficiency of nutrient absorption in the jejenum and upper ileum is enhanced by the occurrence of fingerlike projections called villi, which line the small intestine. These villi increase the surface area of the intestine, thereby increasing the rate of absorption of the nutrients. The villi are lined by a single layer of absorptive epithelial cells that have microvilli that face the lumen of the intestine. As with the villi, microvilli increase the absorptive surface area of the intestine. Each villus is innervated by the vascular and lymphatic system. Individual nutrients are absorbed by passive or active transport. Passive transport of nutrients in-

volves the diffusion of nutrients across the epithelial cell surface. It is a process that requires no additional energy input and is responsible for fatty acid absorption. In contrast, active transport is associated with nutrient transporters and is coupled to the expenditure of metabolic energy. Both amino acids and glucose require active transport. Once absorbed, nutrients that are water soluble will pass into the bloodstream or in the case of water insoluble nutrients such as fatty acids, the lymphatic system. Nutrients that enter the blood are transported by the portal vein to the liver and subsequently distributed throughout the body. Nutrients that are transported in the lymph will pass through the heart before being distributed throughout the body.

> During early postnatal life, the process of pinocytosis supports the engulfment of large particles by the cells lining the intestine. This process accounts for the absorption of immunoglobulins and other intact proteins from the colostrum of some species. This process is only active during the first few hours or days of life for species that rely on colostrum for maternal antibody transfer.

Digestive enzymes of the gastrointestinal tract

Name	Origin	Substrate	Product(s)
amylase	saliva pancreas	carbohydrates	maltose
lactase	small intestine	lactose	glucose + galactose
maltase	small intestine	maltose	glucose
sucrase	small intestine	sucorse	glucose + fructose
lipase	saliva stomach pancreas	triglycerides	diglycerides monoglycerides free fatty acids
pepsin	stomach	proteins	peptides
trypsin	pancreas	proteins peptides	peptides
chymotrypsin	pancreas	protein peptides	peptides
carboxypeptidase	pancreas	peptides	amino acids
aminopeptidase	small intestine	peptides	amino acids

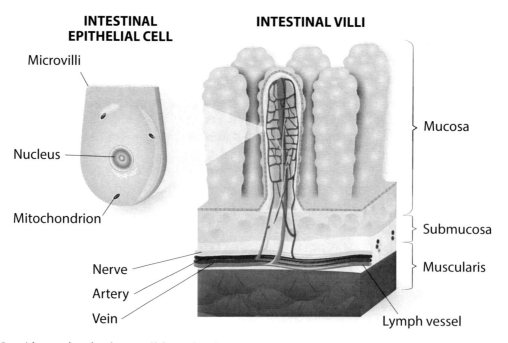

INTESTINAL EPITHELIAL CELL

Microvilli

Nucleus

Mitochondrion

Nerve

Artery

Vein

INTESTINAL VILLI

Mucosa

Submucosa

Muscularis

Lymph vessel

■ **Fig. 4.2.3** Absorption in the small intestine is maximized by the presence of villi, which increases intestinal surface area. Each villus is lined with epithelial cells that further increase surface area for absorptive capacity by the presence of microvilli that protrude from the epithelial cells. The villi are highly vascularized to receive nutrients into the blood system following their absorption. The villi and microvilli effectively increase the transfer rate of nutrients into the body.

The remaining material that has not been absorbed in the small intestine continues to the large intestine that consists of the cecum, colon, and rectum. An appreciable amount of water is absorbed in the large intestine. The cecum and/or colon also support a microbial population that produces amino acids and water soluble vitamins. In omnivores, nutrient production in the large intestine is dependent on dietary fiber. As dietary fiber escapes digestion in the small intestine, it serves as a substrate for nutrient synthesis by the microbial population. However, because the large intestine lacks the digestive secretions and absorptive surface area of the small intestine, water soluble nutrients produced may be of little value to the animal. In rabbits and humans, there is a finger-like extension at the apex of the cecum referred to as the appendix. The function of the appendix remains controversial, however, it is considered dispensable and removal is not associated with adverse outcome.

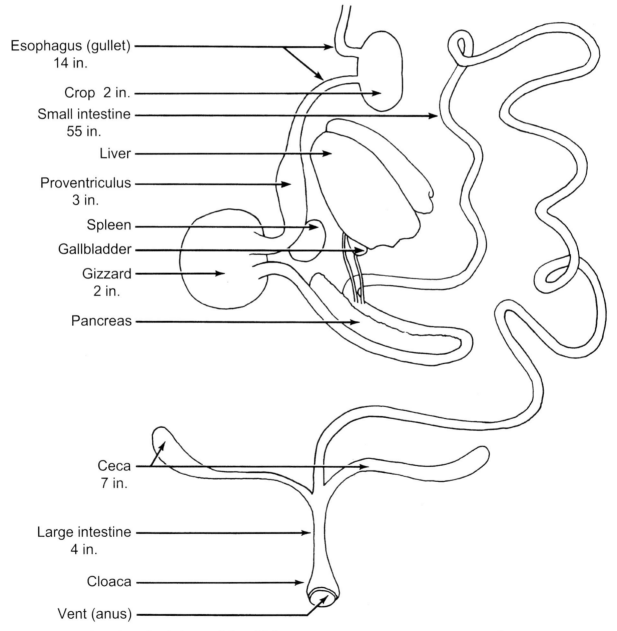

Esophagus (gullet)
14 in.

Crop 2 in.

Small intestine
55 in.

Liver

Proventriculus
3 in.

Spleen

Gallbladder

Gizzard
2 in.

Pancreas

Ceca
7 in.

Large intestine
4 in.

Cloaca

Vent (anus)

■ **Fig. 4.2.4** Digestive physiology of the chicken. (From *Animal Feeding and Nutrition,* 11th Edition, by Marshall H. Jurgens and Kristjan Bredgendahl. Copyright © 2012 by Kendall Hunt Publishing Company. Reprinted by permission.)

Avian Digestive System

The avian digestive system is similar to other non-ruminant systems, though there are some distinctions. First, birds lack teeth and for carnivorous birds, such as the eagle and owl, the beak and claws are important for tearing and reducing the size of the food so that it may be swallowed. Birds have rather poorly developed salivary glands and although the saliva contains traces of salivary amylase, carbohydrate digestion prior to the small intestine is minimal. Due to the anatomy of the mouth, as birds consume feed it is rapidly swallowed and enters an enlarged area of the esophagus called the crop. The crop serves as a holding site for the ingested feed. In select birds that consume high fiber diets, the crop contains an active microbial population to support plant digestion. The crop also is the site of crop milk production, which is a secretion from the crop that is used by some birds to feed their offspring. However, not all avian species have a crop, including some species of penguin.

The stomach of avians is modified. Whereas, chemical and mechanical digestion occurs within a single organ of most nonruminants, these digestive activities are separated into the proventriculus and ventriculus, respectively, of avians. In domesticated fowl, the proventriculus maintains an acidic pH of three and is the site of gastric secretions including hydrochloric acid and pepsin, analogous to other nonruminants. Due to the rapid passage rate of ingested feed through the proventriculus, negligible digestion actually takes place. Instead the feed passes rapidly to the ventriculus, otherwise referred to as the gizzard, which is a highly specialized organ of mechanical digestion. The gizzard has a very thick, muscular wall that acts to physically reduce particle size by muscular contractions and may contain grit (small stones or hard particles) that aids in the digestive process. Enzymes are not directly secreted into the gizzard, though chemical digestion is achieved by the actions of gastric secretions from the proventriculus. In raptors, the gastric secretions are of greater acidity, which supports the digestion of small mammals and corrosion of bone and other animal tissues. Indigestible matter is compacted into a pellet within the gizzard and is subsequently regurgitated.

The acidic digesta that enters into the small intestine is neutralized by pancreatic secretions. Similar to other nonruminants, the small intestine is the site of further digestion. With the exception of lactase, which

birds lack, analogous enzymes of carbohydrate, protein, and lipid digestion are released from the pancreas into the duodenum. Beyond the duodenum there is no anatomical distinction of a jejunum or ileum in avian species. The small intestine serves as the primary site of nutrient absorption; however, the rate of passage through the small intestine is more rapid than in other species. Digesta that passes from the small intestine enters the paired cecum (*pl.* ceca). Although a site of microbial fermentation, nutrients produced do not contribute significantly to the dietary requirements of most avians. Today, chickens are not fed diets containing much fiber, and therefore, the ceca are not well used. Ceca also function in water absorption, water-soluble vitamin synthesis, and the synthesis of vitamin K; however, minimal absorption of the vitamins occur from the large intestine The ceca open into the remainder of the large intestine via muscular ileocecal valves. In avian species, both feces and urine are eliminated through a common opening, the cloaca. In females, egg laying and copulation also occur here.

Hindgut Fermenters

Hindgut fermenters, including the rabbit, rat, horse, and elephant, are nonruminants capable of utilizing fiber due to active microbial populations present in the large intestine. Digestive function in the stomach and small intestine is similar to that of other nonruminants with noted exceptions in the horse including: 1) reduced muscular contractility of the stomach resulting in stratification of the ingesta, and 2) absence of a gall bladder and continual release of bile from the liver into the duodenum. The large intestine is well developed in hindgut fermenters and the microbial population of the digestive tract consists of bacteria, protozoa, and fungi similar to that of ruminant systems, however, the microbial species and percentage of each species differs between the digestive systems. Microbial populations digest fiber by the process of fermentation generating the volatile fatty acids (VFA) acetate, butyrate, and propionate, which following absorption may account for 30 percent of the energy needs in animals such as the horse. The microorganisms synthesize proteins also, but hindgut fermenters are unable to utilize these proteins because the large intestine lacks the enzymes and absorptive capacity for protein absorption. If a hindgut fermenter is fed a low quality diet, the animal may practice coprophagy, which is the act of eating feces. By eating the feces, the ani-

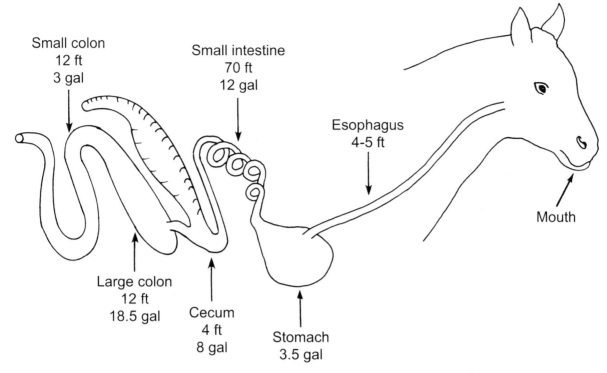

■ **Fig. 4.2.5** Digestive physiology of the horse. (From *Animal Feeding and Nutrition,* 11th Edition, by Marshall H. Jurgens and Kristjan Bredgendahl. Copyright © 2012 by Kendall Hunt Publishing Company. Reprinted by permission.)

mal can obtain the microbial protein that was eliminated. Coprophagy is a common practice among small herbivorous animals. The practice is generally not observed in horses unless insufficient protein is offered in their diet.

Rabbits perform a variation of coprophagy defined as cecotrophy. Rabbits produce two types of feces: one is a pellet type, which is what the owner will commonly see and the other is soft feces that the rabbit consumes. The material found in the soft feces has been fermented in the cecum and rapidly passed through the remainder of the intestine. Soft feces are produced once every twenty-four hours, usually at night. Rabbits eat the soft feces to obtain the protein produced by the microorganisms in the cecum. If rabbits on a rather poor diet are not allowed to practice cecotrophy, death may occur as a consequence of malnutrition due to protein and vitamin B deficiencies.

Ruminants (Pregastric Fermenters)

Ruminants are distinguished from other animals by pregastric fermentation that is aided by the act of rum-

ination, in which the animal regurgitates, remasticates, re-salivates, and re-swallows food that was previously consumed. Ruminants may spend greater than eight hours a day in the process of rumination. When the processes of grazing followed by rumination are considered in the digestive activity of a ruminant, there is nearly continuous motor and secretory activity. The act of rumination is orchestrated by the rhythmic contractions of the ruminant stomach. For advanced ruminants, the stomach is compartmentalized into the reticulum, rumen, omasum, and abomasum. The reticulum, rumen, and omasum are considered the forestomachs. The lining of the organs is non-glandular and does not produce enzymes or mucus, while the abomasum is glandular for these secretions.

The abomasum, or glandular compartment of the stomach, is often referred to as the true-stomach. The designation of true-stomach originates from the analogous roles of the abomasum and the stomach of nonruminants. Here acidic, enzymatic, and mucous secretions promote digestion of dietary or rumen derived compounds.

The complex stomach of the ruminant evolved for microbial fermentation to make use of fiber found within plant cell walls. This microbial digestion is similar to that of hindgut fermenters, though in the ruminant, fermentation is pregastric and the microbial population is responsive to changes in diet, environment, and season. Even differences in microbial populations have been identified in ruminants of different countries of origin. The forestomachs compartmentalize the microbial population of bacteria, protozoa, and fungi into a near neutral or slightly acidic environment (6.5 pH) that is separated from the more acidic abomasum (4 pH). The rumen is the primary site of microbial fermentation. The rumen itself may represent up to 20 percent of the body weight of the animal. The rumen environment is nearly void of oxygen, its contents contain approximately 75 percent water and 25 percent dry matter, the temperature is maintained between 101-103°F, and it is dark—conditions that support an active microbial population. The microbial populations are specialized for the digestion of carbohydrates, proteins, and lipids in an anaerobic environment. Because fermentation in ruminants is pregastric, all dietary carbohydrates undergo fermentation, including the starch and fiber components of the diet. In contrast to nonruminants where glucose is the end product of starch digestion, in the absence of oxygen VFA are produced and are the source of energy for the animal regardless of the carbohydrate type. The VFA are absorbed directly across the rumen, promoted by papillaeted epithelium that increases rumen surface area. In addition, a considerable amount of dietary protein is degraded by microbial species; however, the end products of protein digestion are used by other microbial species to synthesize amino acids. Rumen microorganisms also are capable of using simple nitrogen sources, such as urea, to synthesize amino acids. The net effect is that the microbial population produces the majority of amino acids absorbed, with only 30 percent of protein in a ruminant's diet bypassing microbial degradation. In the rumen, dietary lipid is also altered by the microbial population. Triglycerides are hydrolyzed to free fatty acids, which the microbes modify by changing the position of double bonds, generating trans fatty acids. The rumen microbial population is intolerant of increased dietary fat and for this reason dietary fat in ruminants diet is often less than 5 percent of the total diet.

For the newborn ruminant, the rumen is void of microorganisms and underdeveloped. The reticular or esophageal groove contracts to form a tube in newborn ruminants, allowing milk to bypass the underdeveloped reticulum and rumen and enter directly into the omasum. Direct contact with the dam and the environment inoculates the rumen, which begins functional development around six to eight weeks of age with the onset of eating solid feed.

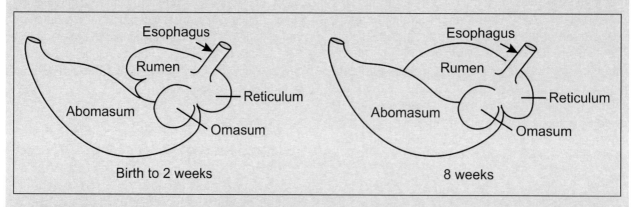

■ **Fig. 4.2.6**   At birth the abomasum accounts for the majority of volume of the ruminant digestive anatomy. The newborn calf is essentially a nonruminant with respect to digestive physiology and milk bypasses the underdeveloped rumen through the reticular groove.

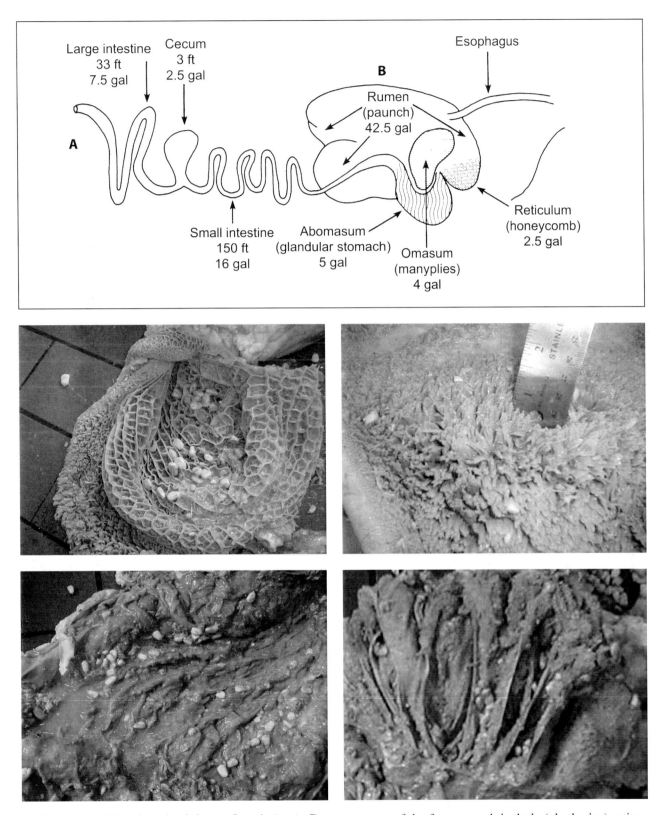

■ **Fig. 4.2.7** Digestive physiology of cattle (top). Compartments of the forestomach include (clockwise) reticulum, rumen, omasum, and abomasum (bottom). (*Top:* From *Animal Feeding and Nutrition,* 11th Edition, by Marshall H. Jurgens and Kristjan Bredgendahl. Copyright © 2012 by Kendall Hunt Publishing Company. Reprinted by permission. *Bottom:* Courtesy of Francis Fluharty, The Ohio State University.)

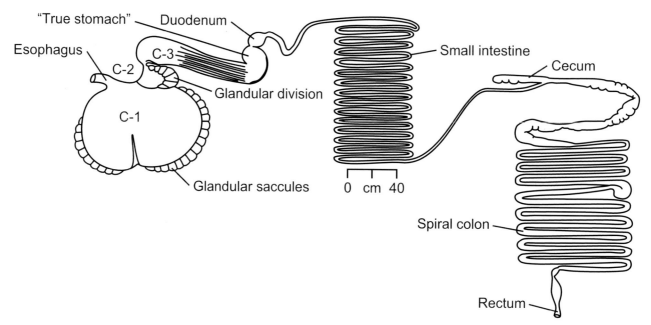

Fig. 4.2.8 Digestive physiology of camelids. (Courtesy of Francis Fluharty, The Ohio State University.)

Microorganisms of the rumen include bacteria, protozoa, and fungi. Numerous bacterial types are found within the rumen and may range from 15-50 billion/mL. Lesser concentrations of protozoa and fungi are found—20,000-500,000/mL. While the microorganisms consume feed particles, protozoa also consume bacteria and therefore may affect feed utilization. There are many different bacterial species alone present (over 6000 species described to date) and they are capable of degrading complex fibers (cellulolytic bacteria), simple sugars (amylolytic bacteria), and proteins (proteolytic bacteria).

The rumen is associated with the reticulum in what is commonly called the reticulo-rumen complex due to the absence of a physical barrier between the two. Microbial fermentation of the rumen is aided by the reticulum, which acts as a pacemaker of rumen contractions. The contractions begin in the reticulum and then spread to the rumen. These contractions act to mix the contents of the rumen and partition rumen contents into three primary layers based on specific gravity. The uppermost region of the rumen will be occupied by gas, the middle region by recently ingested roughages, and grain and digested roughage will partition to the bottom. The gaseous layer is generated by the fermentation process. Adult cattle may produce eight to thirteen gallons of gas per hour. The gasses are eliminated primarily through eructation. The middle, less dense matter of the rumen will be regurgitated with ruminal contractions. Only heavy, digested particles are able to pass through the rumen. The act of rumination aides in reducing larger ingested fiber particles, thereby increasing surface area for microbial access. Dense particulate matter that has settled toward the bottom of the rumen will be pushed toward the omasum. The function of the omasum is less understood, and the structure may be dispensable as normal function is observed in animals following surgical removal. The omasal leaflet structure suggests a role in the regulation of density and volume of the digesta entering the abomasum. The leaves of the omasum retain particulate matter, whereas fluids are allowed to pass rapidly to the abomasum. The abomasum is analogous to the stomach of nonruminants with a similar role in digestion; however, secretion of HCL is greater for ruminants. Greater production of HCl is needed to counteract the effects of the more alkaline rumen contents entering

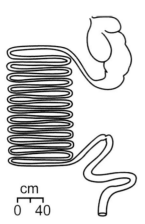

■ **Fig. 4.2.9** Digestive physiology of the hippopotamus

the stomach so that an acidic environment is established for gastric function. The abomasum receives the microbial proteins and modified lipids that were produced in the rumen. Digestion of microbial proteins occurs within the abomasum as does any dietary protein that by-passes the rumen. Any carbohydrate that is not fermented within the rumen will be digested by the enzymes of the pancreas and small intestine, as is the case of starch, or transferred to the large intestine for further hindgut fermentation, as is the case of fiber. Digestion that continues in the large intestine occurs through a microbial population similar to the species identified in the rumen. Quantitatively, VFA production is greatest from pregastric fermentation, which may fulfill more than 50 percent of the ruminants energy needs.

Pseudoruminants

Camelids, which include llamas and alpacas, are considered lesser ruminants or pseudoruminants due to the fact that they have two instead of the three forestomachs when compared to a traditionally defined ruminant. The digestive process is very similar to that of ruminants as they also regurgitate and re-masticate food; however, this process is more frequent in camelids resulting in greater passage rate of digesta through the forestomachs. The forestomach of camelids consists of three sections designated C1,

C2, and C3. The C1 compartment is divided into a cranial and caudal region and is the largest of the forestomachs. The smaller C2 region is attached to C1 and precedes the C3 compartment. Contractility of C1 and C2 is responsible for the passage of digesta from the forestomach for rumination or expulsion into C3. The C3 forestomach has a tube like appearance and terminates in the relatively small glandular stomach. The products of microbial digestion in camelids are the same as those produced in the ruminant, though the relative production of VFA is greater and camelids are able to absorb VFA more rapidly and completely. Also, the degree of fermentation that occurs in the large intestine is reduced in camelids relative to ruminants such as cattle.

Non-ruminating Pregastric Fermenters

Not all pregastric fermenters ruminate; in fact peccaries, kangaroos, and hippopotamus are categorized as non-ruminating pregastric fermenters. The common hippo is characterized by unique digestive physiology that includes two blind sacs connected to a larger chamber that precedes the glandular stomach. The large intestine is simple, consisting of a relatively short colon and marked by the absence of a cecum. Microbial activity has been documented in the forestomach, as well as bacterial fermentation and the production of VFA. Fermentation time in the hippo is considered slow relative to ruminants and the ability to digest fiber is reduced. Both observations have been attributed to the fact that the animal does not ruminate, which effectively reduces particle size for increased microbial digestion. Furthermore, in ruminants the flow of ingesta is bidirectional and outflow of digesta to the glandular stomach is a consequence of particle density as more dense particles settle in the rumen and are pushed ultimately to the abomasum. In the hippo, ingesta flow is unidirectional. Without the mixing of the forestomach contents, fluid exits more rapidly while particles are withheld in the folds of the blind sacs. For the kangaroo, the stomach is a tubular structure with internal banding that establishes a barrier between the microbial populations of the forestomach and the acidic environment of the hindstomach.

Genetics

"We've discovered the secret of life."
　　　　　　　—Francis Crick (1916–2004)

This chapter introduces the basic composition and structure of cells, and the functions of the organelles within the cell. You will learn to distinguish between prokaryotic and eukaryotic cells, as well as animal and plant cells. You will develop a fundamental understanding of cellular function and the processes of proliferation and differentiation that lead to the various cell types that coordinate the life processes of an animal. The role of genes as the determinants of the animal's visible characteristics, and the ways in which variation may arise through the expression of traits are highlighted. You will learn how the study of genetics can be applied to an animal population to direct change and the different types of mating systems that are used to obtain a desirable characteristic within an animal population. Lastly, the impact of our knowledge of the cell and the genome contained within toward an understanding of biotechnology, including genetic engineering, cloning and xenotransplantation, are explored.

CELL BIOLOGY

"The key to every biological problem must finally be sought in the cell"
　　　　　　　—Edmund Beehcer Wilson (1856–1939)

Cellular Biology: A Historical Perspective

In 1655 Robert Hooke made one of history's most important microscopical observations. Using a compound microscope, Hooke observed thin slices of cork and noted the appearance of irregular pores. He termed these pores "cells" after the cells he commonly observed in monasteries. Hooke's observations were actually the cell wall of plant cells.

Almost two hundred years following Hooke's initial observations of cells, Theodore Schwann and Jacob Schleiden proposed the cell theory in 1838. The theory stated that all plant and animal tissues were actually aggregates of individual cells and thus the nucleated cell is the most basic unit of structure and function in living organisms.

Modern Cell Theory

Schwann and Schleiden's early observations have evolved into the Modern Cell Theory that states:

1. All living things are composed of cells.
2. The cell is the fundamental unit of structure and function in all living things.
3. All cells come from pre-existing cells.

Cell Classification

Living systems are comprised of more than one cell type and cells are classified as prokaryote and eukaryote.

Organisms, such as bacteria, are prokaryote, single celled organisms. In single celled organisms, an individual cell carries out all of the functions required for the organism's independent survival. In contrast, complex organisms are multicellular and cannot survive without the coordinated actions of various cells. These cells are eukaryotic. Among eukaryotic cells, animal and plant cells can be distinguished by the presence of a cell wall in plant cells. The cell wall surrounds the cell and maintains the

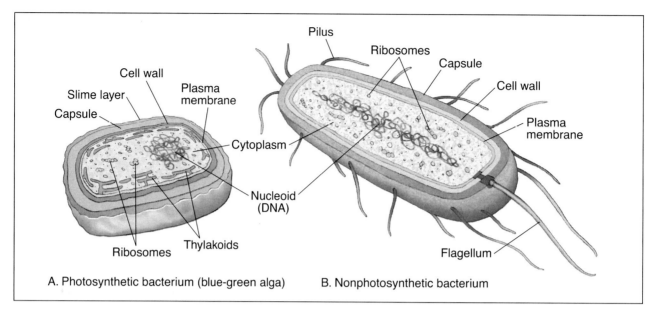

■ **Fig. 5.1.1**   Structure and organization of prokaryotic bacteria (© Kendall/Hunt Publishing Company.)

cell's shape. In animal cells, the cytoskeleton helps maintain shape. Cells range in size from 1 to 100 micrometers and are not visible to the naked eye, thus detailed observation of cells has relied on microscopy.

Cell Structure and Function

In complex animals there are greater than two hundred types of cells known to exist and provide structure and function. This is accomplished by coordinated actions whereby groups of cells regulate a specific function as tissues. In turn, a functional group of associated tissues comprise an organ and organs work together in systems. These specialized cells that function within the hierarchy of living systems all arise from a single cell, a fertilized egg, also known as a zygote.

Shortly following fertilization, the zygote begins to replicate and divide through the process of mitosis. This phase is called cleavage and results in the formation of many smaller cells termed blastomeres, each with its own nucleus. After numerous divisions, the cluster of cells forms a blastocyst. The blastocyst is comprised of an inner cell mass that will give rise to the embryo and an outer cell layer that develops into the placenta. The inner cell mass of the blastocyst lies at the heart of controversial

embryonic stem cells. Stem cells are relatively unspecialized cells that can proliferate through mitotic division giving rise to two identical daughter cells. The daughter cells may remain stem cells or differentiate, beginning the transformation into the various types of cells that comprise the body. Because there is more than one potential cell type that stem cells may differentiate into, they are referred to as being pluripotent.

In vertebrates, this process of differentiation begins during the gastrulation stage of early embryonic development. During this stage, the cells of the blastocyst are rearranged to form a three-cell layered embryo with a primitive gut, called a gastrula. These three layers are embryonic tissues that are called the embryonic germ layers. The ectoderm layer is found on the outside of the gastrula, while the endoderm lines the embryonic digestive tract, and the mesoderm fills the space between these two layers. Every nucleus of every cell in an organism contains the same set of genes. During cell differentiation, certain genes are activated, while others are inactivated. This process is quite intricate and reliant on the cellular environment. Differentiation ultimately allows each cell to develop specific structures and perform certain functions. The various cell types that result from differentiation must also replicate through mitosis. This proliferation of cells is necessary for the organism to grow and tissues and

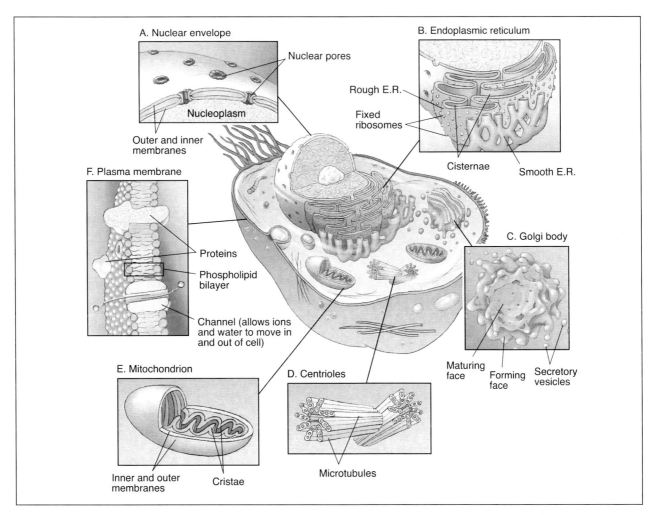

■ **Fig. 5.1.2**   Structures and organization of a eukaryotic animal cell. There are over 200 cell types in the animal body. The above diagram of a cell does not depict the distinguishing shapes or structures of the various cell types but represent common composition and function shared across animal cells. The plasma membrane is a lipid bilayer, consisting of a double layer of phospholipids embedded with various proteins. This is the outer boundary of the cell and functions as a selective barrier, allowing passage of oxygen, nutrients, and wastes in and out of the cell. The cytoplasm contains the cytosol (semi-fluid) and organelles. The nuclear envelope is a double membrane that encloses the nucleus and separates it from the cytoplasm. The nucleolus is a nonmembranous organelle involved in the production of ribosomes that are essential for protein synthesis. A cell may have multiple nucleoli. The nucleus contains the cell's chromosomal DNA. This chromosomal DNA is composed of chromatin, which is a complex of histone proteins and DNA. Chromatin allows for the packaging of DNA into the nucleus. Adjacent to the nucleus is the endoplasmic reticulum. Rough endoplasmic reticulum has ribosomes covering the outer surface of the membrane that function in protein synthesis. Smooth endoplasmic reticulum lacks ribosomes on its outer membrane surface. The smooth ER functions in a variety of metabolic processes including the synthesis of lipids, metabolism of carbohydrates, and detoxification of drugs and poisons. The endoplasmic reticulum works in concert with the golgi. In the golgi complex, products of the endoplasmic reticulum are modified, sorted, and then packaged for export from the cell. Mitochondria are the power plants of eukaryotic cells, using cellular respiration to convert oxygen and food/feed molecules into ATP, thereby providing energy to the cell. Centrioles are usually found in pairs in animal cells and are involved in cell division. Additional cellular organelles include lysosomes, which are digestive organelles where macromolecules are hydrolyzed and peroxisomes, which are membrane-bound vesicles that contain oxidative enzymes that generate and then degrade hydrogen peroxide and produce heat during the oxidative process. (© Kendall/Hunt Publishing Company.)

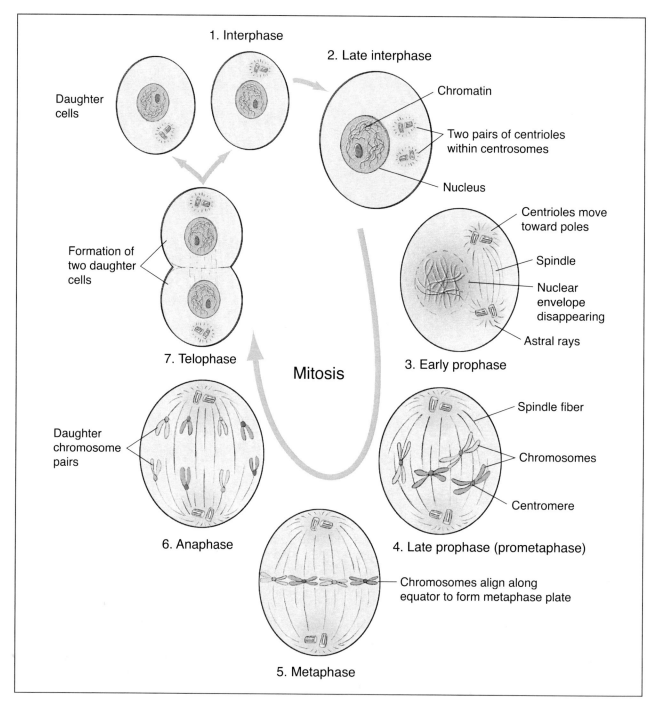

■ **Fig. 5.1.3**   The process of mitosis is a complex and highly regulated event. During interphase, the cell increases in size and the DNA of the chromosomes is replicated. The centrosome is then duplicated. The chromatin fibers condense into distinguishable chromosomes during prophase. These chromosomes consist of two identical sister chromatids. The miotic spindle begins to form during this phase. It is composed of the two centrosomes and the microtubules that extend from each. Centrosomes are nonmembranous organelles that function during the cell cycle to organize the cell's microtubules. As the microtubules lengthen, the centrosomes are propelled away from one another. Prometaphase begins with the breakdown of the nuclear envelope. With this breakdown, the microtubules extend toward the center of the cell. The microtubules then attach to the kinetochore (a structure of proteins located where the two sister chromatids join). Metaphase is the longest phase of mitosis, lasting approximately twenty minutes. The centrosomes are now located at the opposite ends

(continued)

of the cell. The chromosomes are aligned on the metaphase plate, located midway between the spindle poles. The kinetochore microtubules attach the sister chromatids to the opposite poles of the spindle. Anaphase begins when the two sister chromatids of each pair separate, forming two daughter chromosomes. These chromosomes begin to move towards opposite ends of the cell as the microtubules attached to their kinetochore shorten. The nonkinetochore microtubules begin to lengthen, causing the cell to elongate. At the end of anaphase, each end of the cell has a complete set of chromosomes. Anaphase is the shortest stage of mitosis, lasting only a few minutes. During telophase, the two sets of daughter chromosomes arrive at the poles of the cell and decondense. A new nuclear envelope then begins to reassemble around each set of chromosomes from the fragments of the parent cell's envelope. The end result is the formation of two genetically identical nuclei. This is the last phase of mitosis. Following mitosis, a cleavage furrow is formed, which divides the cytoplasm in two. This division results in the formation of two daughter cells, with each cell containing one nucleus. (© Kendall/Hunt Publishing Company.)

organs to maintain themselves, or heal and repair. Proliferation is a controlled cellular event; uncontrolled cellular proliferation is associated with disease states such as cancer.

Because the cell is the fundamental unit of life, the properties of the cell are synonymous with those of life. Succession of life is reliant on the whole organisms ability to grow, reproduce, and maintain physiological processes (digestion, absorption, metabolism, secretion, etc.), properties which are orchestrated at the cellular level and regulated by the chemical composition of the cell.

Cell Composition

Water
All cells have a similar chemical composition. The largest constituent of the cell is water, representing 60–65% of cell volume and the majority of water within a living organism is found within the cell (approximately 40% of body weight). Water acts as a solvent and is an essential constituent of many cellular processes.

Lipids
Lipids are the primary components of the cell membrane and are arranged in a double layer, called a bilayer. The lipid bilayer defines the outer boundaries of the cell and serves as an impermeable barrier to the passage of most water-soluble molecules. The lipid bilayer that defines the cell is considered fluid as the lipids move laterally through the membrane. The membrane also contains proteins, carbohydrates, and other molecules. These dynamic property of the cell membrane is described by the Fluid Mosaic Model.

Proteins
Many cellular processes are ultimately carried out by proteins. The three dimensional conformation of the protein determines the proteins biological function to regulate shape and strength of tissues, movement in muscle cells, and cellular metabolism. Proteins of cellular membranes are important mediators of cell to cell communication as protein hormones, cell receptors, and transporters. The instructions for the production of cellular proteins are contained within the cells genetic material or deoxyribonucleic acid (DNA).

Deoxyribonucleic acid exists as two helical strands. Each of the two helical strands that compose DNA consists of three chemical components: phosphate, deoxyribose (a five carbon sugar), and nitrogenous bases assembled into nucleotides. There are four nitrogenous bases: adenine, guanine, cytosine, and thymine. Adenine and guanine are purines, which are a class of chemicals that have a double-ring structure. Cytosine and thymine belong to a class of chemicals with a single-ring structure called pyrimidines. Nucleotides are often referred to by the first letter of the name of the base they contain; A (adenosine), G (guanosine), C (cytidine), or T (thymidine) and form phosphodiester bonds between the deoxyribose sugar of one nucleotide and the phosphate of a second nucleotide. The two strands of DNA are held together by hydrogen bonds between the nitrogenous bases present on each chain. According to Chargaff's rules there are only two possible pairs of bases, that is: adenine base pairs with thymine, and guanine base pairs with cytosine. The two bases in a pair are said to be complementary, meaning that they will only bond to each other and not to either of the other two bases.

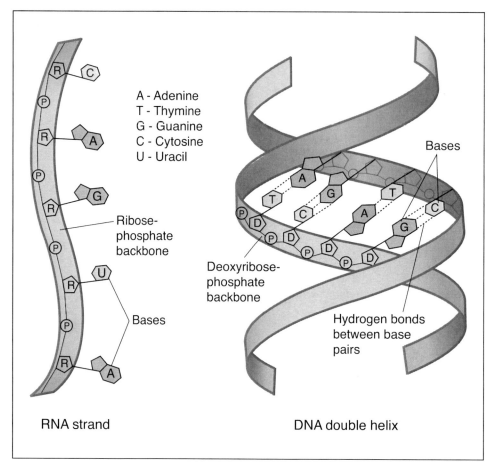

A - Adenine
T - Thymine
G - Guanine
C - Cytosine
U - Uracil

Ribose-
phosphate
backbone

Bases

Deoxyribose-
phosphate
backbone

Bases

Hydrogen bonds
between base
pairs

RNA strand

DNA double helix

■ **Fig. 5.1.4**   General features of RNA and DNA. Although RNA contains the base uracil instead of thymine, they follow the same base pairing rules as DNA. Ribonucleic acid molecules are single standed in many of their biological roles, however, RNA molecules can base pair with other RNA or DNA molecules for regulatory functions.

Deoxyribonucleic acid contains sequences of nucleotides that serve as the precursors of proteins known as genes. Various combinations of nucleotides occur and underlie the diversity of living organisms. The process of DNA serving as the instructions for the synthesis of proteins involves intermediate ribonucleic acids (RNA). There are three main RNA molecules involved in protein synthesis: messenger RNA (mRNA), ribosomal RNA (rRNA), and transfer RNA (tRNA).

While DNA is contained primarily within a cell's nucleus, protein synthesis occurs in the endoplasmic reticulum. The process of getting the message of which protein is to be produced to the endoplasmic reticulum relies on the transfer of information into mRNA. Messenger RNA is transcribed from DNA using the same complementary base pairing language

observed with double stranded DNA; however, in RNA thymine is absent and is replaced by the base uracil. The mRNA that is transcribed is processed and then translated by ribosomes of the endoplasmic reticulum. Translation is the act of decoding the mRNA molecule through directing the synthesis of an amino acid chain that forms a protein. Decoding mRNA involves sequences of triplicate nucleotides termed codons. Codons code for an amino acid. Successful translation requires rRNA as the site of translation. Ribosomal RNA complexes with proteins and must recognize the mRNA and complete the bond formation between the amino acids that are coded for in order for protein synthesis to occur. Amino acids are delivered to the site of translation by tRNA. Transfer RNA is a cloverleaf structure that contains a three base codon that is complementary to

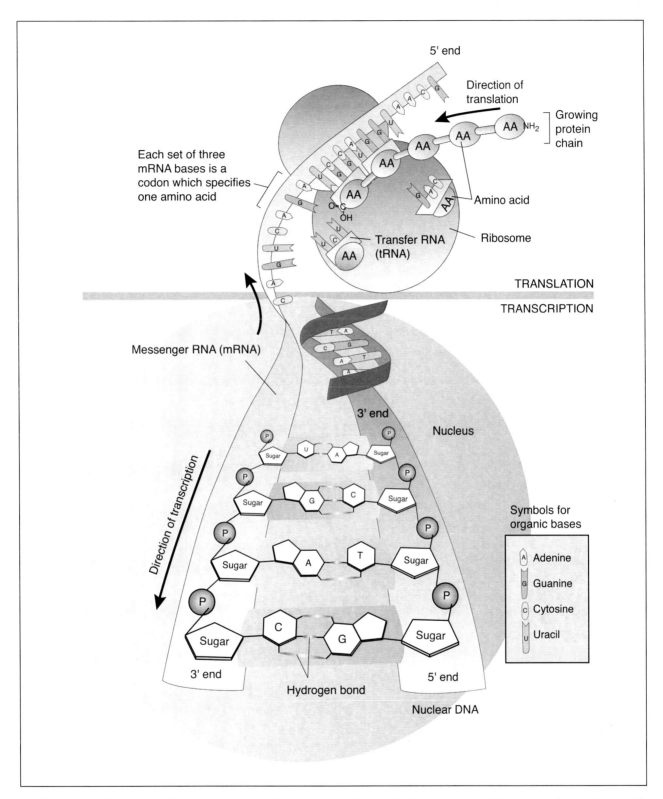

■ Fig. 5.1.5 Processes of transcription and translation. Transcription occurs within the nucleus of the cell. Recently transcribed mRNA leaves the nucleus where it is "fixed" to the ribosome by complementary base pairing between the mRNA and rRNA. Translation occurs at the site of the ribosome and requires delivery of amino acids to this site by tRNA. Peptide bonds are formed between the amino acids through the actions of the ribosome. The resulting peptide chain is released when the ribosome encounters a nucleotide sequence that does not specify a tRNA. (© Kendall/Hunt Publishing Company.)

the mRNA, known as the anti-codon, and an amino acid. Amino acids may be coded by more than one codon, and thus, there are many tRNA molecules.

Carbohydrates

Carbohydrates, which consist primarily of glucose, are major components of the plant cell wall that contribute to strength. They also may be stored in a form known as starch and used to meet the energy needs of the cell. An analogous form of carbohydrate storage, referred to as glycogen, is abundant in animal cells including liver and skeletal muscle cells.

Inorganic Molecules

Inorganic substances are found in the cell in various quantities and are required for a variety of functions in the body as discussed in chapter 4. Calcium, magnesium, potassium, sodium, chloride, and phosphates are required in specific concentrations for normal nerve function and communication, as well as muscle contraction. Calcium and phosphate are both required components of bone tissue and are necessary for bone maintenance. Some minerals also act as catalysts for enzymes. For example, sodium and potassium activate enzymes in carbohydrate metabolism.

GENETICS

"Heredity provides for the modification of its own machinery."

—*James Baldwin (1861–1934)*

■ **Fig. 5.2.1** Domesticated from the wolf, selective breeding has transformed canines into the over four hundred specialized breeds reported today. (© Paunovic, 2009. Under license from Shutterstock, Inc.)

Genetics: A Historical Perspective

The science of genetics arose from the desire of plant and animal breeders to attain a clear understanding of the inheritance of economically important characteristics in their orchards, fields, and flocks. In 1843, Gregor Mendel was recruited to an Augustinian monastery in Moravia to work on the problem of understanding the nature of variation in fruit trees. While fruit trees turned out to be too large and slow growing for this type of study, Mendel did find that pea plants were ideal for his experiments. Today he is famous for his study on these plants and is credited with the discovery of the mechanism of inheritance. It was previously believed that inheritance was the result of the mixing of blood or some other continuous substance, contributed by each parent of an individual. In 1866, after many years of study, Mendel proved that this was not the mode of inheritance; but instead, he found that for each characteristic he studied, an individual carried two particles, which he referred to as factors. He found that one factor is inherited from the male parent and the other factor is inherited from the female parent. These factors kept their individuality in the offspring, that is, they did not blend together. He also found that when offspring produced gametes for reproduction, each of the gametes produced would contain only one of the two factors it had inherited from its parent. Today these factors are referred to as genes. The term was first used by Danish botanist Wilhelm Johannsen in 1909, and was derived from the word *pangenesis* coined by Darwin. *Pangenesis* comes from the Greek words *pan* (meaning "whole" or "encompassing") and *genesis* ("birth") or *genos* ("origin").

The Study of Genes

The study of genes and their heredity is referred to as genetics. When studies focus on the genetic materials RNA and DNA, their structure, and how they control metabolic processes within cells, the term molecular genetics is used. Genetics is a very important field of study, as the knowledge of genes and their actions is necessary to bring about genetic change. Life relies on the transfer of genetic information from parent to offspring. This genetic transfer is referred to as inheritance, and takes place at the time of conception. In animal breeding and production, the goal is to pro-duce animals that excel for desired traits, while at the same time eliminating and/or reducing the occurrence of undesirable traits. Applied genetics, commonly referred to as animal breeding, is the science that aids in the goal of selecting and breeding animals to improve the population. Genetic improvement is the principle goal of nearly all animal breeders. Modern day animals have changed remarkably from their wild ancestors due to the practices of breeding and selection.

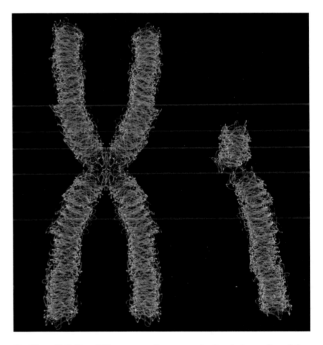

■ **Fig. 5.2.2**   The sex of a zygote is determined by chance. Every somatic cell of an organism contains a pair of sex chromosomes. In mammalian species, the female pair of sex chromosomes is XX. Males on the other hand have one X chromosome like that of the female, but the second chromosome is a Y chromosome. Females (XX) are referred to as homogametic since both of their sex chromosomes are the same, while males are referred to as heterogametic since their sex chromosomes differ. In poultry, the sexes are reversed, such that the heterogametic sex is the female and the homogametic sex is the male. As opposed to the XX and XY designations used in mammalian species, ZZ is used to designate males, while ZW is used to designate females. Since the avian female is the heterogametic sex, the female will determine the sex of the offspring, whereas in mammals, the male is the one who determines the sex. (© MichaelTaylor, 2009. Under license from Shutterstock, Inc.)

Genes Are Organized in the Genome

The complete genetic material of an animal is referred to as its genome. In most plants and animals, the somatic cells contain two copies of the organism's genome. These types of organisms are referred to as diploid (2n). On the other hand, organisms that have only one copy of their genome are called haploid (n). Haploid organisms tend to be rather simplistic and include organisms such as fungi, algae, and bacteria.

The genome of an organism serves as the basic blueprint for the organism and contains both the genes that code for protein and non-coding regions that do not give rise to proteins but may serve a regulatory function. The information the genome carries essentially controls all of the biochemical life processes within the body. Deoxyribonucleic acid of the genome is segmented into, long, threadlike strands called chromosomes that are located in the nucleus of every cell. Each gene in an organism will occupy a specific location on a particular chromosome, called its locus.

Chromosomes are present in pairs, with one chromosome of the pair being inherited from the female parent and the other chromosome being inherited from the male parent, which is why somatic cells are diploid. Since chromosomes occur in pairs in somatic cells, then so do genes as well, with paired chromosomes having corresponding gene loci. Autosomes (non-sex chromosomes) are referred to as homologs or homologous chromosomes. Sex chromosomes are an exception to the general pattern of homologous chromosomes where males are characterized by the XY chromosomes in mammals and females the ZW chromosomes in avians.

The number of chromosomes an animal has is dependent on its species. For example, cattle have sixty chromosomes (fifty-eight autosomes and two sex chromosomes), or thirty chromosome pairs. While the number of chromosomes is known for a species, the total number of genes on a pair of chromosomes is not known but is estimated at several hundred to a few thousand. The number of chromosomes for a given species, the overall size of their genome, and the estimated number of genes present reveal nothing about

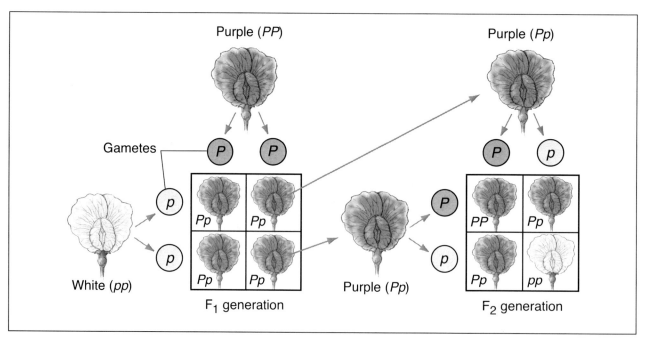

■ **Fig. 5.2.3** Mendel crossed contrasting pea varieties and recorded his observations. Crossing peas with white flowers and peas with purple flowers (referred to as the parental crosses) yielded hybrid offspring (F$_1$ generation) with purple flowers. When the F$_1$ generation were allowed to self pollinate to produce the F$_2$ generation, the white flowers that were absent in the F$_1$ generation reappeared in the F$_2$ generation. (© Kendall/Hunt Publishing Company.)

the complexity of an organism. For example, humans are argued to be one of the most intellectually complex animals, possessing 3.5 billion nucleotide base pairs fixed over forty-six chromosomes. However, marbled lung fish have the largest genome at nearly one hundred thirty-three billion nucleotide base pairs fixed across only thirty-four chromosomes.

The various forms of a given gene at a locus on corresponding homologous chromosomes are termed alleles. While alleles for a particular gene do affect the same trait, different proteins may be produced, and therefore differences in the way the trait is expressed may occur. Alleles of a gene are generally represented by a letter. To differentiate between the different allelic forms of a gene, one form is often noted as an upper case letter, and the other as a lower case letter (i.e., A and a). When the alleles for a given gene are the same, it is said to be homozygous. On the other hand, when the alleles at a given locus differ or contrast one another they are referred to as heterozygous. Using the above lettering system, a gene with contrasting alleles would be represented as Aa, and a gene with the same alleles would be designated as AA or aa. The combination of alleles for a given gene is referred to as the genotype for that locus. In reference to an organism, the term genotype refers to the entire listing of the specific genes (and their alleles) carried on the chromosomes.

Inheritance

The transfer of chromosomes and their alleles from parents to their offspring is the basis of inheritance. Each parent produces reproductive cells called gametes, which are haploid and thus contain only one allele for each gene. Which one of the two parental alleles that ends up in a gamete is determined by the two laws of inheritance derived by Mendel. The first law is the principle of segregation, stating that alleles separate so that only one randomly chosen allele is found in any particular gamete. Mendel's second law is the principle of independent assortment, stating that during the formation of gametes, separation of a pair of genes is independent of the separation of other gene pairs. These two fundamental principles of heredity were conjectured from Mendel's early studies that involved crossing of pea plants.

Mendel's studies disproved the widely accepted notion that inheritance was a result of the blending

Species	Number of Chromosomes	
Humans	46	(23 pairs)
Cattle	60	(30 pairs)
Horses	64	(32 pairs)
Swine	38	(19 pairs)
Sheep	54	(27 pairs)
Chickens	78	(39 pairs)
Goats	60	(30 pairs)
Llamas	74	(37 pairs)
Dogs	78	(39 pairs)
Cats	38	(19 pairs)
Bison	60	(30 pairs)

of traits from parents. Instead, he proposed that parents contained two copies of the factors responsible for inheritance and these factors split, i.e. segregated, during meiosis and the formation of gametes.

Variation in chromosome number may occur due to the many processes that are required to take place for inheritance to be possible. Polyploidy is one of the possible variations and occurs when an individual inherits more than two full sets of chromosomes (one from each parent). This condition is uncommon in vertebrates, but does occur frequently in plants. Another potential variation in chromosome number is aneuploidy, in which an organism has a chromosome number that is not an exact multiple of the haploid number. Monosomy is a form of aneuploidy in which an organism is missing one chromosome from an otherwise diploid cell (2n-1). Similarly, trisomy is another form of aneuploidy where an additional chromosome is present in an otherwise diploid cell (2n+1). Monosomy is often a lethal condition, while trisomy may result in normal, or slightly abnormal development, but usually not death. Down's syndrome is a common example of trisomy in humans and is also referred to as trisomy 21, as it occurs at chromosome number 21. Trisomy is most commonly due to failure of the chromosomes to separate in meiosis, resulting in a gamete carrying two copies of that chromosome. When this gamete combines with another during fertilization, three copies of that chromosome will then be present.

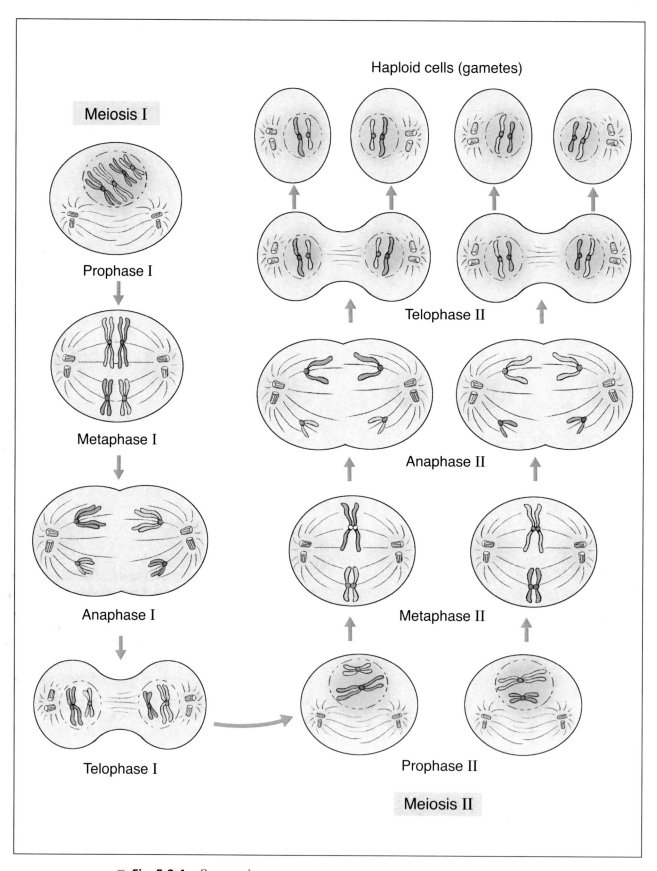

Haploid cells (gametes)

Meiosis I

Prophase I

Metaphase I

Anaphase I

Telophase I

Prophase II

Meiosis II

Metaphase II

Anaphase II

Telophase II

■ **Fig. 5.2.4** See caption on next page. (© Kendall/Hunt Publishing Company.)

■ **Fig. 5.2.4** During meiosis, the chromosome set is reduced from diploid to haploid. This reduction is to compensate for the doubling of the chromosome set that occurs during fertilization. Meiosis is similar to mitosis as it is also preceded by the replication of chromosomes. However, meiosis is different in that this single replication is followed by two consecutive cell divisions referred to as meiosis I and meiosis II, which give rise to haploid cells, each with only half as many chromosomes as the parent cell. During the first phase of meiosis, prophase I, the chromosomes condense. Homologous chromosomes align precisely and are loosely paired. This phase generally occupies 90% of the time required for meiosis. At the end of prophase I, the homologous pairs begin to move towards the metaphase plate. During metaphase I the chromosome pairs align along the metaphase plate. Anaphase I begins with the movement of homologous chromosomes to opposite poles. The sister chromatids (copies of DNA) will remain attached at the centromere as they move together toward the same pole. During telophase I, each chromosome is composed of sister chromatids and each half of the cell contains a haploid set of chromosomes (i.e., there is one chromosome, but two chromatids). Cytokinesis occurs simultaneously with telophase I and forms two haploid daughter cells by dividing the cell in half, marking the end of meiosis I. Meiosis II begins with prophase II, during which time the spindle apparatus forms and the chromatids begin moving towards the metaphase II plate. During metaphase II the chromosomes are lined up along the metaphase plate as in mitosis. In anaphase II, the sister chromatids separate as they more toward opposite poles. During telophase II, the nuclei of the two cells form and the chromosomes begin to decondense. Cytokinesis follows directly, resulting in the production of four haploid daughter cells from the meiotic division of one parent cell.

Hybrid animals may show variation in diploid chromosome number that is intermediate between the parent generations. Species are distinguished by genotypic diversity and hybrids are the result of matings between two species. Some argue that genetic and structural incompatibilities limit the success of hybrids, while others argue the hybridization introduces new genetic material that could lead to the development of new, better adapted species. Hybrids are rare in nature, owing to separate range areas that limit similar species from intersecting. However, there are a number of hybrid animals as the result of human breeding intervention. The most commonly cited hybrid is the mule. A cross between the horse and donkey, the mules genome has 63 chromosome (diploid): 32 chromosomes contributed from the horse dam and 31 chromosomes contributed from the donkey sire. Genetic separation of the horse and donkey results in in-fertility in the hybrid mule as homology between parentally inherited chromosomes is a prerequisite to gamete formation.

Expression of Traits

According to Mendel's second law, the chromosomes of a gamete are a result of any random combination of maternal and paternal chromosomes from the parent generation, contributing to the genetic variability witnessed in a population. Following fertilization and the combining of the alleles from each gamete, the relationship between the alleles will determine which gene is expressed, i.e. the allele inherited with the paternal or maternal chromosome. The ways in which alleles are expressed are of great interest to animal breeders, as they play a major role in determining the visible or measurable characteristics or traits that an animal demonstrates. Dominant alleles are alleles that will be expressed over recessive alleles. Therefore, when a dominant allele is present, the characteristic it codes for will be expressed. On the other hand, a recessive allele will only be expressed when the dominant allele for that gene is not present.

Complete dominance is the form of dominance in which the trait expressed is the same for both the homozygous and heterozygous genotypes. The traits expressed for the homozygous recessive geno-type is different from that of the other two geno-types. An example of this type of dominance is black and red coat color in cattle, where the allele that codes for a black coat is dominant, and the al-lele that codes for a red coat is recessive. Therefore, if an animal receives a dominant allele from each parent, it will have a black coat; likewise, an animal that receives both a dominant allele from one parent and a recessive allele from the other parent will also have a black coat. Only when the animal inherits a recessive allele from each parent will it have a red coat. This is best illustrated using a Punnett square.

Dominant alleles are argued to offer a competi-tive advantage in the offspring, whereas recessive alleles are considered deleterious to the animals survival. Several genetic disorders are associated with homozygous inheritance of recessive alleles. Often recessive alleles are associated with enzyme function. In many instances, only one copy of an enzyme is needed for normal function. In the case of a homozygous recessive genotype, there is either complete loss-of enzyme function or reduction in enzyme efficiency due to an improperly formed en-zyme. Because enzymes are specific for reactions within the body, the loss of the enzyme can be detrimental and lethal. Chondrodysplasia in sheep, otherwise known as spider lamb, is an inherited condition in the offspring of sheep that carry a mu-tation in a cell receptor with enzymatic activity. When the gene is inherited as homozygous reces-sive, the offspring develop a lethal disorder of the muscoskeletal system. Not all recessive alleles are lethal with homozygous genotypes. For example, the red coat color of Angus cattle is a result of a re-cessive gene. Desired and selected for by Red Angus cattle breeders, the gene responsible for red coat color is not associated with lethality in the breed. Similarly, not all dominant traits are benefi-cial. The Manx breed of cat is characterized as be-ing tailless. The tailless trait is a result of the short-ening of the spine associated with a dominant gene and is found in cats with the heterozygous geno-type. The homozygous dominant genotype, how-ever, is considered an embryonic lethal genotype; whereas, homozygous recessive animals are born with tails. In mating two Manx cats, one can predict that 25% of the offspring would have tails, 50% of the offspring would be tailless, and 25% of the off-spring would die in-utero. A similar mode of inher-itance is noted for dwarfism in Dexter cattle, which is also associated with embryonic lethality for ho-mozygous dominant animals.

In humans, lactose intolerance shows recessive inheritance. Lactose intolerance is the result of a reduction-of-function of the lactase enzyme. Many animals do not consume milk after in-fancy and the lactase enzyme is inactivated. Early Europeans that carried a mutation that al-lowed continued production of lactase beyond infancy were able to consume milk into adult-hood. The mutation for lactose tolerance be-came dominant over intolerance.

The classical expression of complete domi-nance is simple in predicting the expression of a trait in offspring; however, rarely are the patterns of gene expression as simple as complete dominance. Co-dominance or no dominance can occur when neither allele masks the other, and both are ex-pressed when they are present in the heterozygous state. The animal will have characteristics interme-diate of the two homozygous genotypes; however, it will always more closely resemble the homozygous dominant trait. An example of this is roan coat color in shorthorn cattle. This occurs when an animal in-herits one allele coding for red coat color, and one allele coding for white coat color. Instead of one al-lele being dominant over the other, they are both ex-pressed, resulting in a roan color pattern that is both red and white.

Classical Mendelian genetics considers that each gene has only two possible alleles. In fact, many genes may have more than two alleles within a population, but only two alleles will be found within an individual, one on each homolo-gous chromosome. The base coat color of cattle is actually controlled by three alleles: black, red, and white. While black and red show classic dominance, the white gene shows incomplete dominance with the black and red allele.

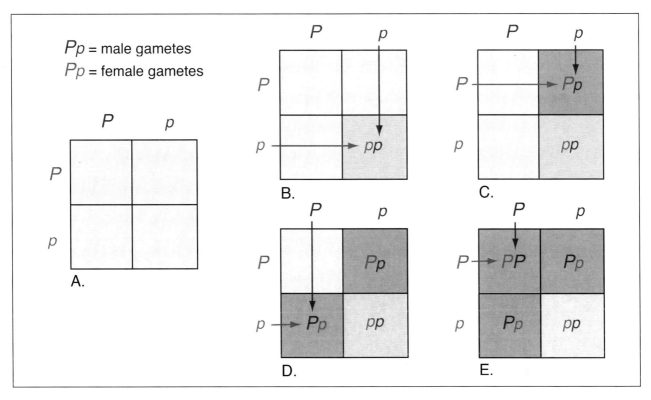

Pp = male gametes
Pp = female gametes

■ **Fig. 5.2.5**   Punnett square. (© Kendall/Hunt Publishing Company.)

In-complete or partial dominance occurs when an allele is expressed in a dose dependant manner. This results in an offspring that is intermediate between the two alleles for a given trait. For example, if one were to cross a red flower with a white flower, incomplete dominance would result in pink flowers. Likewise, palomino coat color in horses is due to incomplete dominance in the offspring of chestnut and cremello mated horses. The chestnut coat color is determined by the genotype CC and the cremello coat color is determined by the genotype $C^{Cr}C^{Cr}$. The offspring of a chestnut and cremello horse results in the genotype CC^{Cr}. With only one C^{Cr} allele present, the C^{Cr} allele is only partially dominant and full suppression of the chestnut color is not achieved, and instead the diluted palomino intermediate is produced.

The example of coat color in palominos further highlights that characteristics of an animal are not due solely to the genes at a single locus, but are also influenced by genes across multiple loci. Epistasis is one type of gene interaction that occurs among genes at different loci to determine gene expression. In this case, the expression of a gene at one locus will depend on the alleles present at one or

■ **Fig. 5.2.6**   A salient feature of the Manx cat is the absence of the tail, which results from an autosomal dominant gene. (© Cheryl Kunde, 2009. Under license from Shutterstock, Inc.)

more other loci. Epistasis may result in gene expression that appears quite different than what the principles of Mendelian inheritance would predict. In the example of palomino coat color, C^{Cr} is a modifier gene responsible for diluting base coat colors in a full ($C^{Cr}C^{Cr}$) or partial (CC^{Cr}) state. The

Predicted Offspring of Black Lab Matings

Black Dam (BbEe)				
	BE	**Be**	**bE**	**be**
B **E**	BBEE black	BBEe black	BbEE black	BbEe black
B **e**	BBEe black	BBee yellow	BbEe black	Bbee yellow
b **E**	BeEE black	BbEe black	bbEE chocolate	bbEe chocolate
b **e**	BeEe black	Bbee yellow	bbEe chocolate	bbee chocolate

(Row labels at left read: Black Sire (BbEe))

■ **Fig. 5.2.7** Coat color of tortoise shell cats is an example of X-linked traits. (© Melissa Ann Kilhenny, 2009. Under license from Shutterstock, Inc.)

base coat color is controlled by a gene at a different locus. Epistasis plays a role in determining coat color in many species, including Labrador Retrievers. Labs can either be yellow, chocolate, or black. These colors are determined by genes at two loci, the black (B) locus and the extension of pigmentation (E) locus. Black coat color is determined by the genotype BXEX , chocolate bbEX , and yellow XXee; where X indicates that either dominant or recessive alleles could be substituted without altering the phenotype.

Note that the expression of the genes at the black locus is dependent on the alleles present at the extension locus. As long as there is the dominant E allele, then the lab will be black or chocolate, depending on the alleles at the black locus; however, if there are two recessive ee alleles at the extension locus, no matter what alleles are present at the black locus, the lab will be yellow.

Sex-related Inheritance

Not all traits are confined to autosomes. Sex-linked inheritance refers to traits that are present on the X chromosome. While males have a Y chromosome, it is argued that it does not contribute to sex-linked inheritance as the few genes located on this chromosome are not expressed in observable traits. As stated earlier, sex chromosomes are an exception to the rule that chromosomes exist as homologous pairs. In mammalian males with XY chromosomes, there is only one copy of the majority of genes as there are no corresponding loci on the Y chromosome for the genes of the X chromosome. Sex-linked genes are those genes associated with regions of the X chromosome that do not have corresponding loci on the Y chromosome. Genes that are X-linked can be passed on to either male or female offspring, since both will inherit at least one X chromosome.

An example of an X-linked trait is the orange and black coloration in tortoiseshell and calico cats. Tortoiseshell cats have a mixture of colors in their coat appearing in patches, always with some orange, and generally with white and black coloration. If there is a substantial amount of white, the cat is considered to be calico. Tortoiseshell cats are females, except for very rare exceptions. This is due to the fact that the locus for orange and black coloration is present on the X chromosome. Black coloration occurs with the genotype $X^B X^B$ for females and $X^B Y$ for males. Orange coloration occurs with the genotypes $X^O X^O$ for females and $X^O Y$ for males. The coat of the tortoiseshell cat occurs with the genotype $X^B X^O$ and results when one X chromosome is randomly inactivated through the formation of a bar body. Inactivation of the X chromosome occurs during early embryonic development and since it is a random event either the X^B or X^O allele will be inactivated. This allele will remain inactive during mitosis, thus as cells divide, they will carry the allele for either black or orange giving rise to patches of color.

Sex-linked should not be mistaken for sex-limited, which are traits that are limited to only one sex. Examples include milk production in females and cryptorchidism in males. While both sexes carry genes for sex limited traits, only one sex is capable of expression of the trait. Furthermore, traits may be sex-influenced, where reciprocal expression of the trait occurs in males compared to females. For example, in breeds of sheep the allele for horned is dominant to the recessive form for polled (absence of horns) in males. In females, polled is dominant to the recessive allele for horned. Therefore, when the genotype is heterozygous males will have horns, while females will be polled.

Qualitative and Quantitative Traits and the Phenotype

Ultimately, an animal's value is associated with the expression of the genotype into traits of economic importance. These visible or measurable traits that an animal demonstrates are referred to as the phenotype and may represent either qualitative or quantitative measurements. Qualitative (or categorical) traits are those that can be described, but can only be subjectively measured. These traits are generally expressed in categories. Examples of qualitative traits include coat color and presence of horns. These types of traits are usually controlled by a relatively few genes and are called simply-inherited traits. In comparison, quantitative traits are traits that can be numerically measured as they show continuous (numerical) expression. These traits are often controlled by many different genes (polygenic), making it difficult to pinpoint the contribution of a specific gene to the animal's phenotype. Examples of quantitative traits include: milk yield; calving ease; weaning, yearling, and mature weight; litter size; and backfat thickness.

The phenotype is not based solely on the genotype, but also is reliant on the environment; a relationship that is represented as: genotype + environment = phenotype. The extent to which environment influences the phenotype is related to the type of trait. Qualitative traits are less influenced by the environment in comparison to quantitative traits. For example, environmental factors such as climate, health, management practices, etc. will have an effect on the animal's ability to reach its full genetic potential. If an animal possesses the genetic potential for enhanced growth, but nutrition is withheld, the genetic potential for growth will not be realized.

Heritability

While phenotype is determined by both genetics and environment, only the genetic effects are inherited. Heritability is a measure of the strength of the relationship between phenotypic values (performance) and breeding values (value of genes to the progeny) for a trait in a population. Simply put, heritability is the proportion of phenotypic variation that can be passed from parent to offspring. In an attempt to determine how much progress can be made in traits from generation to generation, heritability estimates have been made for traits of importance in animal species. Heritability is not constant and may vary from one population of animals to the next and even within a population over time.

Heritability values range from zero to one and can be thought of as a percentage or proportion. Traits ranging in value from 0–0.2 are considered to be lowly heritable. Reproductive traits are generally lowly heritable traits. Moderately heritable traits, such as growth traits are those with values ranging from 0.2–0.4. Highly heritable traits range from 0.4–0.6 and include carcass merit traits. Lowly heritable traits do not demonstrate much change from generation to generation with the use of selection. However, traits that are highly heritable can be selected for and will result in a greater degree of change in less time.

Heritability is used to estimate the value of an individual as a parent through transmission of desirable genes to the offspring and can be used to determine the predicted rate for genetic progress. Genetic progress from one generation to the next is predicted by: *selection differential x heritability,* where selection differential represents the phenotypic advantage of the animals chosen to be parents in relation to the average for the population. Accordingly, the annual rate of genetic progress is predicted by: (*selection differential x heritability)/generation interval,* where generation interval represents the average period of time between the birth of one generation and the birth of the subsequent generation. Genetic progress can therefore be achieved more rapidly in animals that have shorter generation intervals compared to animals with longer generation intervals. This has contributed to the use of rodents and chickens in multi-generational studies to examine the role of genes in disease processes.

Population Genetics

To make genetic progress in animal populations, one must know how often particular genes of interest occur in a population. Within a population, the proportion of loci that contain a particular allele is referred to as gene frequency. In turn, the relative occurrence of a particular genotype within a population is referred to as genotypic frequency and the relative occurrence of a particular phenotype is referred to as the phenotypic frequency. Population genetics is the study of how allele and genotypic frequencies change, and thus influence genetic merit within a population. In population genetics, large numbers of observations are involved and genes are viewed collectively.

If no evolutionary forces are at work to change genotypic frequencies, and a population is randomly mating, then it would be expected that gene and genotypic frequencies would remain constant from generation to succeeding generation. A state of constant gene and genotypic frequencies such as this is referred to as Hardy-Weinberg equilibrium. For a population to be in Hardy-Weinberg equilibrium, it must meet five requirements: large population size, random mating, no migration, no mutation, and no selection. The formula for the Hardy-Weinberg equilibrium equation is as follows:

$$p + q = 1$$

or, it can be extended as:

$$p^2 + 2pq + q^2 = 1$$

Here, p represents the frequency of dominant alleles, while q represents the frequency of recessive alleles. This being the case, p^2 represents the frequency of homozygous dominant genotypes, 2pq represents the frequency of heterozygote genotypes, and q^2 represents the frequency of homozygous recessive genotypes. Due to the requirements of Hardy-Weinberg equilibrium, essentially no populations will exist in Hardy-Weinberg equilibrium.

A major reason that domesticated animals are not in Hardy-Weinberg equilibrium is due to the fact that they experience a great deal of selection. Whereas natural selection is the differential success in the reproduction of different phenotypes due to the interaction of an organism and its environment and is independent of deliberate human control, artificial selection is under human control. Under the premise of natural selection, animals with favorable phenotypes are more suited and have greater opportunity to mate than others; thus, favorable traits become more common over successive generations. Artificial selection is based on management decisions of breeders. Animals with undesirable phenotypes are culled, while superior replacements are chosen to alter the gene frequency of a herd to achieve the desired phenotype.

Gene and genotypic frequencies also are altered by mutations, changes in the chemical composition of a gene that alters DNA. Mutations result in the production of new alleles and while they are considered to occur rarely within a population, they prevent Hardy-Weinberg equilibrium from being obtained. A variety of mutations may occur, including nucleotide base substitutions, deletions, and insertions. Substitutions may involve point mutations, which occur when only one nucleotide base pair is changed. Although point mutations may be silent mutations, in which the nucleotide base substitution still codes for the same amino acid due to the redundancy of the genetic code, missense and nonsense mutations may occur as well. If the resulting codon codes for a different amino acid than the original gene sequence, the mutation is then referred to as a missense mutation. Missense mutations are called such because the code still makes sense (gives rise to an amino acid), although not necessarily the right sense. When a point mutation results in a codon that normally encodes an amino acid becoming a stop codon, this is called a nonsense mutation. A nonsense mutation causes translation to be terminated prematurely, and the resulting polypeptide will be shorter than the polypeptide encoded by the normal gene.

Insertions and deletions are another type of mutation in which nucleotide base pairs are added or lost within a gene. When base pairs are not added or deleted in multiples of three, the result is a frameshift mutation. As mRNA is read as a series of nucleotide triplets (codons) during translation, the insertion or deletion of nucleotides may alter the reading frame. All of the nucleotides downstream from the mutation will be improperly grouped, and therefore will code for the wrong amino acid sequence. This will likely result in a nonfunctional protein unless the frameshift is near the end of the gene.

While mutations can ultimately improve protein function, they also may lead to nonfunctional proteins. Genetic, or hereditary, disorders often result when a mutation has an adverse effect on the phenotype of an organism. An example of a genetic disorder due to a mutation is muscular hypertrophy in callipyge sheep. A point mutation, as a result of an adenine to guanine substitution on chromosome 18, results in extreme muscle growth and compromised tenderness. Similarly, nucleotide deletions in the myostatin gene have contributed to the double muscling phenotype in cattle (11 nucleotide base pair deletion) and to the bully phenotype (two nucleotide base pair deletion) in whippets.

Mating Systems for Genetic Change

There are two tools used by animal breeders to make genetic change in a herd. They are selection and mating. Selection is the process of determining which animals will become parents, as well as how many offspring or litters they produce and how long they will remain in the breeding population. Mating is the process of determining which selected males will be bred to which selected females. The genetic variation among animals allows breeders to select superior animals to be the parents of the next generation. There are a variety of ways to select and mate these animals, referred to as mating systems.

Random mating is a mating system in which selected animals are allowed to mate at random. Random mating does not mean mating without selection; it simply means that the mating of selected breeding stock is not controlled. All of the males and females are combined and kept together so that each male has an equal opportunity to mate with each female. Since an entire herd or flock can be handled as a single unit during the breeding season, random mating demands less labor than other mating systems. When breed registration is required, as with purebred animals, random mating cannot be used with more than one male. This is due to the fact that both the sire and dam of an animal must be

recorded for an animal to be registered. Registration is usually not as important in commercial herds and therefore random mating is more commonly used.

Inbreeding is the mating of related animals within a breed. The degree of inbreeding increases with the mating of more closely related animals. Nearly all purebred animals are inbred to some extent. Inbreeding is used to concentrate the desired genes known to be present in a superior animal, and in turn increase homozygosity and thus predictability for desired traits in offspring. This increase in homozygosity is due to the fact that relatives are more likely to carry the same alleles than animals selected at random from the species. When this is practiced in a population, variation in the genes existing in the population will decrease. Unfortunately, as homozygosity increases, it will increase the chances of detrimental recessive genes being expressed and decrease fitness or performance, known as inbreeding depression. Since inbreeding can lead to the expression of detrimental genes, it is important to carefully control inbreeding to allow the desired genes to be expressed, while, at the same time, minimizing the expression of undesirable genes and inbreeding depression. Inbreeding is not generally recommended for most animal production systems due to the risk of inbreeding depression.

A less restrictive form of inbreeding is linebreeding. It is the mating of animals that have one

■ **Fig. 5.2.8** Artificial selection is responsible for the distinct features of breeds evidenced in animal populations and ensures that desired characteristics are continued in subsequent generations. (© Eric Isselée, 2009. Under license from Shutterstock, Inc.)

common ancestor appearing multiple times, at least three to four generations removed in the pedigree. Every breed of animal is linebred to some extent due to the fact that animals of each breed trace back to common ancestors. In comparing inbreeding to linebreeding, an example of inbreeding would be the mating of a sire to a daughter, while an example of linebreeding would be the mating of a grandsire to a granddaughter. Knowledge of mating systems and detailed pedigrees are required to prevent linebreeding from becoming inbreeding.

Outbreeding is the mating of less closely related individuals when compared to the average of the population and can be used to introduce new genetic material into a population. This mating system is considered the opposite of inbreeding, as the individuals mated are generally unrelated for four to six generations. Outbreeding results in an increase in heterozygous gene pairs. This increase in heterozygosity reduces the chances of the animal expressing undesirable recessive genes and increases the performance of an animal relative to the performance of its parents. This superiority of an outbred animal is referred to as heterosis or hybrid vigor. A type of outbreeding in which unrelated animals within a breed are mated is termed outcrossing. Crossbreeding is another form of outbreeding in which two different breeds of animals are mated. Crossbreeding achieves the greatest level of heterosis. Because animals within a breed have a greater degree of homozygosity, a cross between two breeds will increase heterozygosity in the offspring. The combining of different alleles from different breeds is what contributes to the greater degree of heterosis. Traits that are lowly heritable, such as reproductive traits, demonstrate the greatest levels of heterosis. Moderately heritable traits, including growth traits, only show moderate levels of heterosis. Highly heritable traits demonstrate low levels of heterosis. Heterosis also declines with subsequent generations. It is greatest in the first generation and reduced in subsequent generations, due to a recombining of more similar alleles. Crossbred mating systems strive to improve the overall performance of an animal due to the mating of individuals with different but complementary breeding values, referred to as complementarity.

In the mating systems listed above, selection of the animals precedes the choice of mating system. An exception to this is the corrective mating system in which the goal is to correct the faults of one or both parents in their offspring. In this system, a mating system is chosen first, and a mate is selected in accordance with the desired mating system. For example, if a breeder has a herd of rather small females, but wants to produce offspring of intermediate size, the strategy would be to breed the females to larger males. This is a type of corrective mating, and should result in the production of offspring that are intermediate in size.

Marker Assisted Selection

Selection is primarily determined from an animal's observable phenotype, without direct knowledge of the underlying genes being selected. However, with the advent of DNA-based markers in the 1970's, the possibility of selecting for traits of interest based on genetic material was realized.

It is estimated that less than 10% of the genome actually codes for biological proteins. The function of the genetic material in the remainder of genome is not fully known. However, DNA sequences that do not code for proteins are located throughout the non-coding regions of the genome and are transmitted according to the laws of inheritance. These DNA sequences serve as the markers to identify traits of interest. Knowledge of the genome has allowed for generation of DNA marker maps (genome maps), which identify the marker relative to the chromosome. Location of the marker is indicative of location of the genes contributing to the trait. Marker maps have been developed for a variety of animals including chicken, cattle, pigs, and sheep. With this knowledge, variants (alleles) of the marker associated with the desired traits may be selected. There are many DNA markers used, and they arise from mutations that include point mutations, insertions and deletions, or nucleotide repeats. One of the more widely used DNA markers in the animal industries is the SNP (single nucleotide polymorphism). Single nucleotide polymorphisms occur frequently within the genome. Panels, also known as chips, that contain 50 to 800 thousand discrete SNPs are used to screen animals to identify genomic variants associated with genetic traits of interest. The presence of SNPs for given traits correlates to the genetic value of an animal. With this technology, young animals that have not reached reproductive age can be more accurately identified for their future breeding potential. The use of marker assisted selection does not replace phenotypic performance data, but improves accuracy in selecting animals.

MOLECULAR GENETICS

"Almost all aspects of life are engineered at the molecular level, and without understanding molecules we can only have a very sketchy understanding of life itself."
—Francis Crick (1916–2004)

Molecular genetics involves the purposeful manipulation of an organism's DNA and allows for the production of offspring with genomes that natural breeding is unable to create. Prior to molecular genetics, alteration to an organism could only occur indirectly, on the level of altering the whole organism.

Molecular Genetics: A Historical Perspective

In the 1920s, Hermann Muller speculated that genes controlled evolution by their ability to spontaneously mutate and introduce new genes into populations; however, the specific makeup of the genes was still unknown. In 1944, Oswald Avery attributed nucleic acid as an active component and in 1954, Alfred Hershey and Martha Chase demonstrated that DNA is the carrier of genetic information, not protein which was previously believed to have assumed that role. In the early 1970s it was revealed that the human and primate genomes differed by less than 1.6% and in 2000 the complete map of the human genome was constructed.

Even though the concept of DNA and the theory of inheritance were not fully understood, the first cloned animals were produced in the 1800s. This was accomplished by Hans Driesch in an effort to prove that genetic material was not lost during cell division. In his experiment, he separated a two celled embryo of a sea urchin and allowed each cell to grow independently resulting in two, identical sea urchins. In 1951, another breakthrough was made when scientists in Philadelphia cloned a frog embryo. Unlike previous methods of cloning, the multi-celled embryo was not simply split into two; but, the nucleus of the frog embryo was transferred

to the embryo of a non-fertilized frog egg. This was the first time that nuclear transfer had been used and it is still a process that is used today. In the 1960s, the concept of gene transfer, or transgenesis, in bacterial cells was beginning to be understood. It was not until the 1980s, however, that this knowledge was transitioned into accepted practices and standards of gene transfer into mammalian cells and the process gained efficiency.

Biotechnology

Cloning is the process of producing a genetic copy of a DNA segment, gene, embryo, or animal. The cloning of DNA, also known as molecular or gene cloning, is a routine laboratory procedure that generates multiple, identical copies of DNA of interest.

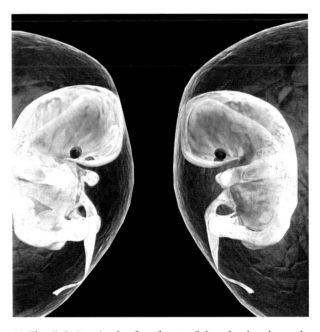

■ **Fig. 5.3.1** At the forefront of the cloning issue is the concern of human reproductive cloning. The American Medical Association and the American Association for the Advancement of Science have issued formal public statements advising against human reproductive cloning. The same problems that have occurred in animal experiments (inefficiency and abnormal gene function) would be anticipated for human cloning. The idea of human cloning is widely viewed as ethically irresponsible. (© Christian Darkin, 2009. Under license from Shutterstock, Inc.)

This procedure has been instrumental in sequencing the genomes of a variety of animals. Embryo cloning, also known as therapeutic cloning, is the production of embryos for research using cloning techniques. The goal of therapeutic cloning is to generate stem cells that can be studied and potentially used in the treatment of disease. Reproductive cloning is the cloning process that produces a "twin" of the animal by harvesting a cell from the animal, and transplanting the cell's nucleus into an egg that has had its nucleus removed, thereby replacing the egg's own genetic material with that of the animal to be cloned. The egg containing this new genetic material is implanted in a female and allowed to develop just as a naturally fertilized egg would develop. The animal produced will be genetically identical to the donor of the nucleus, but will bear no resemblance to the donor of the egg. The genetic material used for cloning can come from an embryo or an adult animal. Dolly the sheep is the most well-known case of animal cloning and represents the first animal cloned from the genetic material of an adult sheep. The prior cases of cloning had only been successful by using an undefined embryonic cell. The successful cloning from a nonembryonic, or somatic cell, was a major advancement as it required the deprogramming of a differentiated body cell, in Dolly's case an udder cell, and the reprogramming to allow it to develop into a separate and fully functioning organism.

Since Dolly, many other animals have been cloned including horses, bulls, and in 2005, a dog. Attempts have been made to clone extinct animals whose cells had been preserved before their death. In January 2009, at the Centre of Food Technology and Research of Aragon in Spain, the first extinct animal was born. It was a Pyrenean ibex, a form of wild mountain goat, which was officially declared extinct in 2000. Scientists were able to preserve skin samples in liquid nitrogen taken from the species before its extinction and clone a female. The kid was carried to term and born alive; however, it died only minutes after its birth due to a lung defect. While advantages of cloning include the ability to reliably produce desired animals and repopulate endangered species, the process remains expensive and inefficient. Furthermore, it will not be able to solve the lack in genetic diversity or influence the environmental factors that may have contributed to the animals' extinction. Lastly, studies in mice sug-

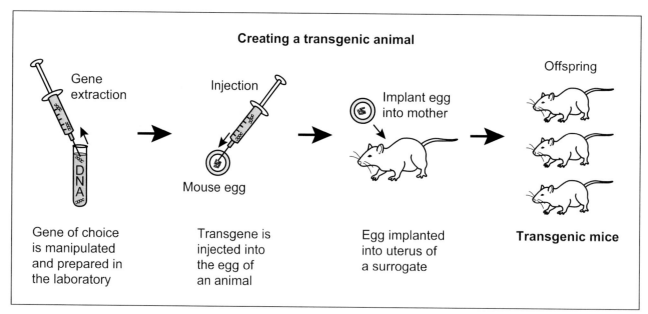

Creating a transgenic animal

Gene extraction

Gene of choice is manipulated and prepared in the laboratory

Injection

Mouse egg

Transgene is injected into the egg of an animal

Implant egg into mother

Egg implanted into uterus of a surrogate

Offspring

Transgenic mice

■ **Fig. 5.3.2** The first step in the process of creating a transgenic animal is to inject the foreign gene into a single-celled embryo. The DNA that is inserted will contain the gene of interest as well as the genetic information that will allow the gene to be transcribed. The DNA is packaged into a viral vector for integration into the host genome. If the foreign DNA is integrated into the genome of the embryo (less than 4% success) a transgenic embryo is created. This embryo will then be implanted into a recipient female. If successful, the female will give birth to a transgenic animal that is able to express the inserted gene.

gest that up to 4% of the genes in the genome of cloned animals function abnormally and the implications of this finding are not known.

Transgenesis

All organisms that are classified as transgenic contain a foreign DNA segment, or gene, which has been inserted into its genome. Transgenic animals may be used in a variety of ways, one of which is the production of pharmaceuticals from livestock in a process that has been termed pharming. Transgenic cows, goats, and sheep can be used to produce bioactive molecules that are secreted into their blood, milk, or urine. Transgenic bacteria also have been used to produce bioactive molecules such as insulin for the treatment of diabetes. The most common use of transgenic animals is for research to study and treat human diseases including cardiovascular disease, autoimmune diseases, sickle-cell anemia, and neurological diseases. These animals, especially rodents, have allowed researchers to study these ailments and investigate possible treatment plans. Other uses include the development of disease resistant animals and animals engineered as a source of transplant organs.

Careful consideration should be given when determining an appropriate transgenic host. In the last twenty years sheep, cows, pigs, rabbits, mice, and rats have all been transgenically altered, but not all are equally suited to help achieve a desired outcome. Take for instance the interest in the use of transgenes to produce human blood clotting protein in milk. Establishing a line of proficient transgenic animals in pigs would be much easier because of their short gestation period and large litters when compared to a single bearing species that requires greater time to reproduce such as cattle. However, milking pigs is considerably more difficult than the well understood milking operation of cows.

Transgenic practices also show promise for direct benefit to the animal and to address environmental concerns. Species of fish have been engineered to attain increased growth rates, disease resistance, temperature tolerance, and product improvement. Chickens and turkeys have been developed to withstand diseases such as avian influenza, and in 2000 a Jersey cow resistant to mastitis was transgenically engineered to produce an enzyme called lysostaphin that destroys a bacterial strain that contributes to mastitis. This application of transgenesis would lessen the 1.7 billion dollar economic loss annually caused by the disease, decrease the number of cows culled from the dairy herd, and benefit the public by reducing antibiotic usage. Similarly, in 2001 a strain of pig, known as the enviropig, was engineered to produce the enzyme phytase within their saliva so that plant phosphorus may be utilized, reducing the dependence on supplemental phosphorus and reducing the amount of phosphorous expelled in the manure, thus lessening the pollutant in the environment. Currently transgenic animals are not permitted to enter the human food supply; while, the Food and Drug Administration is currently reviewing guidelines for their future use, societal discomfort regarding genetically altered animals may pose the greatest barrier to their approval for commercial use.

Xenotransplantation

As of 2002 it was estimated that 250,000 people owed their life to allotransplantation, the transplantation of a human organ to another human. The demand for organs for human transplant increases annually, with a four fold increase in waitlisted patients occurring between 1991 and 2001. Yet the

■ **Fig. 5.3.3**   In 2003, the first transgenic pet was marketed. Transgenic *Danio rerio*, marketed as GloFish™, contain a fluorescent gene harvested from sea anemone. The impetus behind the development of this fish was to aid in the detection of environmental pollutants. Scientists are still hoping to create a strain of fish that will remain non-fluorescent under proper water conditions, but will turn neon when exposed to environmental toxins. Strains of green and orange florescent GloFish have recently been introduced and more varieties are likely to follow. (© David Dohnal, 2009. Under license from Shutterstock, Inc.)

number of transplants as a result of available organs remained steady during this same time frame. To meet this growing demand for organs, xenotransplantation (the implantation of organs, tissues, or cells from one species to another species) has received considerable attention. Pigs have been viewed as ideal models for transplantation for the following reasons: 1) organs of similar size, 2) similarities in anatomy and physiology 3) relatively short reproductive cycles and large litters, 4) rapid growth rates, 5) hygienically maintained at reduced costs, and 6) transgenic techniques for modifying organ systems are established. Indeed, heart valve transplants from pigs have been in routine practice for over thirty years, although it should be noted that the valves are treated prior to transplantation to destroy porcine cells.

Of concern with xenotransplantation is the transmission of zoonoses, whereby an infectious disease is transmitted from the donor animal to the human recipient. Although no infections have been reported from people that have received various living porcine tissues over the past twelve years, the concern is still real as certain porcine viruses (retroviruses) have been shown to infect human cells in the laboratory. A second area of concern is compatibility of donor organs. Although size compatibility is known, the ability to sustain physiological and biochemical function for sustained time is not known. An additional area of concern is immunological rejection. Rejection may be acute (occuring within minutes) or chronic (occurring over time). Proteins produced on the surface of cells of the donor organ are recognized by the host's immune system, ultimately leading to a cascade of events that stimulates organ rejection. The goal in xenotransplantation is to reduce the instance of rejection for the procedure to be a success. While immunosuppressive drugs can be taken to eliminate

■ **Fig. 5.3.4** Recent advances in xenogenic therapy, the transplant of cells from one species to another to restore biological function, has shown promise in treatment of diseases including diabetes. Pig islet cells transplanted into diabetic monkeys reversed the disease process by restoring the pancreas ability to produce insulin. (© Erìc Isselée, 2009. Under license from Shutterstock, Inc.)

certain types of rejection responses that occurs over days, they do not work against hyper-acute rejection that occurs within minutes. To circumvent this issue transgenic pigs that express human regulatory proteins on the surface of pig tissues is under investigation. Studies in primates have shown increased survival rates with transgenesis (seventy to ninety days versus minutes for non-transgenic organs). The immediate goal of these strategies is to provide extension of life using xenotransplantation until a suitable human organ can be obtained.

Reproduction

"It is not a wonder that reproduction sometimes fails, but rather a miracle that so many pregnancies terminate successfully."

—*William Hansel*

This chapter introduces the structural components of male and female reproductive anatomy. Upon completion you will have an understanding of reproductive morphology and physiology and the differences in mammalian and avian reproductive systems. The hormonal underpinnings of puberty, estrus, and gestation are introduced. Lastly, the role of assisted reproductive technologies including estrus synchronization, artificial insemination, and embryo transfer for advancing reproductive success in domestic animals are emphasized.

Reproduction: A Historical Perspective

The first scientific theories of how populations reproduce were postulated by Aristotle (384–322 B.C.) Aristotle formulated two theories governing early embryogenesis. The first, regarded as preformation, stated that the embryo was preformed and grew or enlarged during development. The second, later termed epigenesis, stated that man arose from the successive differentiation of a formless being. Although Aristotle believed the latter one was correct, describing the embryo as being organized from the mother's menstrual blood after being acted upon by semen, this theory was ultimately rejected in favor of preformation until the nineteenth century.

Between Aristotle's initial theories in embryogenesis and the acceptance of epigenesis, several discoveries were made regarding reproductive strategies. In 1562 Fallopius, an Italian anatomist and botanist first described the oviduct (Fallopian tube in humans).

This was succeeded by the description of the corpus luteum in 1573 by Coiter, the ovarian follicle in 1672 by De Graaf and sperm by Hamm and Leeuwenhoek in 1677; with the advent of the first microscopes. However, the realization that sperm fertilized oocytes would not occur until 1825. The final proof that life originates from single cells was provided in 1900 by Driesh, who demonstrated that cells isolated from fertilized ovum were capable of developing into an embryo. By 1907 artificial insemination was being studied in farm animals, dogs, foxes, rabbits, and poultry. The first written English text on artificial insemination was published in 1933 and studies concerning infertility in animals soon followed. Advancements in domestic animal reproductive technology far outpaced progress made in humans. For example, the first successful in vitro fertilization procedure occurred in rabbits in 1959, but almost twenty years would elapse before documented success in humans.

Female Reproductive Anatomy

The female reproductive system and its functions are more complex than that of the male reproductive system. It consists of two ovaries and the portion derived from the embryonic Mullerian ducts (oviducts, uterus, cervix and anterior vagina) and the portion that differentiates from primordial external genitalia (posterior vagina and vulva). The major functions of the female reproductive anatomy are to produce a female gamete, deliver it to a site where it can be fertilized by a male gamete, provide an environment for growth of the embryo/fetus, and deliver the fetus.

Ovaries
The ovaries are the primary reproductive organs in females. They produce female gametes (oocytes) as

well as female sex hormones (e.g. estrogens and progesterone). Ovaries are highly vascularized, which ensures continual nutrient and oxygen delivery. The ovaries serve an endocrine function and the vascular network of the ovaries ensures delivery of hormones to and from this organ to target tissues.

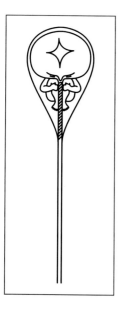

■ **Fig. 6.1** In the seventeenth century it was thought that semen contained animalcules or miniature animals. Sperm was viewed as a homunculus or fully formed miniature man that was placed in the womb of the female to be nourished until birth.

Tubular Portion of the Female Reproductive System

The oviducts, often referred to as fallopian tubes in humans, are a pair of convoluted tubes adjacent to the ovaries and extending to the uterus. Oviducts are the site of fertilization and early cell divisions of the embryo. They function to transport ova and sperm, which must be moved in opposite directions towards each other. The oviduct is a muscular tissue with contractile properties that is generally divided into three sections. The segment of the oviduct adjacent to the ovaries is the infindibulum. This lace-like structure envelops the ovaries and is responsible for capturing the released ocyte and directing its transport. In the, sow, cow, and ewe the infundibulum remains separate from the ovary. In the rat, mouse, and hamster the infundibulum forms a bursa that surrounds the ovary the ovary while in the mare it is attached to the ovary. The middle section, or ampulla, is highly invaginated. These invaginations increase the surface area and are covered with many cilia to aid in the movement of the ocyte down the ampulla. The isthmus is the final section of the oviduct. The site where the ampulla and isthmus join is known as the ampullary-isthmic junction and is the site of fertilization of the oocytes by sperm.

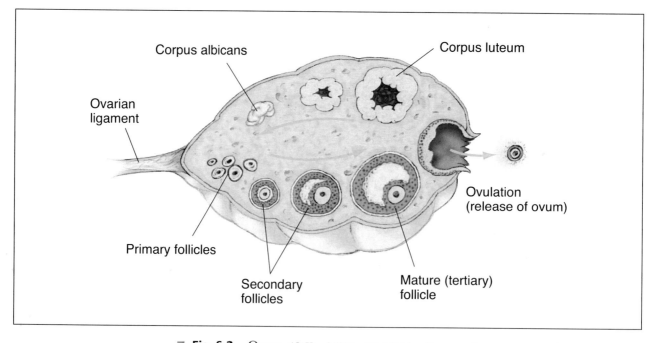

■ **Fig. 6.2** Ovary. (© Kendall/Hunt Publishing Company.)

This junction delays the transport of the ocytes for several hours to increase the chances of it becoming fertilized. Once fertilization occurs, the zygote will then undergo mitotic division as it travels for three to six days through the oviduct towards the uterus. The isthmus is morphologically similar to the ampulla, however, it has decreased surface area and the cilia beat to transport oocytes and embryos towards the utero-tubal junction. Oviductal contractions move sperm in both directions within the oviduct so only a portion of the sperm advance towards the ovary. At the uterotubal junction the isthmus connects to the uterus. The uterus consists of two uterine horns and/or a uterine body and is composed of three major layers: 1) an outer covering, 2) the myometrium, an intermediate smooth muscle layer responsible for uterine contractions, and 3) the endometrium, which is the mucosal lining of the uterus. The endometrium is the site of either embryo implantation or attachment of the extra-embryonic membranes (depending upon the species) for formation of the placenta. The uterine environment is protected by the cervix, a thick-walled, cartilaginous, elongated, smooth muscle sphincter that remains tightly closed except during estrus and parturition. It is located between the uterus and the vagina. The cervix in the cow and the ewe have transverse, interlocking ridges usually referred to as annular rings. The cervix serves a variety of functions, which includes: 1) a passageway and reservoir for sperm, filtering non-viable sperm and consistently releasing viable sperm into the uterus for fertilization of ova;

2) a barrier to prevent bacteria from reaching the uterus via the annular rings and the antibacterial properties of cervical mucous, which forms a mucous plug during pregnancy; and 3) a passageway for the fetus during parturition. The vagina is the organ of copulation and site of semen deposition for cattle, sheep and horses. In swine, the penis of the boar enters the cervix, and semen is deposited at this location. The vagina is tubular in shape, thin-walled, and very elastic. The vulva is the female external genitalia. The visible swelling of the vulva that occurs during estrus can aid in heat detection.

Male Reproductive Anatomy

The male reproductive system functions to produce and deliver gametes to the female reproductive system. The system consists of the testes, the male ducts (Wolffian system) including the epididymis, vas deferens, and the urethra, accessory glands, penis, and prepuce.

Testes

The testes (testicles) are the primary organs of reproduction in males. They function to produce male gametes (spermatozoa) and male reroductive hormones (e.g. androgens). The testes continually produce gametes throughout the animal's life once puberty has occurred. This differs from the female reproductive system, in which all potential gametes are present in the ovaries at birth. The male

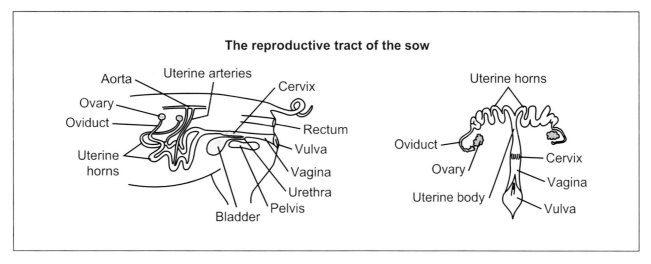

■ Fig. 6.3 Reproductive anatomy of the sow

reproductive system also differs from the female reproductive system in that the testes are located outside of the body cavity in most species, whereas in the female the ovaries are located within the body. The testes are typically oval-shaped and include lobes separated by connective tissue. The seminiferous tubules are small convoluted tubules located within these lobes. They represent 90% of the total mass of the testes. The seminiferous tubules contain germ cells and are the site of spermatogenesis (sperm production). Leydig cells are located in-between the seminiferous tubules and function to produce testosterone and other androgens when stimulated by luteinizing hormone. Sertoli cells, or nurse cells, located within the seminiferous tubules surround developing sperm, nourishing the sperm and mediating the effects of follicle stimulating hormone and testosterone on the germ cells. The testes are suspended within the scrotum by the spermatic cord. The scrotum is a two-lobed sac that conforms to the shape and size of the testes and is divided into two compartments by the scrotal septum. The spermatic cord is composed of blood vessels, nerves, muscle fibers, connective tissue, and a portion of the vas deferens. The spermatic cord and the scrotum both act to physically support the testes, and also work together in regulating the temperature of the testes.

Wolffian Ducts

The epididymis is a long, convoluted tube that functions to store, concentrate, and transport sperm. Conditions within the epididymis are optimal for the storage of sperm and include a low pH and a high carbon dioxide concentration. Concentration of the sperm also occurs here as epithelial cells lining the epididymis absorb some of the fluid in which the sperm is suspended. The transport of semen through the epididymis is affected by three factors: 1) pressure from the production of more sperm, 2) external pressure created by normal movement and exercise, and 3) negative pressure caused by ejaculation. The negative pressure (drawing action) is due to peristaltic contractions of the vas deferens and the urethra. This pressure, in addition to peristaltic contractions of the smooth muscle of the epididymis during ejaculation, transports the sperm from the epididymis into the vas deferens and the urethra. Maturation of the sperm also occurs as it passes through the epididymis, gaining motility and fertility. This maturation process takes approximately 10 to 15 days. From the epididymis, sperm are transported through ducts named the vasa differentia (vas deferens, singular). The vas deferens follow along the spermatic cord, passes through the inguinal canal to the pelvic region. Here they join the urethra near the opening of the bladder. The only function of the vas deferens is to transport sperm. The transport is

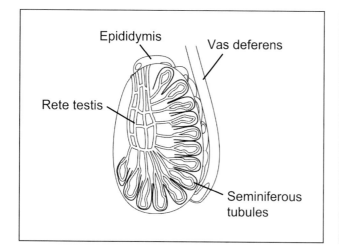

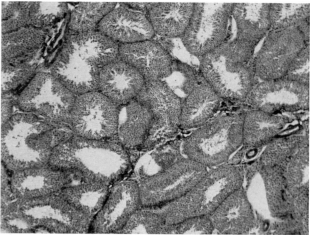

■ **Fig. 6.4** The seminiferous tubules represent 90% of the total mass of the testes and are the site of spermatogenesis and testosterone production. The magnification of the cross section of the seminiferous tubules on the right reveals the tube like structure. Sperm enter the lumen of the seminiferous tubules following meiosis, but must undergo further transformation in the epididymis for motility. (*Right image:* © Jubal Harshaw, 2009. Under license from Shutterstock, Inc.)

facilitated by contractions of smooth muscle in the walls of the vas deferens. The urethra is a single excretory duct extending from the opening of the urinary bladder to the end of the penis. The urethra functions to expel urine and semen.

Accessory Sex Glands

Male accessory sex glands are located along the pelvic portion of the urethra and produce the majority of the ejaculate, or semen, which is the medium for the transport of the sperm. The secretions produced by these glands include buffers, nutrients, and inorganic ions required to assure optimal motility and fertility of the sperm. Semen protects the sperm by aiding in the neutralization of the acidic environment present in the female genital tract. These glands have secretory ducts that open into the urethra. The accessory sex glands include the ampullae, vesicular glands, prostate gland, and bulbourethral glands. The ampullae are enlargements of the vas deferens located just before the urethra. They are the first glands to add fluid to the ejaculate. Ampullae are not present in all species, including swine. The vesicular glands (seminal vesicles) are a pair of lobular glands with a grape cluster appearance. These glands are located near the bladder. The vesicular glands greatly contribute to the fluid volume of semen. The fluid produced by these glands contains

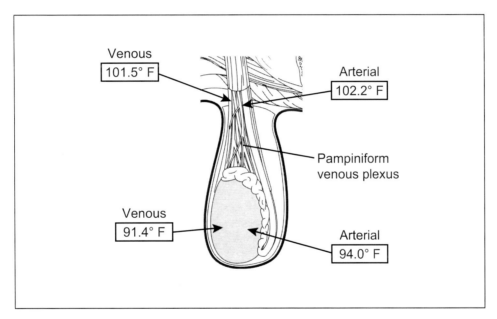

Venous
101.5° F

Arterial
102.2° F

Pampiniform
venous plexus

Venous
91.4° F

Arterial
94.0° F

■ **Fig. 6.5** It is important that the scrotum is kept cool, as high temperatures can cause the degeneration of cells lining the seminiferous tubules, resulting in infertility. If the normal temperature is restored before total degeneration of the cells, then fertility can be restored as well, though this may take a few weeks. Ideally, the testes should be 4–8° F cooler than the normal body temperature of the animal. Thermosensors in the scrotum can detect outside temperature and then initiate the appropriate physiological reactions. One of these reactions is to either draw the testes closer to the body as temperatures fall or to let them be farther away from the body as temperatures rise. There are two muscles involved in this process. One is a smooth muscle called the tunica dartos which lines the scrotum. The other is the external cremaster, a striated muscle that is located around the spermatic cord. The scrotum also contains a specialized vascular system referred to as the pampiniform plexus. The pampiniform plexus is a countercurrent blood supply in which cooler venous blood leaving the testes cools the warmer arterial blood entering the testes.

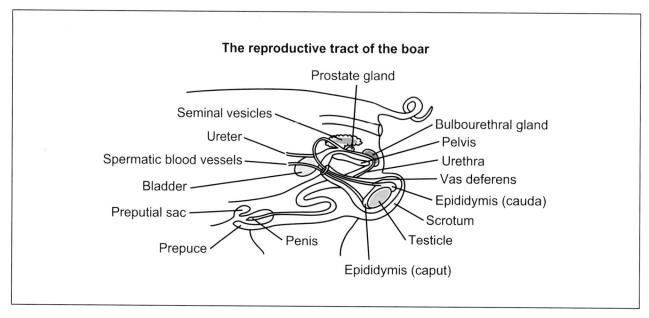

■ Fig. 6.6   Reproductive anatomy of the boar

fructose and sorbitol, which are not found in any of the other glandular secretions. Both fructose and sorbitol are major sources of energy for sperm. The prostate gland is a single gland located where the vas deferens and the urethra converge in some species, including the horse. In other species the prostate gland is embedded in the muscular wall surrounding the pelvic urethra. The prostate gland contributes little of the fluid volume of the semen, but the fluid secreted does contain high levels of inorganic ions including calcium, chlorine, sodium, and magnesium. The prostate may become enlarged in older males and can interfere with urination. Sometimes referred to as Cowper's glands, the bulbourethral glands are paired glands located along the urethra near its exit from the pelvis. In bulls, fluid from these glands is secreted and expelled through the urethra prior to copulation and acts to flush out urine residue. These secretions in boars account for the gel-like portion of the semen which seals the cervix and prevents semen from flowing back into the vagina during natural service.

Penis

The penis is the male organ of copulation and deposits semen in the vagina or cervix (depending on the species) of the female. It can be divided into three general sections: 1) the glans, the free extremity which is well supplied with sensory nerves; 2) the body, the main portion; and 3) two crura (roots), which attach to the ischial arch of the pelvis. The glans varies from species to species. Bulls and rams have helmet-shaped glans. The glans of the boar is relatively small and corkscrew-shaped, allowing it to engage with the corkscrew-shaped cervix of the sow. The semen is deposited in the cervix of the sow instead of the vagina.

The penis may be one of two types: vascular or fibroelastic. Stallions have a vascular penis while the penis of bulls, rams, and boars are fibroelastic. The vascular penis enlarges by retaining blood in erectile tissue during periods of sexual excitement, causing an erection. The blood leaves the penis following ejaculation and thereby decreases the blood pressure and volume in the penis. Fibroelastic penises are firm when not erect, and only contain small amounts of erectile tissue. Therefore, a fibroelastic penis requires a very small amount of blood for erection and does not increase much in diameter. Bulls, rams, and boars have the ability to retract the penis completely into the body by the sigmoid flexure. The sigmoid flexure is an S-shaped bend in the penis that straightens during erection, lengthening the penis. On the other hand, stallions have a pair of smooth muscles which, when relaxed, allow the penis to be ex-

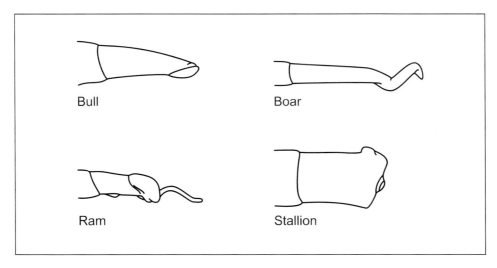

■ **Fig. 6.7**   Glans penis of the bull, boar, stallion, and ram. Note the corkscrew-shaped penis of the boar and the urethral process (filiform appendage) in the ram.

tended, and when contracted, draw the penis back in to the prepuce, an invaginated fold of tissue that completely surrounds the free end of the penis.

Sterility and Castration

In both the male and female fetus, the gonads develop behind the kidneys. The ovaries in females remain in the same location but in the male, the testes descend from the site of origin down through the inguinal canals into the scrotum. The descent of the testes is usually complete by or shortly following birth. If there is a defect in development, one or both of the testes may fail to descend. This may occur in all species of farm animals, but is most common in stallions. If only one testis descends, the animal is referred to as a unilateral cryptorchid and may be fertile. If neither testis descends, the animal is called a bilateral cryptorchid and will likely be sterile. The sterility is due to the high temperature of the abdomen, which inhibits the production of sperm, but does not interfere with testosterone production. Since testosterone production is not affected, the animal will develop secondary sex characteristics and will otherwise appear to be normal. Cryptorchidism can be surgically corrected; however, because this condition can be inherited, it is not desirable to allow these animals to propagate, as it would allow for the perpetuation of the condition. Cryptorchids also develop

testicular tumors at significantly higher rates than normal males; therefore, castration is recommended.

Castration is the process of removing the testes. In livestock, this procedure prevents inferior quality males from reproducing. Since castration removes the source of testosterone, aggressive behavior, which is one component of male sexual behavior (libido), is reduced. This is another purpose of castration. However, a primary purpose of castration is to influence the meat harvested from castrated males. Castration early in life prevents secondary sex characteristics from developing which improves the quality of the meat obtained at harvest. This process also ensures that meat from males will be absent of sex-related odor, especially in the case of swine. When non-surgical castration is performed, the spermatic cord is clamped to sever blood supply to the testes, or a band is placed around the neck of the scrotum to prevent blood supply to the testes. Both procedures result in regression of the testes. With surgical castration, the scrotum is opened and the testes are removed. In horses, castration generally improves the animal's performance and reduces fighting with other horses. Without libido, males are generally much calmer and easier to handle. Castration should be done early in an animal's life to minimize stress. Sterile males can be produced by either vasectomy or epididectomy, which is the surgical removal of a section of either the vas deferens or the epididymis, respectively. This procedure does not affect the production of male hormones and the animal will still behave and appear as an intact male. Vasectomized males are sometimes used to aid in identifying females in estrus.

The Reproductive Process

Reproduction is an essential process required by all species for their continuation and propagation. Reproduction is the most economically important trait in farm animal production that determines the critical endpoints in production such as number of saleable animals, the number of replacement females or the number of eggs produced, and is essential to initiate lactation in milk producing species. For reproduction to occur, the animal must be anatomically sound and exhibit the physiological desire to mate (heat or estrus in females). Successful reproduction relies on coordination of the endocrine system and the reproductive system for the production of hormones required to accomplish germ cell production, fertilization, pregnancy, and parturition. This process is only active during certain phases of the animal's life and is generally age-dependent. Other factors affecting reproduction include photoperiod, the presence of the opposite gender, and the nutritional status of the animal.

Puberty
When an animal reaches a level of physiological maturity that permits conception, this is termed puberty.

Average Age and Weight at Puberty

	Age (mo)	Weight (kg)
Gilt	4–7	68–90
Ewe Lamb	7–10	27–34
Filly	15–24	—
Dairy Heifer	8–13	160–260
Beef Heifer	10–15	225–310

On average, females will reach puberty when 45-55% of mature breed weight is attained; however, considerable variation within and between breeds has been documented and is influenced by factors including season and nutrition.

The signals that induce puberty vary among the different species, but two major influencing factors are age and weight. The age at which an animal reaches puberty is affected by both genetic makeup and environmental factors including nutrition, climate, and in some species, the proximity to a mature male. The weight at which an individual animal reaches puberty is determined by its genetic makeup, but the age at which this weight is attained is greatly influenced by nutritional plane during rearing. Evidence is emerging that suggest that even the nutrition of the mother will influence the time of puberty in her offspring.

In females, puberty is noted by the first estrus accompanied by ovulation. This depends on production of follicle stimulating hormone (FSH), luteinizing hormone (LH) and estrogen at circulating concentrations high enough to induce follicle growth, oocyte maturation, and ovulation. Phenotypically, puberty is characterized by the development of secondary sex characteristics that are generally associated with each sex. Common secondary sex characteristics in males include the hump on the necks of bulls and increased muscling. Characteristics in females include an increase in body fat and mammary development.

The Estrous Cycle
Following puberty, ovarian function will occur in a cyclical manner. This cycle is referred to as the estrous cycle and represents the period of time from one estrus to the next. The production of the female germ cells or oocytes is called oogenesis. Production of the oocytes is accomplished through the process of meiosis, resulting in a gamete (germ cell) containing half the number of chromosomes that are found in a somatic cell (body cell). Surrounding each oocyte is a fluid-filled follicle that grow in size on the surface of the ovary as the time of estrus approaches. During this time, cells within the follicle are producing estrogens. The estrogen produced by this follicle is responsible for inducing estrous behavior. This follicle (or in polytocous animals, several follicles) will eventually rupture, thereby releasing the oocyte(s) into the oviducts. The rupture of the follicle(s) and release of the oocyte(s) is referred to as ovulation. Ovulation will occur either during or near the end of estrus, or after estrus, depending upon the species. The two to five day period preceding and including estrus and ovulation is referred to as the follicular phase.

Following ovulation, the cells of the ruptured follicle wall will transform to become the corpus luteum (CL), representing the initiation of the luteal phase. During this transitional phase from a ruptured follicle to a CL, this structure is referred to as a corpus hemmorhagicum. The CL produces progesterone (the progestational hormone) during the luteal phase. The CL becomes fully functional during the five to seven days following ovulation and is a firm structure; part of which protrudes from the surface of the ovary. The CL will be retained if the animal is pregnant and will continue to produce progesterone during gestation. The corpus luteum and the progesterone it produces are required for maintenance of pregnancy in most species. If pregnancy does not occur, the CL will regress and progesterone concentrations will decrease. Regression of the corpus luteum occurs in response to prostaglandin $F_2\alpha$ (PGF), a hormone secreted from the uterus when a viable conceptus is not present. The regression of the CL will lead to estrus in the next two to five days and the restarting of a new estrous cycle. When the CL regresses the only evidence of its existence is a very small white structure called the corpus albicans. However, if there is a viable embryo in the uterus, regression of the CL is prevented by the embryo through a process termed maternal recognition of pregnancy. Proteins or hormones (depending upon the species) produced by embryos act to prevent production and/or release of PGF; leading to maintenance of the CL throughout gestation.

The sequence and timing of the events of the estrous cycle are controlled precisely by the hormonal interplay of the hypothalamic-pituitary-ovarian axis. The hypothalamus is a small region of the brain responsible for initiating the endocrine functions that control the estrous cycle. The hypothalamus releases gonadotropin-releasing hormone (GnRH), which is the first step in a cascade of hormonal events that coordinate ovarian function for reproductive success. Once released, GnRH acts on the anterior pituitary gland, which is located directly below the hypothalamus. In response to the GnRH, the anterior pituitary releases luteinizing hormone (LH) and follicle-stimulating hormone (FSH) into the blood stream. These hormones stimulate gonadal function and are therefore referred to as gonadotropins. In the female, FSH initiates the early growth of developing follicles. If adequate FSH is

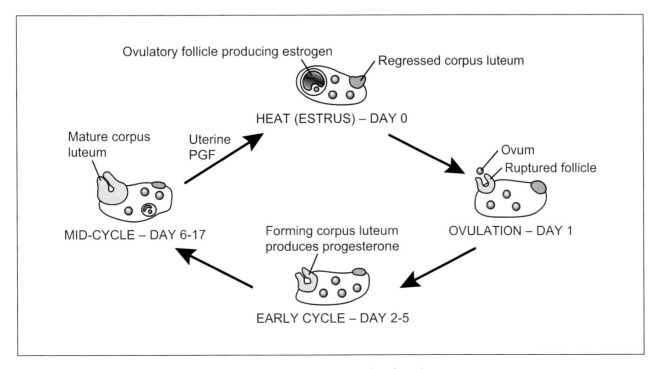

Ovulatory follicle producing estrogen

Regressed corpus luteum

HEAT (ESTRUS) – DAY 0

Mature corpus luteum

Uterine PGF

Ovum

Ruptured follicle

MID-CYCLE – DAY 6-17

Forming corpus luteum produces progesterone

OVULATION – DAY 1

EARLY CYCLE – DAY 2-5

■ **Fig. 6.8** Estrous cycle of cattle.

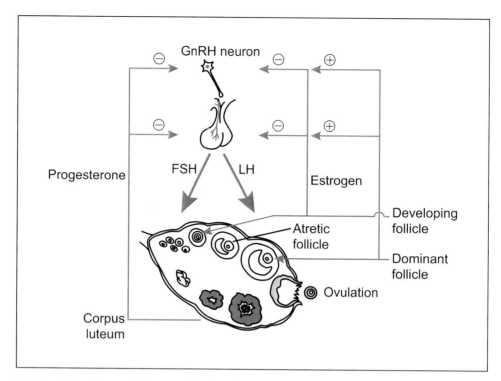

■ **Fig. 6.9** Successful reproduction involves the coordinated interactions of the hypothalamic-pituitary-ovarian axis. Estrogen released from the developing folli-cle is involved in negative feedback of GnRH secretion from the hypothalamus and FSH secretion from the anterior pituitary. This negative feedback is switched to positive feedback regulation when increased concentrations of estrogen pro-duced by the preovulatory follicle initiates an LH surge, whereas progesterone is involved in negative feedback and inhibition of LH release.

not present, smaller follicles will regress through a process termed follicular atresia. Luteinizing hor-mone is critical for the final growth of follicles des-tined to ovulate (preovulatory follicles) and pro-vides the final signal for ovulation via a massive release of LH termed the LH surge. Luteinizing hor-mone is also responsible for the transition of follic-ular cells to become luteal cells of the CL, through a process called luteinization. The release of GnRH from the hypothalamus occurs in a pulsatile manner; hence, the gonadotropins, and especially LH, are re-leased in a pulsatile manner as well.

The secretion of LH and FSH is regulated by ovarian hormones that feed back on the hypothala-mus to control GnRH secretion. Secretion of FSH is inhibited by estrogen, and another hormone called inhibin, that is produced by large ovarian follicles. When large follicles ovulate, or undergo atresia, a surge of FSH initiates growth of a new group of fol-

licles. Progesterone inhibits LH secretion, therefore when a CL is present, secretion of LH is low. When the CL regresses, and progesterone declines, the se-cretion of LH increases and causes the preovulatory follicle to develop. The high concentrations of estro-gen produced by the preovulatory follicle acts at the hypothalamus to initiate the LH surge that causes ovulation.

The estrous cycle can also be described in four phases. Estrus is the first phase of the estrous cycle and represents the time during which the fe-male is sexually receptive (day 0 of the cycle). High estrogen concentrations from the preovulatory folli-cle are responsible for sexual receptivity and for in-duction of the LH surge that initiates the process of ovulation. Estrus is followed by the second phase which is referred to as metestrus. This phase starts with ovulation on day 1, and continues through day 4 to 5 of the estrous cycle. In addition to ovulation,

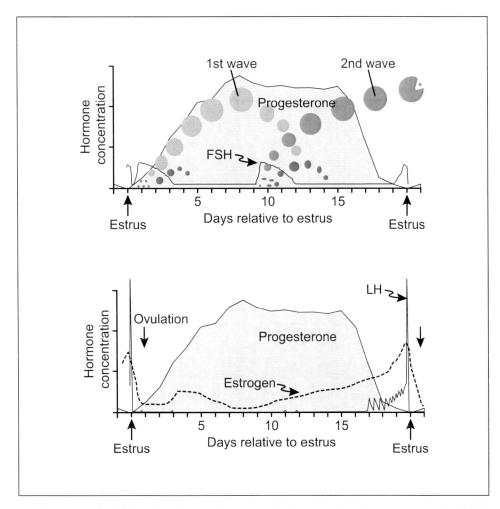

■ **Fig. 6.10** Follicle stimulating hormone initiates early development of a follicle. In the absence of adequate FSH and LH, the follicle regresses and does not reach the ovulatory stage (Top). As LH concentrations increase toward the end of the cycle due to decreasing progesterone, follicular development continues until peak estrogen concentrations initiate a surge in LH and ovulation (Bottom).

the other key event that occurs during this time is the formation of the CL and its increasing production of the hormone, progesterone. Diestrus is the third phase of the estrous cycle and spans the period from approximately day 5 to day 17 in animals with a twenty-one day estrous cycle. During this period, the predominant hormone is progesterone that is being produced by the CL. If the female becomes pregnant, the CL will remain functional until parturition. If not pregnant, PGF will cause luteal regression at the end of diestrus, leading to the final phase of the estrous cycle. The final phase is referred to as proestrus and is typically two to five days in length.

During this interval between regression of the CL and the next estrus, LH secretion is increasing, leading to increased growth of preovulatory follicles and the increasing production of estrogen to cause the next estrus.

Gestation

Gestation is the period of pregnancy for viviparous species that begins with fertilization of the oocyte and ends with parturition. Fertilization results in the formation of the zygote, which will undergo cellular division to produce a 16-cell embryo referred to as a morula. Continuing development transforms the morula into a blastocyst, which moves from the oviduct into the uterus at this stage of development. As the embryo continues to develop, tissues further differentiate to form either the fetus or the placenta. Attachment to the uterus occurs via the placenta. This process, referred to as placentation, allows for exchanges between the maternal and fetal blood supply. It is through the placenta that the fetus receives oxygen and nutrients from the dam and eliminates waste products and

■ **Fig. 6.11** Females of some species will only show estrus on a regular interval during certain seasons of the year (seasonal breeders). The mare and most breeds of sheep and goats are seasonal breeders. Sheep and goats respond positively to decreasing daylength, therefore females of most breeds are most likely to become pregnant in the autumn as daylength shortens after the summer solstice, which is the longest day of the year in the northern hemisphere (June 21). The signal for increasing sexual activity is the hormone melatonin, which is secreted by the pineal gland. Secretion of melatonin increases in darkness, therefore as days get shorter, this hormone increases and stimulates the hypothalmic-pituitary-ovarian axis. On the other hand, mares are most likely to become pregnant in early spring. Their reproductive axis is stimulated by the progressive lengthening of days that occurs after the shortest day of the year on December 21 (winter solstice). Due to the length of gestation in horses and sheep, this seasonal pattern of reproduction helps ensure that parturition will occur in spring. (© ncn18, 2009. Under license from Shutterstock, Inc.)

carbon dioxide through the dam. In livestock species, there is no direct exchange of maternal and fetal blood, and the dam does not pass immunity to the fetus through the placenta.

Placental types differ between species and are classified by distribution of sites of exchange for nutrients and waste products between the fetal and maternal environments. In the diffuse placenta of the sow and mare, sites of exchange are distributed over the entire placenta. In the cotyledonary placenta of the cow and ewe exchange takes place at distinct structures termed placentomes, which are formed between button-like projections called caruncles from the endometrium and cotyledons of the fetal membranes. In non-livestock species such as dogs, the placenta is described as a zonary placenta and is represented by a band of attachment. In primates exchange occurs with-in a disk shaped area of implantation and the placenta is termed discoidal. Due to the different placental structures in dogs and primates, there is immunity passed from the dam to the fetus.

The placenta is also a source of the hormones progesterone and estrogen. In early gestation, estrogen concentrations are low, but increase during middle and late gestation. Estrogens are produced by the placenta and work together with progesterone to develop and prepare the mammary glands for milk synthesis following parturition.

Parturition

Females preparing for birth often show behavioral changes such as the nesting behavior in which the female becomes more active; either building a nest

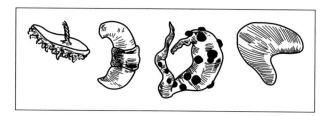

■ **Fig. 6.12** Comparative anatomy of placentation (Left to right): discoidal placenta of primates, zonary placenta of the bitch, cotyledonary placenta of the cow and ewe and diffuse placenta of the sow and mare.

or separating themselves from the herd. Prolactin is responsible for nesting behavior and also stimulates milk synthesis. During this time, females generally will show discomfort and the mammary glands will become swollen as they fill with colostrum.

Parturition is the process of giving birth to offspring. The fetus initiates parturition by secreting the hormone cortisol approximately forty-eight hours prior to birth. Fetal cortisol triggers a variety of hormonal changes within the mother. High concentrations of progesterone that were present throughout pregnancy decline rapidly, while at the same time, estrogens, oxytocin, prostaglandins, prolactin, and relaxin all increase in concentration. The corpus luteum also regresses. Relaxin stimulates the pelvic muscles and ligaments to relax in preparation for the passage of the fetus through the birth canal. Oxytocin causes the gradually increasing uterine contractions required for birth.

There are three main stages in the process of parturition. The first stage includes the dilation of the cervix, which is caused by relaxin and estrogen. These hormones work to soften the cervix and stimulate the epithelial cells to secrete mucous. During this stage, uterine contractions move the fetus into the pelvic canal. During the second stage of parturition, the outer membrane of the placenta (allantochorionic sac) is expelled through the vulva, is typically ruptured in the process, and leads to the release of the fluids that surround the fetus during gestation. It is during this stage that the fetus will subsequently be expelled through the increasingly more intense uterine and abdominal contractions. During parturition, the umbilical cord becomes separated from the placenta thereby breaking the oxygen supply from the dam to the fetus. It is essential that parturition progresses rapidly after this time to ensure that the fetus/neonate receives adequate oxygen. The expulsion of the placenta represents the third and final stage of parturition. If the placenta is not passed, it is referred to as a retained placenta, and, in some species, is considered an emergency that requires immediate intervention.

Difficulty during parturition is referred to as dystocia. The most common cause of dystocia is that the fetus is too large to pass through the birth canal. Another common cause is that the fetus is presented in an abnormal birth position. Dystocia is more common with male fetuses as they are typically larger at birth than females. The normal birth position for most animals is front feet first, with the head between the front legs. In cases of dystocia, assistance is frequently required for the successful delivery of the fetus.

Fig. 6.13 Strong uterine contractions to expel the fetus are initiated by the pressure generated by the fetus entering the birth canal. In cows, mares, and ewes the position of the fetus is usually front feet first. In sows either tail or head-first presentation is often observed. (© Margo Harrison, 2009. Under license from Shutterstock, Inc.)

Reproductive Processes in Avians

Female Reproductive Anatomy
While the goal of reproduction in poultry is the same as that of mammalian species, the process is quite different. The major difference in poultry is that the species is oviparous. The development of young does not occur within the body, but instead eggs are fertilized and then laid outside of the body to continue their development. The eggs must be incubated either naturally by the female, or artificially until development is complete and the young are mature enough to hatch.

Unlike mammals, only the left ovary is functional in poultry. The right ovary regresses and the right oviduct atrophies, and therefore neither are involved in reproduction. A female chick hatches with approximately four thousand ova, each enclosed in

Reproductive Characteristics of Females

	Estrous Cycle Length (days)	Estrous Duration	Ovulation	Gestation (days)
Sow	21	60 h	18–60 h	114
Ewe	16	30 h	1 h	150
Mare	21	4–10 d	24–48 h	336
Cow	21	14 h	10–14 h	280

In the sow ovulation is relative to the onset of estrus, or receptivity of the female to the male. In the mare and ewe ovulation occurs prior to the end of estrus, whereas in the cow ovulation occurs after the end of estrus.

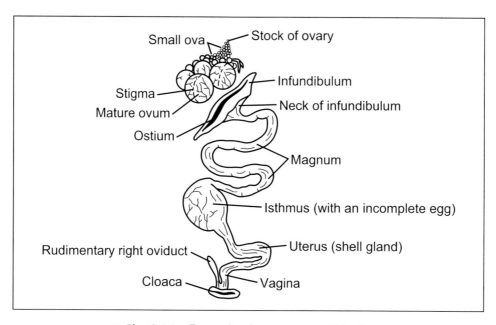

■ **Fig. 6.14** Reproductive anatomy of the hen.

separate follicles, attached to the left functional ovary. Maturation of these ova begins at puberty. The ova actually contain the yolks of the egg and are much larger than mammalian ova. Before each ovum is ovulated, it grows as yellow and white yolk granules are deposited. The granules are high in nutrients, especially fat and protein as the yolk provides a major source of nutrients for the developing embryo. This process is stimulated by FSH. Once the yolk has been fully formed, LH levels in the blood increase, stimulating ovulation along a line of the follicular wall termed stigma. The surge in LH that initiates ovulation relies on positive feed back of progesterone, this is in contrast to the stimulatory effects of estrogen on LH release in mammals. Furthermore, avians lack the formation of a corpus luteum. Following ovulation, the ovum is captured by the infundibulum and enters the oviduct. If sperm are present, fertilization will occur in the infundibulum. Sperm can remain viable for four to six days in the infundibulum, depending on the species. However, the yolk (ovum) will only spend approximately thirty minutes in the infundibulum and fertilization must occur within fifteen minutes. Whether

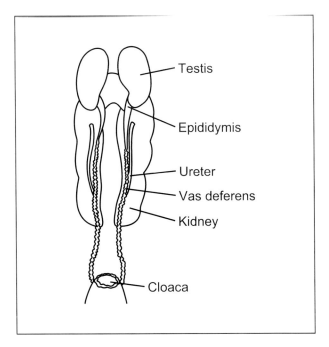

■ **Fig. 6.15** Reproductive anatomy of the rooster.

the ovum is fertilized or not does not affect the remaining processes of egg formation.

The ovum moves from the infundibulum to the second section of the oviduct, the magnum, where it will stay for two to three hours. Here the thick portion of the albumen, the egg white, is deposited around the yolk. In the third part of the oviduct, the isthmus, two thin shell membranes are secreted and surround the albumen. These are the membranes that lie just inside the egg shell. The ovum also takes up water and mineral salts during its one and a half hour stay in the isthmus. The egg then moves into the uterus, which is sometimes referred to as the shell gland. The egg will spend eighteen to twenty-one hours in the uterus. Here, the rest of the albumen is added to the egg by diffusion. The egg also undergoes plumping, in which water and minerals diffuse through the membranes that were previously formed in the isthmus. Following plumping, calcification of the shell occurs and pigment is added. The eggshell is primarily composed of calcium carbonate ($CaCO_3$). There is no cervix present in the female reproductive system, though there is a sphincter muscle between the uterus and the vagina. Within the vagina, the cuticle or bloom is added, which is the very thin outer layer consisting primarily of protein. When the cuticle dries, it seals the

small pores present in the eggshell. The egg then passes through the vagina and into the cloaca and is expelled from the cloaca to the outside of the body. Oxytocin, a pituitary hormone, stimulates muscle contractions of the uterus, inducing laying of the egg. The cloaca is a common orifice for copulation, defecation, and expulsion of the egg.

The total time from ovulation to laying is about 25.5 hours on average in chickens, but varies with poultry species and is approximately forty-eight hours in ostriches. Poultry do not have an estrous cycle; instead, they are continual ovulators, laying eggs in succession once they reach sexual maturity. Approximately thirty minutes after an egg is laid, another one is ovulated. Eggs are usually laid in the morning, and are laid about thirty minutes later each day. Since birds are sensitive to photoperiod and laying of the egg becomes later each day, at some point the laying cycle will reset. Selection for egg production and improved management has resulted in birds that lay three hundred to three hundred and fifty days eggs consecutively, depending on species. In comparison, wild birds as well as emus and ostriches reproduce seasonally and tend to only lay eggs in the spring.

Male Reproductive Anatomy

The reproductive system of the male bird is much more simplistic than that of the mammalian species, consisting of two large testes; each with an epididymis and vas deferens that lead to a rudimentary phallus (penis) that is erectile in ducks and geese. The location of the testes differs from the reproductive system of the male mammals. The testes of the rooster are located high within the abdominal cavity, along the backbone, and near the front of the kidneys. As in mammals, sperm are produced and mature within the seminiferous tubules; however, spermatogenesis in birds occurs at core body temperature, on average 106.7° F, and is aided by the nightly cooling of the body to 104° F. The epididymis of the rooster is relatively small compared to mammals and has reduced capacity for sperm storage. Instead, the vas deferens, leading from the testicles to the cloaca, acts as the major storage site for sperm. Roosters lack accessory sex glands as well and as a consequence the volume of the seminal fluid is reduced. In poultry, the seminal fluid produced by the testicles, carries sperm from the testicles to two papillae present on the cloacal wall that deliver the sperm at mating. During copulation, sperm is transferred from the male to the female through cloacal contact. This type of copulation is inefficient and unsuccessful in turkeys due to intense genetic selection for increased breast size, therefore, artificial insemination is used extensively.

Reproductive Technology

Artificial Insemination

Artificial insemination (AI) is a procedure in which semen is collected from the male and placed into the female's reproductive tract, using artificial means. Artificial insemination provides breeders and producers with opportunities to increase performance and profitability of their herds or flocks. Currently, AI is practiced in approximately 80% of dairy cattle, 95% of swine, 15% of beef cattle, and 95% of turkeys. The use of AI is limited in sheep due to cervical anatomy and to some degree in horses as a result of stringent breed registry requirements or occlusions. There are three major advantages of AI including: 1) an increased rate of genetic improvement by use of superior sires, 2) the reduction or elimination of the cost and risk of owning and maintaining sires, and 3) the reduced risk of reproductive diseases. The process of AI requires the collection of semen from the male, the detection of estrus or synchronization of ovulation in the female, and the proper placement of semen in the female. Following good management practices, the reproductive efficiency using AI is comparable or improved compared to natural mating systems.

The most common method of semen collection in cattle and horses is causing the male to ejaculate into an artificial vagina (AV) which has a collection receptacle. This is commonly referred to as the AV method. In this method of collection, a teaser-mount animal or a dummy mount is used. In cattle, cows, bulls, or steers can be used as teaser-mounts. As the bull mounts the teaser, the penis is guided into the artificial vagina and maintained until the bull dismounts. Alternatively, semen may be collected through the use of an electroejaculator. This method is typically used with males that are unable to mount a female as a result of injury or age, or with males who are not trained to mount an animal or dummy for this purpose. An electrical probe is placed in the rectum of the male and a very low electrical current

stimulates the accessory sex glands and contraction of muscles that causes ejaculation. The method of semen collection used with boars is referred to as the gloved hand technique. Boars are usually trained to mount a dummy structure. Once the boar mounts, the tip of the penis is grasped firmly and pressure applied to simulate the cervix of the female until ejaculation is complete. A pre-warmed thermos is used to collect the semen. Semen is collected from poultry by using a stroking and milking technique. Stroking of the male bird from the pelvic arch to the pubic bones will cause the male to raise his tail and invert the cloaca. The enlarged portion of the vas deferens that stores semen, located near the entrance to the cloaca, is squeezed by the pressure of the thumb and forefinger and the semen is ejaculated.

Once semen is collected from the male, it is evaluated to determine concentration, motility and morphology. Since a single ejaculate of semen contains more sperm than needed to impregnate a female with AI, the semen is diluted with an extender solution containing nutrient, buffers, and antimicrobials and aliquoted into storage vials. Aliquoted semen is either rapidly frozen in liquid nitrogen or cooled to prolong the life of the sperm. Semen from bulls can be frozen and stored indefinitely. On the other hand, semen of boars and stallions is not frozen as efficiently; therefore, it is often stored fresh at 40° F. Poultry semen is usually used within two hours of collection.

One of the most important, yet difficult tasks in the process of artificial insemination is the detection of estrus in the female. The insertion of semen into the female must be coordinated with ovulation to ensure fertilization of the oocyte. Most species demonstrate specific behaviors that are characteristic of estrus, and aid in its detection. General behaviors demonstrated by females in estrus, regardless of the species, include restlessness, irritability, and excitability. Females may exhibit increased vocalization and show interest in other animals present. Cows in estrus will spend more time than usual walking and spend less time resting or eating. They will mount other cows, and allow female or male cattle to mount them, beginning approximately twenty-eight hours before ovulation. When the cow shows standing heat (standing to be mounted by other cattle) she is normally inseminated approximately twelve hours later. Sows in estrus will stand when pressure is applied to their back and will typically have a swollen, red vulva. Sows should be artificially inseminated each day they are in standing estrus. Ewes do not exhibit any obvious signs of estrus if a ram is not present, making detection more difficult. In the presence of a ram, a ewe in estrus will roam around the ram and will rub his neck and body, while vigorously shaking her tail. Mares in estrus will allow the stallion to smell and bite at her. She will stand with her hind legs extended, lift her tail to the side, and lower her hindquarters. The clitoris will be exposed by frequent contractions (winking) of the labia. For accurate detection of estrus, the mare should be teased by a stallion. If the mare is aggressive toward the stallion, it is an indication that she is not in estrus, despite other signs being present.

Estrous Synchronization

Estrus synchronization is a process in which estrus and or ovulation is induced in all the females in a herd or group at the same time, resulting in ovulation in the group of animals occurring within a short window of time. This is done to help eliminate the problem of estrus detection. Estrus synchronization also simplifies the use of AI by having all or a select portion of the herd ready to be bred by AI at the same time. Overall, synchronization reduces labor and allows for a more organized and efficient production system. It also may allow the producer to shift the parturition season to more closely coincide with the most favorable marketing patterns or seasonal forage production. There are several different approaches to synchronization in each species. These systems employ the use of PGF, GnRH and/or progesterone (or progesterone-like compounds) to control the estrous cycle.

Superovulation and Embryo Transfer

The process of embryo transfer involves the collection of fertilized embryos from one female to be placed in another female for further development. This technology became available in the livestock industry in the 1970's. Today it is most widely used in the cattle industry. The value of embryo transfer is the opportunity to produce more offspring from genetically outstanding females, just as AI provides the opportunity to produce more offspring from superior males.

The first step of the process is to induce the female donating the embryos to superovulate. Cows generally only ovulate one oocyte per cycle.

Superovulation allows a superior cow to produce several oocytes at one time for embryo transfer. This is accomplished by giving successive injections of FSH over three to four days to the cow to induce growth of multiple preovulatory follicles. Prostaglandin $F_2\alpha$ is injected causing the CL to regress and the cow will exhibit signs of estrus thirty-six to sixty hours later. FSH prevents the death of follicles that would normally undergo atresia and recruits new follicles into the existing pool, allowing more follicles than usual to reach the preovulatory stage. The donor female is inseminated twelve hours following onset of estrus and again after another twelve hours to fertilize the oocytes. These resulting embryos are collected from the donor female seven days later through a process of flushing the uterus with fluid. The embryos are identified under a microscope and are examined for normal morphology. After being evaluated each embryo is frozen for storage and later use or placed immediately into a recipient female synchronized to be at the same stage of the reproductive cycle as the donor female. The embryo, barring any complications, will develop into a fetus within the recipient. The offspring will carry the genetics of the sire and the donor female.

The most common method for transferring embryos to recipients is non-surgically, through a procedure very similar to that described for artificial insemination. The embryo is placed into an AI straw,

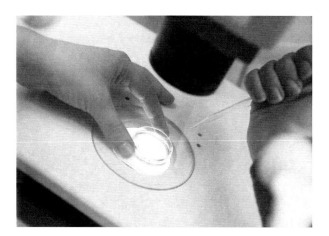

■ **Fig. 6.16** Embryos produced by the donor female are collected and examined for normal morphology under a microscope before being frozen for storage or transferred to a recipient female. (© Monkey Business Images, 2009. Under license from Shutterstock, Inc.)

the straw is placed in a transfer gun and carefully passed into the uterine horn that is adjacent to the ovary containing the functional CL. The pregnancy rates in cattle average 60% for commercial embryo transfer companies. Embryo transfer is typically limited to use with elite females. It is also mostly used in cattle. There is less incentive with swine since they are litter producers and have much shorter generation intervals. The procedure is difficult and costly to perform in sheep and goats.

In Vitro Fertilization

In vitro fertilization is the fertilization of oocytes outside of the body in a culture dish. The embryo is allowed to develop for approximately seven days in culture until it is transferred to a recipient. Oocytes can be collected from the ovary using a procedure referred to as ultrasound-guided follicular aspiration. These oocytes are then fertilized as described above. In vitro fertilization provides some advantages over other methods of embryo production. One advantage is that it allows for more frequent collections of oocytes as well as the collection of oocytes during pregnancy. Pregnancy rates with embryos produced in vitro are approaching those that can be achieved by traditional embryo transfer procedures. As technology advances, these rates will likely increase.

Lactation

"Man is the only creature that consumes without producing. He does not give milk, he does not lay eggs, he is too weak to pull the plough, he cannot run fast enough to catch rabbits. Yet he is lord of all the animals."

—George Orwell (1903–1950)

This chapter describes the process of lactation and the hormones involved in the control of milk production and letdown. You will gain an understanding of basic mammary anatomy and the anatomical difference of the mammary glands among various mammalian species. The lactation cycle and the processes involved in milk synthesis are emphasized. Current societal concerns, such as the use of the bovine somatotropin hormone, and industry concerns, including the incidence, detection, and treatment of mastitis, are discussed.

Lactation: A Historical Perspective

The origins of lactation trace to over two hundred million years ago but remain obscure. It was originally suggested that following the beginnings of live birth, prolonged contact between the mother and offspring was necessary to maintain temperature of the young. Food originally brought to the young, a nutritional strategy maintained in the care of birds posthatch, was gradually replaced by maternal secretions derived from the mobilization of the female's body reserves. Indeed, the mammary gland is simply a highly evolved skin gland. More recently, it was proposed that lactation evolved as a mechanism to maintain the moisture of eggs, but became an important alternative food source that led to the loss of egg yolk for nourishment and the rise of mammals. Regardless of origin, lactation is a universal feature of all mammals that affords continued maternal contact between the female and her offspring that extends beyond the in-utero environment. Lactation enables the young to be born at a relatively immature stage and permits continued growth after birth. For many mammals, the developing fetus only represents a fraction of mature weight of the adult. The newborn of elephants, chimpanzees, and cattle are only 4–7% of maternal weight. Limiting the weight of the fetus in-utero allows the female to remain highly mobile during gestation. This ensure the animals ability to travel over long distances to secure food and flee predation when necessary. The prolonged investment in lactation and care of the young at the expense of the mother provides both nutritional support and protection through milk.

Overview of Lactation Strategies

A major characteristic of all mammals is that they employ lactation as a part of their reproductive strategy to feed and nourish their young. The complexity of the mammary system is coincident to the complexity of placentation and differs between monotremes, marsupials, and eutharians. In monotremes, which are egg laying mammals, primitive mammary glands are characterized by the absence of teats. For the duck-billed platypus, newly hatched immature young are reliant on milk for three to four months post hatch. On each side of the midline of the abdominal wall milk extrudes from one hundred to one hundred fifty paired tubal glands that open at the base of a stiff mammary hair. Secretory tubes have two cell layers, an inner secretory layer and an outer contractile layer. There is no internal storage of milk and as milk is produced it is secreted and subsequently licked by the young.

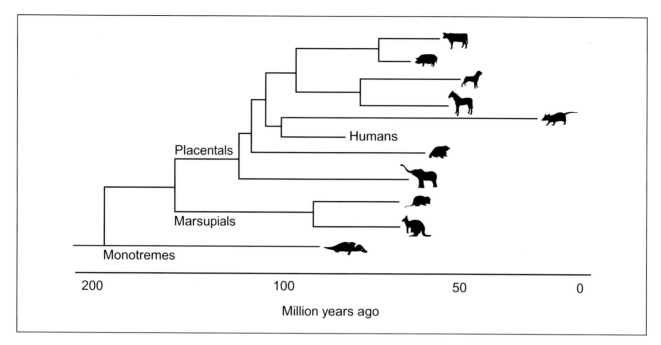

■ **Fig. 7.1**   Lactation as a strategy to nourish the offspring evolved with the origins of mammals. It is evident in the egg bearing monotremes and remained with the divergence of placental mammals. (Animal images © Goran J, 2009. Under license from Shutterstock, Inc.)

Marsupials, including the koala bear, opossum, and kangaroo, give birth to live young after a relatively short gestation. The majority of the physiological development of the young occurs after birth and is coincident with an extensive lactation period relative to gestation. The placenta that develops is simple and only supports embryonic development for thirteen to forty days, depending on species. Generally referred to as pouched animals (although not all have a pouch), premature young that are characterized by limited neural, kidney, and lower limb development at birth will move from the uterus to a ventral depression or pouch. For the kangaroo, teats are located within the pouch and each are associated with a simple mammary gland. There are four mammary glands and the development of the gland and teat corresponds to the suckling stimulus of the young. As the immature joey reaches the pouch it will seek out and latch on to a single teat. As the joey suckles, the mammary gland will be stimulated to initiate lactation. Only the teat and mammary gland stimulated by the joey will initiate lactation, the remaining three mammary glands will regress. The teat develops in size as the joey grows and the joey remains with the teat during the entire lactation. Once the joey reaches a stage to intermittently

leave the pouch, it will return to suckle from the same teat. During this time, the female will conceive to support the development of another neonate. As the first joey leaves the pouch, the second is born, travels to the pouch, and begins to suckle from a different teat than the first joey. The two teats that are not stimulated by suckling will regress. Thus, two mammary glands are stimulated to provide milk to support the unique needs of different aged offspring at the same time. Greater investment in lactation with minimal input into gestation allows the female to terminate the young under unfavorable conditions without reproductive costs, as there was little time invested into the gestation phase. For example, the female may eject the young from her pouch if there are not adequate resources to support lactation and postnatal growth; however, there will be limited delay in the time required to conceive a subsequent joey when resources become available. For the opossum, an excess of twenty-five young are born, however, only fifteen teats are available to nourish the young. Excess young are expendable and postnatal mortality rates are high, however maximal investment in lactation is ensured.

Placental mammals account for over 90% of all mammalian species and are distinguished by a com-

plex placenta. In contrast to marsupials, a greater investment is placed in gestation and the fetus develops in-utero, receiving maternal nutrients through the placenta. The young are classified as altricial or precocious depending on the degree of maturity at birth. For example, altricial animals are considered immature at birth and are more reliant on maternal care than precocious neonates. Altricial young are incapable of coordinated movement, eyes are frequently closed, and there is an absence of hair and teeth at birth. Mammals in this group include many rodent species. In contrast, precocious young, including many species of agricultural significance, are relatively mature at birth and senses of sight and hearing are fully developed. Precocial young are born with hair and teeth and often within minutes after birth, are capable of standing and fleeing from danger or predation. Although there is a greater investment in gestation, milk remains of vital importance for early nourishment and immune protection.

Mammary Gland Structure

The mammary glands of animals including goats, cattle, horses, and giraffes are located in the pelvic, or the inguinal region; whereas the mammary glands of the sow, dog, and cat develop in two parallel rows along the abdomen. Humans, other primates, and elephants have mammary glands located in the pectoral or breast region. The number of mammary glands varies among species and may be referred to as simple or complex. In the simple mammary gland, all milk secreting tissues empty through a single teat opening. In complex mammary glands, there are multiple openings within the teat, each corresponding to a simple mammary gland. Humans have two complex mammary glands, one in each breast, with each complex gland containing ten to twenty simple glands. Cows have four simple mammary glands, each with a corresponding teat. Ewes have similar anatomy to the cow, except there are only two simple mammary glands and two teats. Mares have two complex mammary glands, but they actually have four simple mammary glands, with two streak canals, two teat cisterns, and two glands cisterns per teat. The sow has two rows of complex mammary glands, each row usually having six to eight teats for a total of twelve to sixteen, although

■ **Fig. 7.4**   Placental mammals first appeared approximately 101 million years ago. The young are classified as altricial (left) or precocious (right) according to maturity at birth. Extensive gestational development occurs relative to other mammals. In livestock species, lactation affords the newborn immunological protection against environmental pathogens until active immunity is established. (*Newborn mice:* © max blain, 2009. Under license from Shutterstock, Inc. *Newborn horse:* © Stephanie Coffman, 2009. Under license from Shutterstock, Inc.)

greater and lesser teat numbers are present in some animals. Each teat is served by two streak canals and two teat cisterns continuous with the glands cisterns. The teats may or may not be spatially paired; therefore, the number on each side of the midline may differ.

External Structure

The mammary gland is commonly referred to as the udder in cattle, goats, sheep, and horses. Each mammary gland is supplied by its own nervous and lymphatic systems and the right and left halves of the udder receive their own blood supply. There is no direct exchange of milk or blood between any of the glands; however, as blood recirculates throughout the body, it will pass through the other three mammary glands. Therefore, if one gland receives drug treatment, all milk produced must be discarded. If one mammary gland is injured; however, the remaining glands will continue to function since they are separate from one another.

In dairy cattle, the front mammary glands of the udder are smaller and produce and store approximately 40% of the milk, whereas the back two quarters are larger, producing and storing the remaining 60% of the milk. On average, a cow's udder will hold twenty to forty pounds of milk, although some high producing cows may have udders holding up to sixty or seventy pounds of milk. In addition to milk weight, there is the weight of the blood present in the udder. For one gallon of milk to be produced, one hundred thirty gallons of blood must pass through the udder, thereby providing a continuous nutrient supply to the secretory cells. When considering both milk and blood, the udder holds a total of approximately one hundred pounds. It is important that the udder is well supported since it is located completely outside the body and is required to hold substantial weight.

The udder is supported by the suspensory system, consisting of a medial suspensory ligament and two lateral ligaments. The medial suspensory ligament provides the majority of support. This ligament is an elastic tissue that originates from the animal's midline and separates the two halves of the udder. The two lateral ligaments also provide support for the udder. These ligaments are non-elastic and fibrous. Originating about the udder, they are located on each side of the udder and meet with the medial ligament across the bottom of the udder.

Internal Structure

Alveoli are the primary functional units in the mammary gland responsible for milk synthesis and storage. Spherical and arranged in clusters within lobules, alveoli are lined with a single layer of epithelial (secretory) cells. As milk is synthesized, it is transferred to the lumen of alveoli and for some species, including water buffalo and dairy cattle, the alveoli are a major site of milk storage. The epithelial cells of alveoli

■ **Fig. 7.5** The gross anatomy of the mammary gland differs considerably across species. Whereas the cow has one simple mammary gland corresponding to each of four teats, sows have between twelve and fourteen teats each with two simple mammary glands associated.. (*First image:* © Carsten Erler, 2009. Under license from Shutterstock, Inc. *Last image:* © RestonImages, 2009. Under license from Shutterstock, Inc.)

have three major functions: 1) absorb nutrients and other precursors of milk from the bloodstream, 2) synthesize nutrients specific to milk, and 3) secrete milk components into the lumen of the alveolus. Each alveolus is vascularized for the delivery of nutrients and each is surrounded by specialized muscle cells termed myoepithelial cells. Contraction of the myoepithelial cells forces milk from the lumen of alveoli into a ductal network that transports milk to the gland and teat cisterns. In some species, the gland and teat cisterns are the primary site of milk storage and are demarcated by the annular ring. Milk exits from the teat cistern through the streak canal of the teat. At the junction of the teat cistern and streak canal is Fürstenberg's rosette, folds of tissue which compress under milk pressure to prevent milk leakage and which contain bactericidal proteins to protect the mammary gland from infection. The main barrier against infection; however, is the streak canal. The canal is lined with keratin, which has bacteriostatic properties and helps to seal the streak canal between milkings. The canal is maintained closed between milking through the teat sphincter.

■ **Fig. 7.6** The median suspensory ligament divides the udder into right and left halves and can be viewed from the rear of the cow. (© Chris Turner, 2009. Under license from Shutterstock, Inc.)

Lactation Cycle

Mammals reproduce more than once, and therefore, lactate more than once. The mammary gland is one of a relatively few structures of the body that undergoes repeated cycles of structural development, functional differentiation, and regression. Structural development, or mammogenesis, begins the lactation cycle and is greatest during pregnancy. Functional differentiation represents active mammary tissues and is marked by lactogenesis, milk synthesis and secretion, and galactopoesis, or maintenance of lactation. Lactation is an event that coincides with the formation of colostrum, in coordination with parturition, and is

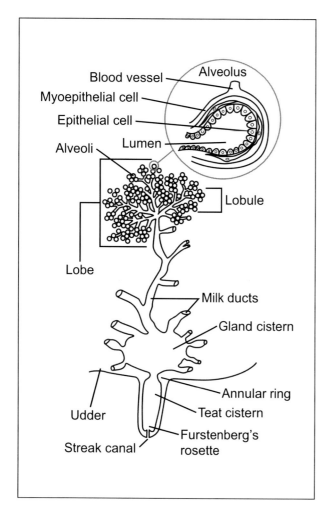

■ Fig. 7.7 Milk synthesis occurs in the alveoli, which are comprised of a single layer of epithelial cells. Each alveolus is surrounded by myoepithelial cells. When the myoepithelial cells contract, milk is secreted into the lumen of the alveolus and subsequently into the milk ducts and then the gland cistern.

maintained until the young no longer needs milk or milk is no longer removed from the gland. The mammary gland then undergoes regression, or involution, which is characterized by a return of the mammary gland to a non-lactating state.

Development and Growth of the Mammary Glands

In all mammals, the mammary gland is a highly evolved organ that arises from thickening of the epidermis during embryonic development. It consists of functional secretory tissue, also referred to as the

parenchymal tissue, and a non-secretory framework known as the stroma. In the lactating mammary gland, the parenchymal tissue is composed of epithelial structures, such as alveoli and ducts, and the associated stroma consists of connective tissue as well as blood vessels. There also is considerable white adipose tissue that exists in a mammary gland from the early phases of fetal development extending through much of pregnancy.

The development of the mammary system is similar among lactating species and occurs during the prenatal, prepubertal, peripubertal, and postconception stages of development; however, the duration of each stage differs and is influenced by gestational strategy. In cattle, the mammary gland begins formation at day twenty-five of embryonic development. A single layer of cells associates on each side of the midline to form the mammary streak. Along the mammary streak, distinct areas differentiate into mammary buds. In rodents, formation of the mammary buds occurs around embryonic day ten. The number and location of mammary bud formation determines the number and location of external mammary glands observed in the adult. Each bud will give rise to a primary spout, which is the precursor of the teat and gland cisterns that are evident by day ninety of gestation in cattle. Several secondary sprouts will arise from each primary sprout. These secondary sprouts represent future mammary ducts and will become canalized shortly before birth. Additional events in prenatal mammary development include the formation of the mammary fat pad. First observed at day eighty of gestation in cattle and day sixteen of the rodent, the mammary fat pad supports late fetal and postnatal ductal development. By six months of gestation the mammary tissue of the bovine fetus is well developed and further structural development does not occur. Immediately after birth, the immature ductal system will regress while the mammary gland experiences isometric growth, in which the rate of growth parallels that of the body. Prior to the onset of puberty, the isometric growth of the mammary gland is replaced by allometric growth, in which the rate of growth exceeds that of the body. Signaled by ovarian function and estrogen production, allometric growth corresponds to proliferation and expansion of the ductal system. Interestingly, excess energy intake during the period of allometric growth can impair mammary development and reduce milk production during the subse-

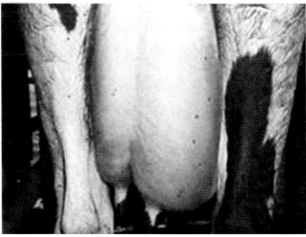

■ **Fig. 7.8** Following milk release in cattle, the teat sphincter will remain open for fifteen to sixty minutes, thus leaving the mammary gland open and susceptible to microbe entry during the hour following milking. A concern in lactating animals is mastitis, inflammation of the mammary gland that is caused by a bacterial infection. It occurs most often in dairy cattle, but it can affect all other domestic animals, especially swine, as well as humans. Symptoms of clinical mastitis include flakes or clots appearing in the milk (left) and the infected quarter of the udder is usually inflamed: swollen, red, hot and sensitive to the touch (right). A persistent low level of infection, referred to as subclinical infection, occurs more frequently in dairy herds than clinical infection, and significantly reduces both the production and quality of milk. Low level infections are generally not evident to the milker-operator, but can be detected using certain laboratory tests. The somatic cell count (SCC) is one such test that provides an indication of milk quality as well as the severity of subclinical mastitis. Mastitis must be treated with antibiotics. Milk from cows being treated for mastitis cannot be sold for human consumption due to zero tolerance laws that prevent milk that contains any antibiotic residues from being sold for human consumption. Therefore, mastitis leads to a decrease in profitability. If not treated promptly or appropriately, mastitis can cause cellular damage to the mammary gland, thereby causing a decrease in milk production and severe cases of mastitis can result in death. In efforts to control the incidence and severity of mastitis, milk producers often take the following precautions: ensuring proper cleaning and sanitation of the milking machines; using correct milking procedures that minimize the chance of infection; providing feed for the cows as they leave the parlor to encourage them to remain standing until their teat canals close; carefully monitoring somatic cell counts; and treating clinical cases promptly. (Courtesy of Joe Hogan, The Ohio State University)

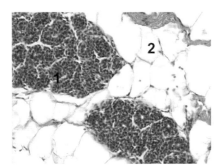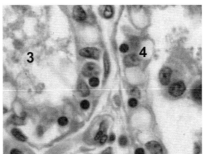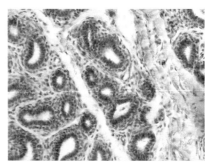

■ **Fig. 7.9** Haematoxylin and eosin stained bovine mammary tissue. Note the lobules (1) of the nonlactating developing mammary tissue and the presence of adipose tissue (2). Milk (3) is observed in the lumen of the alveoli in the lactating mammary tissue and lipid droplets are observed prior to being secreted from the epithelial cells (4). In the mammary gland undergoing involution, the alveoli (5) are reduced in size and there is considerable non-lactating stromal tissue (6) between alveoli. The excretory milk ducts also are observed (7). (Courtesy Ann C. Ottobre, The Ohio State University)

quent lactation in dairy cattle. The allometric growth ceases with subsequent estrous cycles, however, fluctuations in estrogen and progesterone that occur during the estrous cycle are important toward complete ductal development and establishment of the lobule-alveolar system.

The postconception stage of mammary development is characterized by increased progesterone concentrations and extensive lobule-alveolar development. Although alveolus budding occurs prior to conception, it is not until pregnancy that the secretory cells develop. There is a significant increase in the number of cells, and thus the amount of DNA, of mammary tissue during pregnancy with the greatest increases corresponding to the last stages of gestation when rapid fetal development also occurs. Gestation also corresponds to increased vascularization of the mammary gland. Structural development of the mammary gland during gestation requires both estrogen and progesterone.

Milk Synthesis and Secretion

The increase in size of the mammary gland just prior to parturition is due to the accumulation of milk in the alveoli. As parturition approaches, prolactin from the anterior pituitary initiates and maintains lactation. During parturition prolactin and oxytocin, in addition to other hormones, peak to stimulate the mammary glands to come into a full lactational state. This allows for the production of colostrum for the first few days, and then provides an increasing amount of milk for subsequent days. It is not until several weeks following parturition that milk production will reach its peak. The reason for this delay in maximum output is not fully known, but may be attributed to the many physiological and biochemical changes that the mammary gland must undergo. On average, dairy cattle achieve maximal milk production between fifty and seventy days postcalving and may produce in excess of one hundred pounds of milk per day during peak lactation. In contrast, dairy goats achieve peak milk production between thirty and forty-five days postkidding and produce an average of 9.4 pounds of milk per day during peak lactation. Once peak production has been reached, milk production will begin to decline as the secretory cells lose their functional abilities. The prolonged and steady production following the peak is referred to as persistency. Persistency is necessary for high production rates over the entire

lactation period. Many factors may affect persistency including genetics, nutrition, disease, as well as the frequency and completeness of milking.

Oxytocin is the most notable hormone responsible for stimulating milk letdown (release). Produced in the hypothalamus and released from the posterior pituitary gland, oxytocin reaches the mammary gland via the bloodstream. At the mammary gland, oxytocin acts on myoepithelial cells, causing them to contract. As stated, these cells surround the secretory cells and the contraction of these myoepithelial cells forces the milk that is stored in the lumen into the ducts that lead towards the gland and test cisterns. The initiation of milk letdown increases the pressure in the mammary gland, thereby promoting

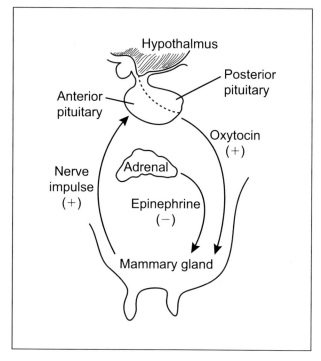

■ **Fig. 7.10** Milk ejection is stimulated by oxytocin release from the anterior pituitary. Direct stimulation of the teats by the milker's hands or suckling by the young can initiate a nerve impulse to trigger oxytocin release. In addition, cows associate the milking routine with certain environmental cues and will begin to eject milk in response to these cues, which may include the sounds of the milking equipment or the presence of feed. It is important to minimize stress as the release of epinephrine by the adrenal glands works in opposition to oxytocin to inhibit milk ejection.

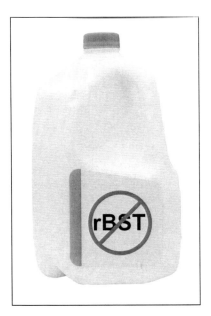

Fig. 7.11 Treatment of dairy cows with the naturally produced hormone, bovine somatotropin (BST; also known as growth hormone) increases milk production by decreasing the rate of decline in milk production following peak lactation; however, it will not prevent involution or eliminate cessation of lactation. In the 1980's, following years of research, pharmaceutical companies developed a recombinant bovine somatotropin (rBST) product that could be given to dairy cows to complement their natural concentrations of somatotropin. Both the Food and Drug Administration (FDA) and National Institute of Health (NIH) concluded that dairy products produced from the milk of cows treated with rBST were safe for human consumption. In 1993, based on the results of numerous studies, the FDA approved the use of BST. In 2009, following pressure from consumers who demanded milk products free of rBST, its use diminished substantially in the United States dairy industry.
(© Jaimie Duplass, 2009. Under license from Shutterstock, Inc.)

milk letdown. In dairy cattle, one to two minutes following the initial stimulation of the udder, oxytocin and mammary pressures are the greatest. Six to seven minutes following initial stimulation, the liver and kidney remove the oxytocin from the bloodstream, causing milk letdown to cease.

Oxytocin release is dependent on the female recognizing a stimulus. If the young are with the female, milk letdown is initiated by stimulation of the teats by the female's offspring beginning to nurse. Females also are conditioned to release milk by other stimuli associated with milk removal, such as the visual appearance of the offspring; or in the case of dairy animals, the sights and sounds associated with the milking parlor.

Epinephrine (adrenaline) and norepinephrine are hormones that also play a role in lactation; however, they work in opposition to oxytocin. These hormones are released when an animal becomes stressed, frightened, or nervous. Epinephrine and norepinephrine inhibit milk letdown by causing the blood that usually bathes the myoepithelial fibers

Milk Composition

	Water (%)	Lactose (%)	Fat (%)	Protein (%)	Energy (kcal/100g)
Cow	87	4.6	3.9	3.2	74
Human	87	7.1	4.5	0.9	72
Buffalo	83	4.8	7.4	3.8	101
Goat	87	4.3	4.5	3.2	70
Donkey	88	7.4	1.4	2	44
Elephant	78	4.7	11.6	5	143
Rat	79	2.6	10.3	8.4	137
Bat	60	3.4	18	12.1	223
Seal	35	0.1	53	9	516

and contains oxytocin, to be shunted to the body's extremities. Milk letdown will thereby decrease as the fibers relax due to the lack of oxytocin. During stress, the hypothalamus may fail to stimulate adequate release of oxytocin from the posterior pituitary or the myoepithelial cells may fail to respond to the oxytocin. Due to these effects, it is important to minimize stress on animals during milking time so as to not interfere with milk letdown.

Involution

During this period following peak production, the mammary gland will undergo a gradual involution; decreasing in weight, volume, and productivity. Involution may involve a reduction in alveoli size and their synthetic capacity, as occurs in dairy cattle; or extensive tissue degeneration and an almost complete loss of alveoli, which has been observed in mice. The loss of the epithelial cells of alveoli that occurs in rodents is considered programmed cell death, or apoptosis. As the female approaches her next parturition, structural development of the mammary tissue is once again initiated. Redevelopment of the alveolar system will support the ensuing lactation. The increase and decrease in milk production during lactation is referred to as the lactation curve. The lactation curve is a plot of milk production over the duration of the lactation period and each species has a characteristic lactation curve. Involution and the gradual decline in milk production demonstrated over the lactation curve are inevitable and occur despite continued milk removal from the mammary glands. Cessation of milk removal, however, will accelerate involution at any point along the lactation curve and promote the relatively rapid return of the mammary gland to a non-lactating state.

Milk Composition

Milk is the primary source of nutrition for mammalian infants during early postnatal life and is comprised of water, triglycerides, lactose, protein, minerals, and vitamins. Milk production in most females corresponds to the nutrient needs of the young. For example, the milk of marine mammals that occupy cold environments is greater in lipid and negligible in lactose as these young do not readily digest carbohydrates, yet deposit considerable lipid after birth for insulation. In addition, most females produce only enough milk to feed their young, however, dairy animals have been bred and developed to maximize milk production so excess is available for human consumption.

Milk synthesis occurs in the alveoli, where epithelial cells receive a continuous supply of nutrients from the blood. The epithelial cells of the alveoli acquire glucose, medium and short chain fatty acids, vitamins, and minerals from the circulation. In addition, milk specific nutrients including lactose and casein are synthesized locally.

Carbohydrates

The primary carbohydrate found in the milk of nearly all species is lactose. Only the epithelial or secretory cells produce lactose, which is a disaccharide of glucose and galactose. All glucose required by the mammary tissue is provided by the bloodstream, as the mammary gland is incapable of synthesizing glucose. The glucose may be used directly by the epithelial cells for energy, converted to glycerol for triglyceride synthesis, or converted to galactose and used for the production of lactose. The inability to synthesize glucose makes glucose availability an important limiting factor in milk

■ **Fig. 7.12** There is a greater than two fold increase in blood flow to the mammary glands prior to parturition. It is estimated that each liter of milk requires the circulation of five hundred liters of blood through the mammary glands. The abdominal vein, otherwise known as the milk vein, is highly visible under the skin of the lactating cow. (© Luca Flor, 2009. Under license from Shutterstock, Inc.)

component synthesis. A reduction in circulating glucose limits the amount of lactose synthesized by mammary tissue. Furthermore, lactose production contributes to milk volume. Lactose secretion into the lumen of the alveoli increases the concentrations of dissolved solutes relative to the bloodstream. As this occurs, water is drawn into the lumen of the alveoli to maintain osmotic pressure. This results in a high concentration of water in milk, generally greater than 80% in most species. As lactose production declines, the amount of milk produced per day also declines.

Since lactose is a disaccharide, it must be broken down in the small intestine by the enzyme lactase. Individuals that lack adequate amounts of lactase are therefore unable to degrade lactose in a condition referred to as lactose intolerance. Lactose intolerant individuals experience gastrointestinal distress, including diarrhea, intestinal gas, and bloating following milk consumption. The incidence of lactose intolerance varies widely with approximately 90% of the population of Asian ancestry affected, 70% of the African-American population, and only 2–8% of those of Scandinavian and Western European descent affected. Individuals with lactose intolerance can alleviate the condition by consuming lactose free products or consuming cultured products that have undergone fermentation of the lactose.

Lactose intolerance is frequently misperceived as a milk allergy since the two conditions share many symptoms. Milk allergy experienced by humans is in response to proteins found in milk. The direct protein responsible for the allergenic response remains unknown and both casein and β-lactoglobulin are suspected milk allergens. The incidence of milk protein allergy in human infants is estimated to be less than 1%.

Protein

There are many proteins found in milk, though the primary protein is casein, comprising over 80% of the total milk protein. The mammary gland is the only tissue capable of synthesizing casein. There are many forms of casein and all carry negative charges as a result of the phosphate groups held in associations with the proteins. The epithelial cells sequester calcium from the bloodstream in order to secrete these negatively charged proteins, making casein responsible for the calcium content of milk. Milk serum proteins make up approximately 18% of the

protein in milk. These milk serum proteins provide amino acids for the synthesis of other proteins by the body when digested.

Colostrum, the first milk produced following parturition, contains increased concentrations of proteins, as well as fat, minerals, and vitamins A and C. The increased protein is due to the transfer of immunoglobulins (antibodies) into the milk. These immunoglobulins provide the offspring with increased resistance to diseases and infections. This type of immunity is referred to as passive immunity since the animal receives the antibodies from the mother's milk. The functionality of these immunoglobins depends on their absorption from the small intestine.

Mammals may be classified by the mechanism of acquired immunity. For pigs, horses, goats, and cattle, immunity is acquired passively through the

■ **Fig. 7.13** Calcium required to neutralize the protein casein is removed from the bloodstream of the female. In early lactation, the demand for calcium in milk production and the removal of calcium from the blood can cause a rapid decline in circulating calcium that exceeds the ability to mobilize calcium from body reserves. As a consequence, hypocalcemia may occur and in dairy cattle, the reduced blood calcium may result in milk fever, which is characterized by muscle weakness, loss of appetite, and eventual heart failure if left unattended. This condition, also referred to as parturient paresis, is treatable by an intravenous dose of calcium.

consumption of colostrum. For dogs, rats, mice, and hamsters, transfer of antibodies to the young occurs both in-utero through placental transport and postnatal through colostrum. In the last classification, which includes humans and rabbits, placental transfer of antibodies is the only source for passive immunity. The intestine of a neonate is able to absorb large macromolecules such as immunoglobulins as a consequence of its immaturity. Immunoglobulins absorb to the surface of the enterocyte and are internalized by pinocytosis. The absorbed immunoglobulins are subsequently transferred into the bloodstream. As the intestine matures, it is no longer able to absorb immunoglobulins, a process known as gut closure. Gut closure is considered complete at twenty-four hours in the calf, thirty-six to forty-eight hours in the pig, twenty-four to forty-eight hours in the foal, and twenty-four to forty-eight hours in the dog and cat; therefore, it is essential that the offspring receive colostrum as soon as possible following birth. Studies in pigs show that the time of gut closure can be extended if the neonate is fasted or fails to suckle immediately following birth, increasing the likelihood of receiving passive immunity. The production of colostrum by the female exceeds the window of gut closure in the neonate as colostrum is produced during the initial three to four days postpartum; however the quality of colostrum decreases during this transition. The most important antibody in the colostrum is IgG, which acts as a nonspecific antibody.

Lipids

Over 90% of the lipids found in milk are triglycerides, while the remaining 10% are cholesterol and phospholipids. Triglycerides of varying chain length and saturation are found in milk. The sources of milk lipids are: 1) circulation, originating from diet or mobilization of body fat, and 2) production by the mammary tissue. The characteristics of milk fat depend on the source of the fatty acids. Long chain fatty acids in milk are absorbed from the bloodstream, while short and medium chained fatty acids are synthesized by the mammary tissue.

The majority of dietary unsaturated fatty acids in ruminant species are saturated by the microorganisms present in the rumen. By the time the fats reach the small intestine and are absorbed, they are predominantly saturated; however, lesser amounts of unsaturated fatty acids are absorbed and incorporated into milk as well. Mammary tissue is capable of synthesizing fatty acids. Glucose is the primary precursor for fatty acid synthesis in the mammary gland of nonruminants. In ruminants, which have very limited supplies of glucose, the volatile fatty acids acetate and butyrate are the primary precursors to fatty acid synthesis. Since the amount of milk fat depends on the amount of acetate and butyrate produced, and the amount of acetate and butyrate present depends on the amount of fiber in the diet, the amount of fat in milk can be manipulated to some extent by altering the animal's diet.

Dairy Cattle

"All the good ideas I ever had came to me when I was milking a cow"

—Grant Wood, (1891–1942)

This chapter introduces the different breeds of dairy cattle and their specific contributions to the United States dairy industry. The role and size of the dairy industry is highlighted along with the basic management of dairy farms, facilities associated with dairy farming, and diseases of the dairy industry. In addition, the importance of dairy as a dietary component and the nutritional benefits of milk are explained.

Dairy Cattle: A Historical Perspective

The use of cow's milk for human consumption first occurred in northeast Africa and Asia. Early evidence of dairying can be traced through the written records of the Sumerians of Mesopotamia dating to 6000 B.C. In the Euphrates Valley ancient architectural art depicts cows being milked from behind; however, around 3000 B.C. Egyptian records indicate that cows were milked from the side. In both cultures, calves were placed in front of the cows during milking, which would stimulate milk release in the dam. Records of dairying in Greece date to 1550 B.C. and in Rome, 750 B.C. Milk was an important constituent of the diet; however, lack of refrigeration meant that consumption of milk was primarily by shepherds and farmers, who consumed milk fresh after milking the cow, and the wealthy who could pay to have the fresh milk expressly delivered. Sources of animal milk varied through out the world, as it does today; however, cattle represent the primary species milked today and contribute over 80% of milk consumed worldwide. The first recognition of the benefits of dairy products can be traced to Aristotle. In Aristotle's (384–322 B.C.) writings on the history of animals he documented that the cow was a mammal producing milk in excess of what was required by the calf and could be used to benefit man in processes including tooth development.

In the United States specialized breeds of cattle for dairying did not exist until the 1850s. Although the first cows to provide milk arrived in Jamestown in 1611, these cows were the same animals that supplied labor and meat. Usually, one cow was sufficient for a family and no surplus of milk was produced for sale. Milk was immediately consumed due to its tendency to spoil. However, technological advances following the Civil War allowed for the preservation of excess milk, providing the foundation of today's current dairy industry. The development of condensed milk, mechanical refrigeration, and the process of pasteurization (controlled heating to destroy microorganisms) helped to expedite the development of the industry. Soon, large dairy herds were established by importing significant numbers of cows from Europe to sustain the growing industry. The breeds of imported cows varied and breed associations were formed to track each breed's registry.

There have been many milestones that have marked the advancements in the dairy industry. In 1776 the first ice cream parlor opened and during the years of 1848 to 1873, sixty-nine patents were filed for hand crank ice cream freezers. The 1800s also marked the first cheese factory (1851), a centrifugal cream separator (1878), and the invention of the milk bottle (1884). The Babcock Cream Test developed in 1890 served as the basis for setting milk prices by determining the amount of milk fat contained within the milk. Commercial pasteurization soon followed in 1895 and homogenization in 1919. The first milking machines were developed as early

■ **Fig. 8.1**   The Egyptian goddess of love, Hathor, is often depicted as a heavenly cow. (© Bill McKelvie, 2009. Under license from Shutterstock, Inc.)

as 1903 and in the 1930s, dairy farms began to utilize bulk tanks in replacement of the milk can. The number of dairy cows reached its peak in 1945 and in 1964, plastic milk containers were introduced. In the 1970s nutritional labeling of milk became common and by the late 1980s, the sales of reduced fat and skim milk exceeded whole milk sales. The demands of the consumer for dairy products have been met by a reduction in dairy cows and increased milk production per cow during the last century.

Breeds of Dairy Cattle

Cattle that are considered dairy breeds are those that excel in their ability to produce large quantities of milk for a sustained period of time. Today, dairy cattle can be found in every state in America and consist primarily of six dairy breeds that were imported to the United State. These breeds include the Ayrshire, Guernsey, Holstein, Jersey, Brown Swiss and Milking Shorthorn. The prominent breeds that make up a large majority of the United States dairy industry are the Holstein and Jersey.

The Holstein-Friesian originated from the Netherlands and Northern Germany. Its origins trace to black and white cattle owned by Batavian and Friesian immigrants to the region. Interbreeding of the cattle contributed to the characteristic black and white markings of today's breed. The breed's first importations to the United States occurred in the mid 1600s by Dutch sailors. However, significant importation did not occur until the 1860s and present-day Holsteins are descendents of the cattle imported from 1877 to 1905, after which time importations stopped. In the United States the breed is commonly called Holstein; elsewhere, it is known as Friesian.

The Holstein displays either a dominant black and white spotted pattern, or less commonly, a recessive red and white pattern. Solid black and solid white animals are unable to be registered with the breed. Holsteins are the largest of the dairy cattle breeds in terms of size and are characterized by large udders, and increased milk production relative to the other breeds. Holsteins are ranked first among dairy breeds in average milk production and can produce more than one hundred pounds of milk per day during peak lactation. In 2011, average milk production rates of 23,385 pounds per lactation were reported. Increased production capacity contributes to the Holsteins susceptibility to udder problems due

■ **Fig. 8.2**   Holstein (© Viorel Sima, 2009. Under license from Shutterstock, Inc.)

■ **Fig. 8.3**   Jersey (Courtesy American Jersey Cattle Association)

to stress placed on the udder ligaments. Holsteins rank fifth in milk fat production, with an average milk fat of 3.5 percent. The Holstein is the major dairy breed by far in the United States, accounting for 85 to 90% of all dairy cows. Approximately three hundred sixty thousand Holsteins were registered in 2011.

The Jersey breed originated from the Island of Jersey, located between the coasts of France and England and is considered one of the oldest purebred breeds of dairy cattle, due in part to its isolation by the island. Importation first occurred around 1815, with major importations occurring between 1870 and 1890. Jerseys can range in shades of cream, to light fawn, to nearly black. All Jerseys have darkly pigmented skin and black muzzles. They are the smallest of the dairy breeds and are noted for their large eyes. This breed has well attached udders and is known for its efficient use of feed and excellent grazing abilities. The breed also has a productive life span that exceeds the average of all other dairy breeds by 185 days. Jerseys rank third in average milk production with an average yield of 18,633 pounds of milk per year; however, this breed is first in production of milk fat with a percentage of approximately 4.8. Additionally, milk from Jersey cows yields 20% more cheese and 30% more butter relative to market-average milk. The Jersey is the second most popular dairy breed in the United States, representing 10 to 15% of all dairy cows. Over ninety-six thousand Jersey cattle were registered in 2011.

The Brown Swiss originated from the Alps of Switzerland. Importations first occurred in 1869 and the Brown Swiss cattle of today are descendants from twenty-five imported bulls and one hundred forty imported cows. Brown Swiss vary in solid shades of brown, ranging from light to dark. The nose and tongue of the Brown Swiss are characteristically black with a light colored band around the muzzle. The early breed was strong and had a tendency to be heavy muscled, which lead to their earlier classification as a dual purpose breed. Today, the breed has retained its large frame, but is more refined for the dairy industry. Heifers of this breed mature more slowly than those of some other dairy breeds and are noted for heat tolerance. Brown Swiss rank second in average milk production with annual production rates of 22,252 pounds of milk. The breed ranks third in milk fat production, producing an average of 4.1% milk fat and is a desired breed for cheese production due to the protein content and the fat to protein ratio of the milk. The Brown Swiss is third in registration numbers with an estimated ten thousand seven hundred registrations in 2013.

The Guernsey breed originated on the Island of Guernsey, located in the English Channel between France and England. The breed was developed by Monks through the careful selection for desired traits. Guernsey's were first imported into the United States in 1840; however, major importation did not occur until after 1870. It is estimated that over thirteen thousand head of Guernsey cattle were brought into the United States. The breed may be any shade of fawn and contain clearly defined white markings. The skin of Guernsey is yellow in color compared to the darkly pigmented skin of the Jersey. The Guernsey is an early maturing breed noted for producing milk typically yellow in color and with a high milk fat content of 4.5%. The yellow color is a result of increased concentrations of β-carotene in the milk. The breed ranks second among the dairy breeds in average milk fat and fifth in average milk production with an annual yield of approximately sixteen thousand pounds. Milk production is sustained by consuming 20% to 30% less feed per pound of milk produced compared to larger breeds of dairy cattle. The Guernsey is the fourth most popular dairy breed in the United States.

The Ayrshire has its origins in Ayrshire County in Southwestern Scotland. The first importation of Ayrshires into the United States occurred in 1822. Ayrshires may be solid white or light to deep brown with white coloring. The preferred coloring is a distinctive red and white. The breed is noted for its symmetrical and well attached udders, sturdy legs and feet, and excellent grazing ability. The vigor of the offspring and their ease of care, contribute to the desirability of the Ayrshire breed. Ayrshires rank fourth among the dairy breeds in milk production and average over seventeen thousand pounds of milk per cow annually. The breed ranks fourth in milk fat production, with an average of 3.8% milk fat. Ayrshires are currently ranked fifth in United States dairy registrations.

The Shorthorn was developed in England and first served as a triple purpose breed supplying meat, milk, and draft power to early settlers of the United States in 1783. However, through careful selection, the breed was adapted for either milk or meat production and in 1882 a breed association was established to distinguish both types of Shorthorn cattle. Although it still maintains a status of a dual purpose breed, Milking Shorthorns were recognized as a distinct dairy breed in 1969. Milking Shorthorns can be red, white, or a combination of the two. This breed is known for its adaptability and continues to undergo more refinement in the United States to improve its milk producing ability. The breed averages approximately fifteen thousand pounds of milk annually with an average milk fat content around 3.3%. The Milking Shorthorn is sixth in dairy breed registrations.

United States Dairy Industry

Dairy cattle are found in each of the fifty states. Early dairying relied on dairy farms to be in close proximity to areas of large human populations due to the perishable nature of milk. Today, dairy farms are concentrated in the north and west, with western areas offering the advantage of reduced costs of production relative to other regions of the United States. The ten largest milk producing states in 2012 include: California, Wisconsin, New York, Idaho, Pennsylvania, Texas, Michigan, Minnesota, New Mexico, and Washington. Idaho's proximity to West

■ **Fig. 8.4** Brown Swiss (© nagib, 2009. Under license from Shutterstock, Inc.)

■ **Fig. 8.5** Guernsey (Image courtesy of Cheri Oechsle, the American Guernsey Association.)

Coast states allows excess product to be shipped to this location, whereas, Wisconsin has historically maintained a strong milk production industry and utilizes its many dairy farms and manufacturing plants to produce the majority of the country's manufactured milk products, including over a fourth of the total cheese produced.

Since 1950 milk production per cow has increased over 50%, number of dairy cows has decreased by over 50% and the average herd size has increased. Most dairies in the United States fall within two broad categories: traditional or large scale specialized dairies. The traditional dairy consists of less than two hundred cows and equates to the average herd size a family could accommodate and still retain

satisfactory living standards while maintaining a functioning farm, including growing feed for the cattle. In 2008, 32.5% of total United States milk production was attributed to traditional dairying systems. For states including Pennsylvania and Wisconsin, these traditional dairies dominate the industry contributing up to 74% and 56% of total milk produced, respectively. The second category of specialized dairy systems represents operations dominated by herd sizes greater than two hundred cows. Operations that exceed five hundred cows are classified as concentrated animal operations and are subject to regulation by the Environmental Protection Agency. Herd sizes of two hundred to four hundred ninety-nine contribute to 12.6% of total milk produced, whereas dairies of herd

■ **Fig. 8.6** Ayrshire (© Viorel Sima, 2009. Under license from Shutterstock, Inc.)

Breed Characteristics

Breed	Body Weight (lb)	Milk (lb/305 d)	Fat (%)	Protein (%)	Lactose (%)	Solids (%)
Ayrshire	1200	17,230	3.88	3.17	4.60	12.77
Brown Swiss	1400	22,252	4.10	3.32	4.68	13.08
Guernsey	1300	15,877	4.50	3.37	4.71	14.04
Holstein	1400	23,385	3.50	3.06	4.68	12.16
Jersey	1000	18,633	4.79	3.63	4.83	14.42
Shorthorn	1200	15,000	3.33	3.32	4.89	12.90

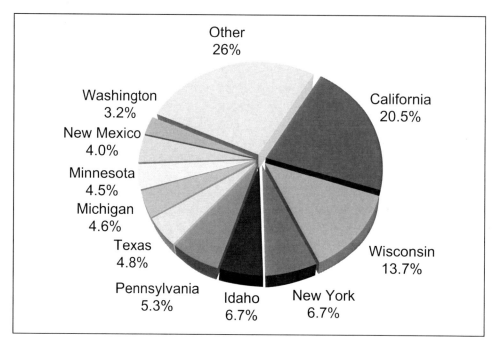

■ **Fig. 8.7** Top ten States for milk production in 2013. (United States Department of Agriculture-National Agricultural Statistics Service)

size of five hundred or greater represent 54.9% of total milk production. These enterprises may still represent family farms or family corporations, but are specialized in the production of milk and often buy the majority or all of the feed necessary to maintain the herd. Specialized operations are usually found in dairy states such as California, New Mexico, Arizona, Texas, Idaho, and Florida. In California and Idaho, operations with greater than five hundred cows contribute to 92% of the states, total milk produced. From the beginning, these larger enterprises were able to adopt business techniques, management strategies, and labor-saving technologies that smaller operations could not afford; resulting in lower milk production costs that have given the larger farms a competitive advantage. For the traditional dairy operation, producers often belong to cooperatives that assemble and distribute the milk to the processors and manufacturers; however, even the cooperatives are experiencing consolidation and seeing a decrease in number with an increase in size. It should be noted, that although the total number of dairy farms continues to decrease, the herd size of both traditional and specialized dairy operations has experienced growth, whereas the average

herd size of traditional dairies was twenty-five cows in 1960 it was eighty-eight in 2000. Currently, traditional dairy enterprises dominate the industry in terms of number of producers, whereas a majority of milk originates from large scale specialized operations. Both traditional and large dairy conglomerates are predominantly family owned and operated, which is likely to continue to restrict ownership and partnerships to family members.

Management Systems

Types of Dairy Systems

Types of dairy operations include traditional confinement and grazing-based systems. Traditional confinement systems rely on cows being housed year round. Grazing systems are defined as those that meet 30% of the animals forage needs by grazing and provide fresh pasture at least once every three days. Currently, approximately 90% of dairy operations are managed as traditional confinement systems. The size of the herd influences the dairying system, only 1% of herds greater than five hun-

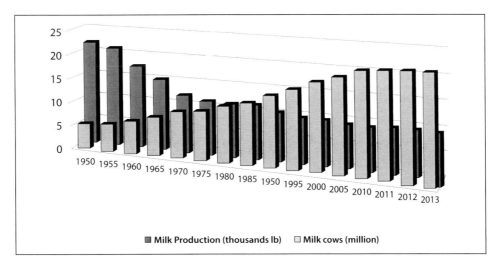

■ **Fig. 8.8**   Since 1950 number of milk cows has decreased whereas amount of milk produced per cow has increased. (United States Department of Agriculture-National Agricultural Statistics Service)

dred practice grazing, whereas approximately 10% of herds less than five hundred practice grazing-based dairying. In traditional confinement systems, females are maintained in various stages of lactation throughout the year, which provides for continual income. The rate of milk production is greater per cow compared to grazing systems; however, the cost of hired labor is greater as well. Because of the need to house animals throughout the year, traditional confinement dairies experience increased capital investment and are sometimes associated with increased feet and leg problems. Nutrition is commonly maintained through the provision of total mixed rations; balanced rations of forages, concentrates, proteins, vitamins, and minerals that meet the nutrient needs of the animal in one feed source. Approximately, 52% of all traditional confinement dairies use total mixed rations, 36% of dairies of herd sizes less than one hundred employ this strategy while 94% of dairies with herd sizes greater than five hundred feed total mixed rations. The grazing-based system is defined by the consumption of forages directly harvested by grazing cows and is reliant on forage availability and management practices. Supplemental feed is required to provide optimal nutrient ratios for milk yield and requires an understanding of the grazing behaviors demonstrated by cattle and seasonal changes in pasture quality. Grazing-based dairying may be seasonal, relying on available pasture and synchronization of breeding and lactation cycles, or year round with the feeding of harvested forage during winter months. Although the average pounds of milk produced per cow is less in a grazing-based system, some costs are less than that of traditional confinement systems.

Dairy Housing Systems

The two most common types of dairy housing in traditional confinement dairies are tie stall/stanchion barns and free stall barns, which represent 49.2% and 32.6%, respectively, of the housing systems employed. Outdoor and indoor lot group housing facilities, as well as individual pen systems, are used to a lesser extent by the industry. Whereas tie stall facilities are commonly used by small herds of less than one hundred, herds greater than five hundred animals rely on freestall systems in 75% of the operations.

In a tie stall/stanchion system, each cow is confined to an individual stall and is held in its stall through the use of ties or stanchions. In a tie stall system, a collar, chain, or strap, is placed around the cow's neck and is fastened to the front of the stall. Stanchion structures rely on restraint of the animal by a metal yoke or pivoting bars. Restraint to individual stalls facilitates better health care through continued observation for signs of disease or illness and ease of treatment through confined space. In addition, animals are more easily restrained for breeding and feeding. However, this system is more labor intensive as effort is needed for the addition and release of each cow from its stall. This system also is more difficult to use in conjunction with the traditional milking parlor because of the manual need to release each cow. In smaller herds where a milking parlor is not used, the milking must take place within the stalls. This requires the producer to assume a stooping position during milking and may require manual transport of the milk to storage.

Freestall barns were first constructed in the 1960s to take advantage of aspects of tie stall facilities and group housing lots. In this system individual resting stalls are provided, but the cow is not restrained to an individual stall. Cows have freedom of movement and may enter or leave any stall upon their choosing. It is common to have 10% more cows than stalls as all cows do not rest at the same time. Separate from the stall area is a milking facility and designated community feeding area. Free stall barns are advantageous for use with milking parlors as cows are easily moved into the milking facility. The communal feeding area allows for easier feeding of a group compared to the labor of individualized feeding. Freestall systems require less bedding than conventional stall barns and are associated with reduced feet, leg, and udder problems. Some cows will not use the provided stalls and must be trained to enter them. The individualized attention afforded to cows in a tie stall barn is lost with the freestall housing system and competition at feeding is increased.

Waste Management Systems

With an increase in size of herds waste management is an increasing challenge. Improper waste management can lead to pollution of air, land, and water supplies. Dairy cattle produce approximately 8% of their body weight in manure and 3% in urine each day, allowing a mature cow to produce one hundred twenty pounds of manure and forty-three pounds of urine daily. Manure must be removed from the barn regularly and can be accomplished through the use of alley scrapers (mechanical or tractor), gutter cleaners, alley flushing (using fresh or recycled water), slotted floor, or manure vacuum. Facility designs are associated with specific manure handling methods. Approximately 83% of tie stall/stanchion operations rely on a gutter cleaner to remove waste from the housing area, whereas, manure is handled by alley scrappers in 72% of freestall facilities. The majority of dairy operators apply removed waste as nutrients to rented or owned land. The waste may be applied daily, which is the lowest-cost method of handling, or stored for future hauling. Daily application is weather dependent. Runoff from frozen or snow-covered ground can contribute to pollution through the loss of fertilizer nutrients, whereas application during wet weather can be detrimental to ground cover, impacting future crop yields. Manure storage can be accomplished through below-ground storage tanks, earthen basins, or above-ground silos.

 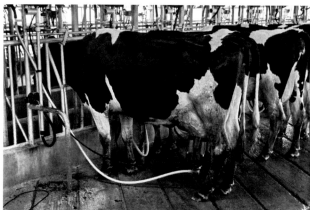

■ **Fig. 8.9**   Stanchion dairying system (Both images © Nitipong Ballapavanich, 2009. Under license from Shutterstock, Inc.)

The size of the storage area is dependent on the number of cows in the herd and the length of time the waste is to be held. Once the storage facility meets its capacity, the manure is applied to land.

Milking Systems

The removal of milk from the lactating female is the primary objective of a dairy operation. Considerable time is invested in the milking process as cows must be milked at regular intervals each day for optimal milk production. A majority of dairy operations milk cows twice daily, however, milking three times daily is practiced and has been shown to increase daily milk production by 10%. The removal of milk from the udder occurs under the application of a vacuum to the teat by the milking unit. Pressure in the udder is created by the ejection of milk from the alveoli; the milking machine is maintained at a lower pressure to promote milk flow. Prior to milking, the cows must be prepped, requiring the milking operator to wash, dry and manually strip the initial milk from the teat to prevent bacterial contamination. This preparation is also important toward reducing environmental pathogens on the teat surface and promoting milk let down. Milking begins after preparation of the udder and continues until the completion of the process, around six to eight minutes later. After the milking is complete, the milking apparatus is removed manually or through the use of an automated system, which responds to a reduction in milk flow rate. Proper removal of the milking apparatus is important toward sustained health of the udder. Teats are subsequently treated with a disinfectant, commonly iodine, which reduces the incidence of pathogens entering the opened teat and thus reduces the incidence of infection. The milking apparatus is connected to a milk pipeline that sends milk to a collection, or bulk tank where it is stored until transport. Ultimately, milking systems are designed to allow for complete removal of milk without damage to the teats or udder, provide for the

■ **Fig. 8.10**   Freestall barns (© Laila Kazakevica, 2009. Under license from Shutterstock, Inc.)

■ **Fig. 8.11** Pail milkers may be used in tie stall/stanchion dairying systems, but have been predominantly replaced by pipeline milkers. (© Vladimir Mucibabic, 2009. Under license from Shutterstock, Inc.)

production of uncontaminated milk, provide comfort for the cows and their handlers, and utilize equipment that can quickly and thoroughly be cleaned and disinfected. Cleanliness in the housing facility reduces likelihood of cows contracting infections and sanitation of the milking facility and equipment minimizes milk contamination.

In tie stall/stanchion facilities, milking typically is linked to the individual stall and may include pail milkers, suspension milkers, or pipeline milkers. Each system relies on a vacuum line positioned above the stall to establish the pressure necessary for milk ejection by the milking apparatus. Pail milkers are transported and positioned next to each individual and stationary cow. The milker is attached to the udder and draws the milk into a pail that must be emptied by hand after the completion of milking. Suspension milkers are similar to pail milkers; however, suspension milkers are not placed on the ground, but rather hung from a strap called a surcingle that is placed over the cow's back. Again, milk is transported manually to the bulk tank. Pipeline milkers also make use of a vacuum line, but rely on a milk line that is also built into each stall. As milk is removed from the udder it is drawn directly into the milk line and transported to the bulk tank, allowing for less handling of the milk, which decreases labor and minimizes milk contamination. However, capital investment into facilities is greater

with pipeline milkers compared to pail or suspension milking systems.

Milking facilities also may be separate from herd housing and almost 80% of cows are milked in parlors. Milking parlors are specially designed to be labor and time efficient. Usually, the milker works within a pit with the cows positioned in an elevated manor to eliminate the need of stooping to attach and detach milking machines. Milking parlors not only reduce labor, but can accommodate herd expansion and require less milk handling, which can reduce the amount of milk contamination. However, milking parlors require greater capital investment due to the costs associated with the need for additional facility space and equipment.

There are several styles of milking parlors; herringbone, parallel (side-by-side), side-opening (tandem), rotary and robotic. The herringbone is the most common milking parlor style used by over 50% of dairy operations. Cows enter and leave the parlor in groups and during milking, stand at a 30° angle to the operator pit with the rear of cow accessible to the milker. Having the cows stand at an angle reduces the distance the operator needs to travel when compared to when cows are standing broadside. Herringbone parlors are classified by the number of individual stations available for cows on each side of the operator pit and can range from as small as a "double-four" to very large parlors. Cows are restrained by gates that are mechanically controlled by the operator and all cows on one side must enter and leave at the same time. This becomes a hindrance as a slow-milking cow can impede the progress of the group. However, sorting slow-milking cows into a separate milking group is one strategy to resolves this problem. The parallel parlor is the second most utilized of the parlor systems. Cows stand facing away and perpendicular to the operator pit allowing access to the udder from behind. As attachment of milkers to the teats occurs between the rear legs, visual access to the front quarters is limited and milking apparatus attachment is more difficult. The side-opening (tandem) parlor is designed for cows to enter parallel to the milking pit. The parlor is equipped with two side opening gates, one for entrance and the second for exit. Cows enter each milking station individually, without relying on the movement of a group of cattle. This is advantageous as slow-milking cows do not hinder the progress of an entire group during milking. However, walking

■ **Fig. 8.12** Herringbone milking parlors (top) are the most common of milking parlors in use, whereas rotary parlors (bottom) are gaining popularity in large dairy herds. (*Top:* © Beth Van Trees, 2009. Under license from Shutterstock, Inc. *Bottom:* © Janet Hastings, 2009. Under license from Shutterstock, Inc.)

distance between cows is considerably greater for the operator, as compared to the herringbone or parallel parlors. In more recent years, use of rotary milking parlors has increased, especially in larger dairy herds. In this system cows enter the milking station, on a platform with multiple stations that is rotating at a very slow rate. In most rotary parlors, cows are positioned to be milked from their outside of the rotating platform. This system optimizes automation of the milking system. The milking apparatus, is attached shortly after cows enter the platform and removed either manually or automatically as the cows complete one revolution on the platform. This parlor style allows for large numbers of

cows to be milked simultaneously in a rather small area, but requires a greater initial investment.

Health Management

One of the most well known health issues in the dairy industry is mastitis. Udder inflammation that underlies mastitis can be caused by physical trauma or bacterial infiltration of the teat, with the latter representing the majority of cases. It is approximated that mastitis costs the United States dairy industry approximately $2 billion annually. Subclinical mastitis and is undetectable by the human eye. Tests, such as the California Mastitis Test (CMT), that test for increased somatic cell counts as an indicator of the disease, should be done at least once a month to screen the herd for the incidence of infection. If the disease progresses past the subclinical stage then, clinical mastitis occurs. Signs of clinical mastitis are readily observed and include a swollen, red, and hot udder that is painful to the touch. Other symptoms include a drop in milk production rates concomitant with the production of abnormal milk that can be lumpy, stringy, yellow, or contain clots or blood. Clinical mastitis can result in permanent udder damage. Microorganisms responsible for mastitis are classified as contagious or environmental. Contagious pathogens are found within the udder and can be spread from infected to non-infected udders during the milking process if proper sanitation procedures are not observed. Pathogens originating in the cows surroundings are referred to as environmental pathogens and are contracted in facilities between milkings. Most treatments involve the injection of an antibiotic into the udder through the teat canal, which renders the milk unsuitable for human consumption until treatment has been stopped and the appropriate withdrawal for the antibiotic has been observed. Good management techniques are the best approach to controlling the outbreak of mastitis and include: administering routine mastitis tests, strict sanitation of milking facilities and equipment, pre- and post-milking teat disinfection, use of teat sealants during the dry or non-lactating period, and separation of the milking of mastitic cows from non-mastitic cows.

As the majority of dairy cattle are housed, feet and leg problems are an issue within the industry. Recent studies indicate that lameness is the second leading health problem in dairy cattle, affecting

■ **Fig. 8.13**   Mastitis in a dairy cow. (Courtesy of Joe Hogan, The Ohio State University)

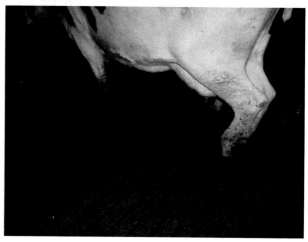

■ **Fig. 8.14**   Signs of lameness in dairy cattle. (Courtesy Jeff Firkins, The Ohio State University)

approximately 15% of dairy cows annually. Underlying causes of lameness are varied and include origins of an infectious or dietary nature, as well as physical trauma. Causes of lameness include physical trauma due to confinement housing, laminitis, digital dermatitis, and foot rot. Rumen acidosis, which is a decrease in the pH of the rumen, is a primary contributor to laminitis in dairy cows. This condition causes inflammation of the hoof and subsequent hemorrhage of the sole. Digital dermatitis, also known as hairy heel warts, has become a routine occurrence in some dairies. Cattle experiencing this condition lie down excessively or avoid applying weight to the affected hoof. The factors contributing to digital dermatitis are not known, but it is suggested to be bacterial in origin. Treatment often involves the use of topical sprays. Medicated foot baths may be used and are somewhat effective in reducing the incidence within a population. Necrotic regions between the claws are commonly indicative of foot rot, a contagious disease that is caused by bacteria of the intestine that is shed in the feces. This condition can also be corrected by antibiotics. Housing cattle in a dry, clean environment, use of medicated foot baths and regular hoof trimming and maintenance are effective preventative measures.

Capacity of Production

The aim of the dairy industry is economical milk production, which is reliant on reproductive success.

Age at first calving is important to the lifetime productivity of the female and is recommended to coincide with approximately twenty-four months of age. Once in production, cows are expected to calve in intervals of twelve to fourteen months. To meet a twelve month calving interval, a cow would need to become pregnant within ninety days of calving to accommodate the two hundred eighty day gestation period. In today's dairy operations, the average calving interval is 13.2 months. A twelve month calving interval provides for a ten month lactation period and two month dry period in which the female is not lactating. Within this lactation period, peak milk production will be obtained within six to eight weeks after calving, therefore, multiple lactations optimize milk production over the cows lifetime. The dry period is integral toward attaining maximal milk production rates in subsequent lactations. This period, lasting between forty and sixty days, allows for the mammary gland to return to a non-lactating state and subsequently regenerate secretory tissue in preparation of the next lactation. The dry period also allows the female to repartition nutrients away from sustaining lactation and toward replenishing body nutrient reserves and fetal development.

As high culling rates are encountered in some dairy operations, reproductive success is integral toward producing sufficient numbers of replacement females for the herd. Currently, the numbers of females born each year exceeds the number of replacements required for culled cows. Advances in the sexing and sorting of semen used for artificial insemination has increased the availability of replacement females and allows for greater selection of superior replacements.

Dairy as a Food Commodity

The United States is the largest producer of cow's milk in the world with an estimated two hundred billion pounds produced in 2013. Approximately 40% of the milk supply is processed into cheese, 30% is processed into fluid milk and cream, and the remaining is processed into butter, frozen dairy products, and powdered milk. Prompt marketing of milk in its fluid form or processing into additional products is necessary due to the perishable nature of fluid milk. The removal of water to create milk powder is important toward increasing the stability for transport and storage. In 2007, per

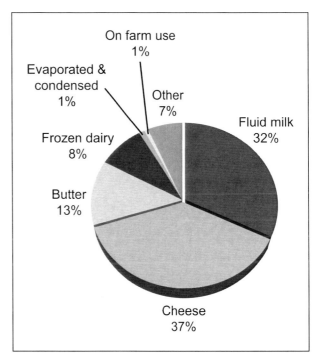

■ **Fig. 8.15** Distribution of the United States milk supply.

■ **Fig. 8.16** Dairy products are recognized as leading functional foods, foods that have health promoting benefits. The most recognized functional dairy products are those enriched with pre- and/or probiotics. Prebiotics are non-digestible food ingredients that beneficially affect the host by promoting the selective growth of healthy bacteria in the colon, whereas, probiotics are defined as beneficial live bacteria that are introduced to the host through food. Yogurt and yogurt drinks are common sources for pre- and probiotics. (© Olga Lyubkina, 2009. Under license from Shutterstock, Inc.)

capita consumption of fluid milk was approximately twenty-two gallons per year, with whole and 2% representing the majority of fluid milk consumed. The consumption of fluid milk has been declining since the 1960s; competition with other beverages that are not perishable and are marketed by vendors outside of retail markets has contributed to this decline. A decline in butter consumption has been felt by the industry as well; however, per capita cheese consumption has increased nearly four fold in the last four decades. Recently, mozzarella surpassed cheddar as the most popular cheese marketed. It is the growing cheese market that has contributed to stabilizing the dairy industry.

Nutritional Value of Milk

Milk is a nutritious food and is recommended, as are other dairy products, by all major health organizations as part of a balanced diet. Consuming milk and milk based products provides numerous health benefits and nutrients that are vital for the sustained health and maintenance of the body. Dairy products are especially important during childhood and adolescence when bone mass is rapidly developing as these products are a rich source of calcium and help build and maintain bone mass throughout life. In addition to calcium, milk contains nutrients such as potassium, vitamin D, and protein. As a source of potassium, dairy products are important to the maintenance of blood pressure, while vitamin D further contributes to bone development through its regulation of calcium and phosphorous concentrations. More recently, the consumption of three servings of low-fat dairy products a day has been reported as an important part of the diet to maintain a healthy body weight. Fortification of dairy products, by the addition of pre- and probiotics, has the potential for even greater health benefits by improving digestive health and immunity.

CHAPTER 9

Beef Cattle

"The feeling of friendship is like that of being comfortably filled with roast beef . . . "
—Samuel Johnson (1709–1784)

This chapter introduces characteristics of beef cattle and select breeds commonly encountered in the industry. You will develop a basic understanding of the role that the beef industry plays in United States agriculture. The structure of the industry is described as well as the practices required for beef cattle management and the products obtained from the industry.

Beef Cattle: A Historical Perspective

Cattle were domesticated from the Auroch, a primitive breed of cattle that survived until 1627. Christopher Columbus is accredited with the introduction of cattle to the western hemisphere during his second voyage in 1493. In 1519, Cortez brought cattle to what is now Mexico. It is thought that Spanish cattle were distributed throughout what is now the southwestern United States by Spanish missionaries as they began to settle the area. These cattle became the feral Longhorn cattle that occupied the plains region and served as the foundation for the United States beef industry. Settlers first imported cattle to North America via the Jamestown colony in 1611. Many cattle importations followed, and as the cattle population became established, cattle migrated westward with the pioneers. These early cattle were not only used for milk, but also for meat and to serve as draft animals. Until the mid-1850's, most cattle were dual purpose in that they were used for both milk and meat. Milk was usually for the producer's own use or for local sale from very small herds.

By the late eighteenth century imports of Shorthorn cattle were changing the face of cattle in the United States. These original imports were crossed with existing Longhorn cattle herds. Longhorn cattle began to disappear from the landscape and the population was further devastated following the Civil War as Longhorn cattle in the southwest were rounded up to feed the growing United States population. Shorthorn progeny did not perform adequately in the harsh conditions of the Western range; therefore, Hereford bulls were imported from England in an effort to improve the hardiness of range cattle. Herefords quickly became the dominant breed of the west. Later, Hereford-based cattle were improved by the infusion of the Angus breed, producing the commonly known black baldy. This cross remains a critical component of many cowherds today.

Many changes occurred in the beef industry following World War II. Prior to this time, cattle were generally raised and fattened on pastures and forages to produce the beef consumed by Americans. As technological advancements led to greatly increased production of grains such as corn, barley and soybeans, grain surpluses developed and beef producers began to feed grain, rather than forage, to increasing numbers of cattle. The beef from grain-fed cattle quickly became popular with consumers despite the increased market price. While the industry continued to depend on pastures and grassland for cow-calf production, a new feedlot sector of the industry was created based upon the practice of feeding grain during a finishing phase directly prior to harvest. This industry included many smaller feedlots throughout the grain belt of the United States as well as larger feedlots in the western states. The large feedlots originated in the 1950s in California and Arizona and later became concentrated in the Plains states in the 1960s. By the 1970's, the feedlot industry and the marketing of grain-fed beef became a driving force behind the United States beef cattle industry.

As the feedlot industry developed, concerns arose that the existing population of cattle, developed primarily for fattening on forages, became too fat when provided grains in the feedlot. While these small-statured animals performed very well on for-ages, they tended to deposit excess fat in feedlots. As a result, the need arose for cattle with enhanced growth potential that when fed grain would produce carcasses with a reduced fat content. This need resulted in an era commonly referred to as the Breeds Revolution in the industry. The industry turned to importation of different breeds of cattle, primarily of European origin, that would provide the neces-

■ **Fig. 9.1** Prehistoric cave painting from Lascaux, France depicts the now extinct wild Auroch. (© DeA Picture Library/Art Resource, NY)

■ **Fig. 9.2** Longhorn cattle (left) were a breed of early significance to the establishment of beef animals in the United States. Today, the Black baldy, crossbred of Angus and Hereford breeding, is of significance in the beef cattle industry. (*left:* Courtesy of Star Creek Ranch, Somerville, TX and the Texas Longhorn Breeders Assocation of America, Fort Worth, TX. *right:* © Ed Phillips, 2009. Under license from Shutterstock, Inc.)

sary genetics for enhanced growth. The first quarantine station for importation of new breeds of cattle into the United States opened in Canada in 1965. This allowed the introduction of breeds of cattle that had previously been prohibited for importation to North America. During the next twenty-five years, the number of cattle breeds increased to more than seventy breeds. Crossbreeding with these newly introduced breeds became an important part of the beef industry and provided cattle that more closely met the needs of the feedlot industry. In 1975, the number of cattle in the United States peaked at 132 million, including cattle and calves in both the beef and dairy industry. This number has slowly decreased to approximately one hundred million. However, the current number of cattle in the industry actually produces more meat annually than was produced during peak population numbers as a result of increased knowledge and improved management practices.

Breeds of Beef Cattle

Over two hundred fifty breeds of cattle are currently recognized. Beef cattle breeds are defined by the pedigree records in their breed registry and purity of ancestry. Animals in this classification are referred to as purebred or seedstock cattle. Purebred cattle are crucial to direct genetic change in the beef industry. Cattle that are not registered or consist of more than one breed are typically described as commercial or crossbred cattle. Commercial beef cattle are typically crossbreeds although some herds of non-registered cattle of a single breed do exist. Over 90% of cattle in the United States are classified as commercial cattle. Purebreeds of beef cattle are categorized into three classes based upon their origin as British (*Bos taurus*), Continental breeds (*Bos taurus*), and Zebu (*Bos indicus*) breeds.

British Breeds

British breeds originated in the British Isles and are generally considered to be maternal breeds due to their characteristic traits of moderate mature size, high fertility, extended longevity, and moderate to high milk production. These characteristics contribute to the use of British breed females in cow-calf production. British breeds typically are early maturing, reaching puberty and mature body size at an earlier age, and are often less muscular in comparison to other breeds. The breeds in this classification that are most prevalent in the United States beef industry are Angus, Hereford, and Shorthorn.

The Angus breed, officially named Aberdeen-Angus, originated in the shires of Aberdeen and Angus in Scotland. While the earliest written records of Angus cattle trace to the 1700s, it was not until the 1800s that farmers began keeping records of purity of ancestry. This breed was developed exclusively for meat and was first imported to the United States in 1873 by George Grant of Victoria, Kansas. Angus cattle are black or red in color, each with a separate breed registry. Red Angus cattle originated in England and Scotland and were recognized under a separate breed registry in 1945. Red is the recessive

■ **Fig. 9.3**   Red Angus. (Courtesy Red Angus Association of America.)

■ **Fig. 9.4**   Hereford (© dcwcreations, 2009. Under license from Shutterstock, Inc.)

■ **Fig. 9.5**   Shorthorn. (Courtesy American Shorthorn Association.)

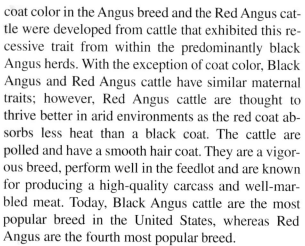

■ **Fig. 9.6**   Simmental (Courtesy American Simmental Association.)

coat color in the Angus breed and the Red Angus cattle were developed from cattle that exhibited this recessive trait from within the predominantly black Angus herds. With the exception of coat color, Black Angus and Red Angus cattle have similar maternal traits; however, Red Angus cattle are thought to thrive better in arid environments as the red coat absorbs less heat than a black coat. The cattle are polled and have a smooth hair coat. They are a vigorous breed, perform well in the feedlot and are known for producing a high-quality carcass and well-marbled meat. Today, Black Angus cattle are the most popular breed in the United States, whereas Red Angus are the fourth most popular breed.

Hereford cattle originated in England, in Hereford County. They were first imported to the United States by Henry Clay of Kentucky. Herefords are red with white faces, and have white on the legs, abdomen, and tail. While Herefords were originally horned, a population of polled Herford cattle was developed in Iowa in 1901 as a result of a genetic mutation for this trait. Warren Gammon is responsible for development of the polled Hereford after identifying and purchasing males and females that carried the genetic mutation and selecting for the trait in offspring. Separate Hereford and Polled Hereford registries existed for several years; however they are both currently represented by the American Hereford Association. Herefords are vigorous, hardy animals that adapt well to range environments in the western United States. They are generally docile and easy to manage. Herefords are often used in crossbreeding programs and the white pattern, especially the white face, generally dominates. Hereford is the second

most popular breed in the United States in terms of number of registrations.

The Shorthorn originated in northern England in the Tees River Valley around 1600. Shorthorn was originally a dual-purpose breed used for both meat and milk. The breed was first imported to the United States in 1783 to Virginia, where early settlers referred to them as Durhams. These cattle were used for milk, meat, and draft. The development of the modern Shorthorn used for beef production began in the late 1700s. Mature Shorthorns may be either horned or polled. They may be red, white, or roan in color. Shorthorns are also fairly heat tolerant and are known for their reproductive efficiency, longevity, and excellent milk production. There is a dairy lineage of Shorthorn cattle that are referred to as Milking Shorthorn. Shorthorns are the tenth most popular breed in the United States.

Continental Breeds

Continental breeds are sometimes referred to as exotic or European breeds. The breed designation is derived from the origin of these breeds tracing to the European continent. Continental breeds may be paternal, or terminal sire, breeds due to their characteristic traits of rapid growth rate, heavy muscling, leanness of carcass, and large mature size. The paternal breeds have characteristics that make them most desirable to be used as sires to produce calves that will be used for production of meat rather than in the reproducing cow herd. Other Continental breeds are referred to as dual-purpose as they possess both paternal traits and most of the maternal traits described above for British breeds. The dual-

■ **Fig. 9.7** Limousin. (Courtesy North American Limousin Foundation (NALF).)

■ **Fig. 9.8** Charolais. (Courtesy of Kori Conley, The Charolais Journal.)

purpose beef breeds are breeds used for both milk and meat production in Europe. The Continental breeds that are most prevalent in the United States beef industry are Simmental, Limousin, Charolais and Gelbvieh.

Simmental cattle originated in the Simmen Valley located in Switzerland and the breed dates to the Middle Ages. They were first used as a dual-purpose breed for both meat and milk production. Nearly one-half of the cattle in Switzerland are Simmental. This breed was first imported to the United States in 1969 from Canada. While the traditional color of Simmental cattle is red (or yellow) and white patterned, registered Simmental cattle in the United States can be white, grey, solid red, or black, with or without white markings and are either polled or horned. Simmentals are typically large animals that are generally docile and adaptable to a variety of climates. They are noted for their rapid growth rate, thick muscling, and lean carcasses as well as their milk production and maternal ability. The Gelbvieh breed, which originated in Germany, and came to the United States in 1971, possess traits similar to those of Simmental cattle. Simmental is the third most popular beef breed and Gelbvieh is the sixth most popular breed in the United States.

Limousin cattle originated approximately seven thousand years ago in the province of Limousin in central France and are recorded as the oldest of domesticated cattle. The breed was first mentioned in the United States by soldiers returning from War World II whom reported on the golden cattle of France. However, it was not until 1968 that the

breed was introduced to the United States following importation of semen from Canada. Limousin cattle can be red, gold, or black in color. Compared to some other Continental breeds, they have shorter necks and a smaller head with a broad forehead. While Limousin cattle are acknowledged for paternal traits in general, their large loin muscle area and the leanness of carcasses are most recognized and contribute to production of carcasses that have excellent yield of lean muscle. The breed is the seventh most popular in the United States

The Charolais breed was developed in central France near Charolles. It is one of the oldest of the French breeds of beef cattle. Charolais were imported to Mexico in 1930, then to the United States when two bulls were imported from Mexico to King Ranch in Texas. Only a limited number of Charolais were imported into the United States due to import restrictions imposed in the 1940s. Charolais have pink skin with a white to light straw color coat. They are a large, heavy-muscled breed that can be either polled or horned. They are noted for their rapid body weight gain in feedlots, lean, red, tender meat, and are often used in crossbreeding programs. In the United States, the Charolais is the fifth most popular breed.

Zebu

Zebu cattle in the United States consist predominantly of cattle that received a percentage of their genetics from the American Brahman breed. Zebu cattle have had the greatest genetic influence world-wide on cattle populations as they predominate in equatorial regions of the world. There are seventy-five types of

Zebu cattle reported, each characterized by the distinctive hump that occurs as a result of a spinal process at the thoracic vertebrate, an extended dewlap, and large drooping ears. These features allow Zebu cattle to adapt to arid environments and contribute to their heat and insect tolerance. The American Brahman is the most influential *Bos indicus* breed in the United States. The Brahman breed originated from bulls and cows of the *Bos indicus* type that were imported from India between 1854 and 1926. The American Brahman Breeders Association was started in 1924. They can vary in color from light gray or red to almost black. As a pure breed, Brahman only rank eleventh in terms of popularity, however, the pure breed has been crucial to creation of crosses of *Bos indicus* and *Bos taurus* breeds that were subsequently used to create pure breeds that are of greater importance in the United States beef industry. Brahman-influence cattle are most prevalent in the southern regions of the country from Florida to southern California. However, they tend to reach puberty at a later age, have a reduced growth rate, and less muscling than the *Bos taurus* breeds described above.

Two of the most important hybrid breeds from the *Bos taurus* and *Bos indicus* are Brangus and Beefmaster. Brangus cattle consist of 5/8 Angus and 3/8 Brahman genetics and are recognized as a pure breed with their own registry. This is the eighth most popular breed in the United States. Beefmaster cattle contain just less than 50% Brahman genetics and just over 25% each of Hereford and Shorthorn genetics. This is the ninth most popular breed of cattle in the United States.

The United States Beef Industry

The beef industry plays a substantial role in the United States economy and accounts for approximately one-fifth of all yearly cash receipts in agriculture, making it the largest money-generating commodity in this industry. The industry grosses an income of approximately $40 billion annually. Animal agriculture accounts for approximately one-half of the annual agricultural cash receipts and beef cattle account for approximately 40% of all livestock cash receipts. There are approximately seven hundred and fifty thousand beef cattle operations in the United States with 7.1% of the world's cattle that produce approximately 21% of the world's beef and veal. For thirty-five of the fifty states, cattle and calves rank in the top five commodities. Eight states record more than $1 billion each in gross income from beef cattle and an additional ten states gross more than $500 million. Cattle are widely distributed throughout the fifty states; however, the population of beef cattle in the northeast is limited primarily due to the greater human populations in these regions.

A major advantage of the beef industry is its ability to utilize resources from which humans cannot directly benefit. In addition to non-tillable land that can only be used for grazing, by products such as corn stalks and wheat straw, cannery waste, beet pulp, brewer's grain and distiller's grain, can be used very effectively as feed for beef cattle. In this way, products of minimal value as food for humans are converted into a nutrient dense food for humans. When cattle are fed grain, the quality and palatabil-

■ **Fig. 9.9** Brahman. (Image courtesy of Meghan Rheiners, the American Brahman Breeders Association.)

■ **Fig. 9.10** Brangus. (Courtesy International Brangus Breeders Association (IBBA).)

ity of the beef is increased, efficiency of production is enhanced, and year-round production of beef is achieved. Major seasonal fluctuations in the availability of beef for human consumption do not occur in the United States. Of all domestic livestock, cattle are the most numerous with 1.4 billion head reported worldwide and it is believed that every country in the world has at least some cattle. It is also estimated that there is one cow for every 4.6 people in the world and there are 26.6 head of cattle per square mile of the land surface on Earth.

Beef Cattle Management Systems

The structure of the beef industry varies from some of the other livestock industries as it is typically divided into more distinct segments. Animals are typically owned by multiple individuals or companies as they move from one stage of production to the next. Also, regions of the country often tend to specialize in only one or two segments of the beef industry. The four major segments of the beef industry are purebred producers, commercial cow-calf producers, yearling or stocker operators, and feedlot finishing operations.

Purebred

In purebred production, the primary objective is to produce breeding stock and primarily bulls to be used to mate with commercial cows. This makes purebred production the only segment of the beef industry where the primary goal is not to produce cattle for the purpose of meat consumption. These animals will ultimately be harvested for meat, but only after they have been used to produce breeding stock or used to mate with cows in commercial production. In the purebred production segment, the greatest demand is for bulls. Bulls are most often purebreds whose genetic information derived from pedigree, individual and sibling performance, progeny performance and genomic evaluation is used to calculate estimates of genetic merit called Expected Progeny Differences (EPDs). These estimates of breeding value are used to aid in the selection process for sires and increase the predictability and efficiency of offspring. Through the marketing of bulls to commercial producers and other purebred producers genetic change in the beef industry is directed by this segment of the industry. Purebred producers that raise exceptionally high-quality cattle also market semen and embryos to other cattle producers or sell or lease bulls to stud.

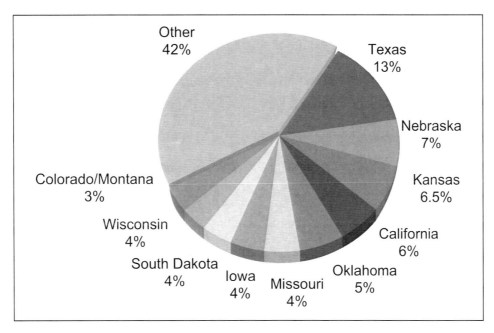

■ **Fig. 9.11**   Top ten States in beef cattle numbers in 2013 (United States Department of Agriculture-National Agricultural Statistics Service)

■ **Fig. 9.12** Commercial cow-calf production is the first phase of beef cattle production, which supplies weaned calves to stocker or feedlot operations. (© Bill McKelvie, 2009. Under license from Shutterstock, Inc.)

■ **Fig. 9.13** In the United States the majority of beef cattle are finished in feedlot operations. (© Thoma, 2009. Under license from Shutterstock, Inc.)

Commercial Cow-Calf

Commercial cow-calf production represents the first phase that produces animals intended for meat consumption. Commercial cow-calf producers typically produce calves that are six to ten months of age and three hundred to seven hundred pounds which are weaned and sold. These calves would typically be crossbred and intended for harvest as beef. The goal of most commercial cow-calf producers is to produce the heaviest calves possible with the least cost. The calving season for these operations is usually in the spring or fall. The calving season is generally chosen based upon the availability of feed, environment, management expertise and options, and marketing avenues for the calves. A majority of calves are born in the spring in order to match the rapid forage production in pastures during this season with the peak lactational demands of the dam. While most producers sell their calves at weaning, some will retain ownership of their calves and add additional weight to the animals. Calves sold at weaning typically go to either a feedlot or to a stocker production system. Most producers will retain the top 20–50% of heifer calves to use to replace mature females in the herd that have reached the end of their productive lifetime. Texas and Missouri are the top two states in beef cow numbers.

Stocker

Yearling or stocker operations are one of the two destinations of calves at weaning. In this phase, calves are grown to heavier weights on low-priced feedstuffs for six to eighteen months before the calves are placed in feedlots to be fattened for harvest. Stocker operations generally purchase their calves from commercial cow-calf producers and grow them on forages or crop residues and eventually sell them to feedlots. The feed utilized in stocker operations varies by region and producer and may include summer pastures and range, small-grain pastures, crop residues, by-products of production of food for humans, by-products of ethanol production, or standing prairie hay or silage. For cattle with less growth potential, this period allows them to grow in stature and muscling before they move into the feedlot and prevents over-fattening of these animals in the feedlot. In cattle with high growth potential, it is often more efficient to place these animals directly into feedlots at weaning.

Feedlots (Finishing or Fattening)

Feedlots represent the last phase of the production system for live animals. Cattle are typically fed to an endpoint based upon their body fat composition, thus this segment of the industry may be referred to as the fattening or finishing phase. Cattle are fed to the desired endpoint on grain-based diets and then sent to be harvested by packers. Both heifers and steers, and recently weaned calves or yearlings from stocker operations are fed in feedlots. These cattle typically enter this phase between six hundred and one thousand pounds of body weight and are harvested between one thousand and fifteen hundred pounds of body weight. Cattle are fed in a feedlot for between sixty to two hundred days and a vast majority of young cattle in the United States are finished on grain in feedlots before being sent to harvest. Most feedlot cattle are harvested between 18 and 24 months of age. Feedlots vary greatly in size from very small operations that either purchase calves or raise them from birth, to very large operations that concentrate tens of thousands of cattle in operations specializing only in feeding cattle for harvest. The top two states in cattle feeding are Texas and Nebraska. Feedlots sell finished cattle to packer/processor operations for harvest and processing. The packer/processor sells carcasses or boxed beef to retailers, distributors, HRI (Hotels, Restaurants, and Institutions), or for export. Retailers and distributors then further process these larger primal cuts for sale to the consumer. Over the last few decades, feedlots and meat packing plants have increased in size, while simultaneously decreasing in number. While the beef industry can be divided into these various segments, they are not always a clear-cut delineation. For example it is not unusual for a producer to own their cattle through more than one phase, by either keeping them on their farm or contracting with other farms to fatten or finish their cattle.

Health Management

The movement of beef cattle from one stage of production to the next contributes to bovine respiratory disease, which is the most costly health concern in the industry. Young animals are susceptible to pneumonia, and economic losses arise from treatment cost, failure to thrive, reduced weight gain, and in some cases death. The disease is multifactorial and both microorganisms and management practices are known to contribute to the onset. Pneumonia is representative of inflammation of the lungs and results in reduced respiratory capacity. Both viruses and bacteria can trigger pneumonia. The severity of respiratory infection and resulting lung damage differs between the microorganisms and often the initial infection leads to secondary infections by other organisms. A reduction in feed intake is one of the earliest observations of the disease; this is followed by coughing, thick nasal discharge, and overall unthrifty appear-

■ **Fig. 9.14**   Pneumonia is a multifactorial disease that mostly affects calves, but can affect cattle of all ages. (© ason, 2009. Under license from Shutterstock, Inc.)

ance. Antibiotic treatment is often successful and management control practices to minimize stress of the calf are important toward reducing the incidence of respiratory disease within the herd.

Capacity of Production

The goal of the beef industry is to satisfy consumer demand for palatable, tender, consistent, edible meat. Animals may be raised for market or retained for breeding. Calves are typically weaned between five and nine months of age. Females in the breeding herd will be bred between thirteen and fifteen months of age. After an approximately two hundred eighty-five day gestation, the heifer will calve for the first time at approximately twenty-four months of age. Females are bred approximately eighty days after giving birth, resulting in the production of one calf each year. This annual cycle will continue until the female is removed from the breeding herd. Cows are removed or culled from the breeding herd for a variety of reasons including failure to become pregnant in a timely manner, feet or leg problems, udder problems that impair the ability of the calf to suckle, age, temperament and productivity. On average, beef cows remain in the herd for six to seven years of age. Once a cow is culled, she is sent to be harvested.

■ **Fig. 9.15** Beef carcasses are graded according to quality and yield. Fat is one primary factor in determining both grades. Quality is assessed primarily by the degree of intramuscular fat and animal age. Yield is determined from the amount of backfat, muscling and internal fat. (© svitlana10, 2009. Under license from Shutterstock, Inc.)

Beef Cattle as a Food Commodity

The primary end products of beef production are derived from grain-fed cattle that are between twelve and thirty months of age. The carcasses of animals that represent this primary product are graded for both quality of the beef and the yield of beef from the carcass. Quality grades used for the primary product are prime, choice, select, standard, and utility, with prime being the most desirable quality grade. Quality grades are based primarily on the amount of intramuscular fat or marbling that is present within the loin muscle of the carcass. Age of the animal and color of the meat are also factors used to calculate quality grade. Yield grades provide an estimate of the percentage of the carcass that will be lean meat relative to content of fat. Yield grades are based upon the amount of backfat on the carcass, the cross-sectional area of the loin or longissumus dorsi muscle at the twelfth rib, carcass weight and the amount of internal (kidney, pelvic and heart) fat. Yield grades are from one to five. A carcass with a yield grade of one is from an animal with minimal backfat, is heavy muscled and will have a high yield of edible beef. A carcass on the other end of the range with a yield grade of five would have excessive backfat, would typically be light muscled and would have a much lower yield of edible beef. The most valuable beef carcass would have a quality grade of prime and a yield grade of one. Meat from animals that are culled from the production herd, such as cows and bulls that are older than thirty months, are a secondary product of beef production, and represents about 20% of the beef produced in the United States. These secondary products are used in ground beef, processed beef products and in lower quality steaks and roasts.

Nutritional Value of Beef

The United States beef cattle population peaked in 1976, this marked a gradual decline in annual per capita beef consumption that would persist for nearly twenty years. In the last ten years, per capita consumption has remained steady. Explanation for earlier decline can be attributed to the concern over cholesterol intake in the late 1970s and early 1980s.

Fig. 9.16   Beef remains an integral part of many diets. (© Gregory Gerber, 2009. Under license from Shutterstock, Inc.)

During this time, researchers tended to over emphasize the negative health effects of cholesterol, explaining the decrease in red meat consumption. However, current research has shown that cholesterol in moderation is not detrimental and the stigma attached to red meat has slowly declined, thus explaining the recent, higher per capita consumption. Beef is a nutritious food, and lean beef can be included as a part of a balanced diet. Concerns by consumers regarding the fat content of beef, especially in comparison to other meat and protein sources, have been addressed by trimming of excess fat before retail sale and by genetic selection for leaner cattle.

Small Ruminants

This chapter introduces the classification of sheep and goat breeds in the United States, including fiber, meat, and dairy types. The prominent breeds that fall under each category are described, as well as the production systems and managerial aspects of successful sheep and goat enterprises. Lastly, uses of sheep and goats for fiber, conservation, and food along with the nutritional benefits of sheep and goat meat are discussed.

SHEEP

"The shepherd always tries to persuade the sheep that their interests and his own are the same."
—*Stendhal (1783–1842)*

Sheep: A Historical Perspective

According to archaeological evidence, domestication of sheep traces to 8000 B.C. in South West Asia, an area now occupied by modern day Iran and Iraq. Sheep are one of the earliest food animals to undergo domestication. Their domestication success is attributed to their relative docility, size, and socially dependent nature. Modern domestic sheep are descendents of Mouflon sheep, which still exists as a primitive breed of Iran and Iraq today. Although domestication of sheep was originally for a source of food, evidence that specialization of sheep for wool occurred as early as 3000 B.C. As a means of cultivating the wool rendered from sheep, weaving was one of the first arts to develop and woolen factories began to appear in Rome during 50 A.D. This period of time also corresponds to the development of the Merino breed specifically for the production of fine wool. Sheep also provided milk and capital. Shepherds and their flocks are routinely mentioned throughout the Bible as sheep were of crucial eco-

nomic importance. As with other livestock, the domesticated sheep was spread throughout the world by migrating human populations.

All ancestors of domesticated sheep found in the United States were introduced. Rocky Mountain sheep, also known as Big Horn sheep, are native to the continent and served no role in the development of the United States sheep industry. Christopher Columbus was the first to introduce sheep into the Western Hemisphere during his second voyage to the New World in 1493. Sheep were brought to the West again by Cortez in 1519. These sheep were used by early Spanish settlers and armies as they moved to conquer and colonize the New World. Subsequently, flocks were acquired by Native Americans and these sheep became the Navajo-Churro sheep breed of today. The introduction of sheep by English colonizers occurred in 1607 and again in 1609 by the Jamestown colony in modern day Virginia. Initial imports were primarily for food and wool quality declined during this era. However, the wool industry would begin to develop coincident with the first woolen mill in Massachusetts during 1662. As wool gained importance in America during the seventeenth century, labor laws were enacted requiring colonists to work in the industry. England, enraged by the growing status of the colonist's wool industry, outlawed the American wool trade, which contributed to events that prompted the Revolutionary War.

The term spinster originally defined a young woman who spun yarn. One of the earliest labor laws enacted to support the United States wool industry required the eldest unmarried woman within a family to spin yarn. Thus, the term spinster was redefined and is used today to reference an older, unmarried woman.

■ **Fig. 10.1.1** Domesticated sheep are descendants of the Mouflon. (© Fernec jCegledi, 2009. Under license from Shutterstock, Inc.)

By the mid 1800s, the sheep industry started drifting westward as expansive, untapped, and cheap land became available. The Civil War contributed to the greatest wool demand in American history. Shortly after, however, the United States wool industry began to face competition from imported wool and cotton. The emphasis of the industry began to shift away from wool production and into raising and marketing lambs for food. The greatest number of sheep coincided with World War II when a record 56 million head were reported. After this time, sheep numbers began to dramatically decline, a trend that continues today in many regions. Peak sheep numbers during War World II are attributed to the supply of mutton as a dietary staple to soldiers during the war. Post World War II, lower returns per investment, scarcity of sheep herders, uncertainties of grazing allotments on public land, susceptibility of flocks to predation, and seasonal production and fluctuations in annual income contributed to the decline of the industry. The National Wool Act of 1950, which demanded increased production of wool as a strategic material for military uniforms, stabilized the industry until the 1960s, when wool was removed as a strategic material and cheaper, synthetically engineered materials began to replace the need of wool as a fabric. Although the industry has attempted to capitalize on meat production, challenges have been faced in part to lack of incorporation of lamb in the consumer's diet. For those who do consume lamb, it is a high priced commodity when compared with other meats available. The decline in sheep numbers

■ **Fig. 10.1.2** In Massachusetts, early law required that every young person learn the art of spinning and weaving wool. (© akva, 2009. Under license from Shutterstock, Inc.)

has occurred to the extent that sheep were labeled as a minor agricultural species by the Federal Drug Administration in 1990. Further insult came in 1995 as the federal government withdrew its subsidies to wool producers. Globalization has further contracted the industry as the United States struggles to compete with product from the world market. The future of the sheep industry will ultimately be determined by consumer demand, which is driven by availability of convenient and user friendly cuts, increased knowledge of preparation and cooking, and improved consistency in retail products. There is optimism for producers that have withstood the challenges to the industry. Increased production efficiency, new technologies that improve product stability, increased profit from culled ewes, and emergence of speciality markets offers promise in stabilizing the industry.

■ **Fig. 10.1.3**   During Woodrow Wilson's term (1912), sheep were kept on the White House lawn. The wool was auctioned off to support charities and the small flock also helped to trim the lawn. (Library of Congress.)

Breeds of Sheep

Approximately two hundred breeds of sheep are documented in the world today. According to the American Sheep Industry Association, over forty breeds of sheep are found in the United States. Breeds of sheep are classified according to their commercial use and include: maternal or wool breeds, paternal or meat breeds, dual-purpose breeds, hair breeds, and dairy breeds. Breeds vary considerably in weight and number of offspring. Weight may range from one hundred to four hundred pounds, whereas, number of offspring may range from one to six per lambing depending on the breed type. Universally, sheep are considered the most timid and least trainable of the domesticated species. These features have prompted the assumption that sheep are the least intelligent of the animals; however, these attributes are due to early selection based on flocking instincts. The sociable and group mindset of the flock is desired by the shepherd, but discourages independent behavior. This has left the sheep more reliant on human care relative to other domesticated animals.

Maternal or Wool Breeds

Maternal, or ewe, breeds are classified according to reproductive efficiency rates, maternal ability, wool production, and milk production. These breeds are typically white faced and include the Merino, Rambouillet, and Finnsheep.

The Merino breed was developed during the reign of the Roman Empire between 41 and 50 A.D. in an area now occupied by Spain. Considered the most influential breed of sheep, the Merino traces its origins to Tarentine and Laodician sheep, breeds of Rome and Asia Minor, respectively. The Merino has played a crucial role as the foundation breed for the development of almost all other fine wool sheep breeds in existence today. During the fall of the Roman Empire, England and Spain rose to become world leaders in Merino production. Spain, outlawed the exportation of Merino sheep to maintain its status as the producer of the greatest quality Merino wool in the world and maintain dominance of the industry. This law remained in effect until 1809, when Spain was invaded by Napolean Bonaparte. William Foster would risk penalty of Spanish laws in 1793 to smuggle the Merino breed to America. The Merino is medium bodied, produces a high quality white fleece, is able to thrive on relatively poor grazing land, and is known for its strong flocking instincts. There are multiple types of Merino sheep developed throughout the world. In the United States, Merinos are characterized by their skin surface area, which coincides with wool yield. Types are classified as A, B, or C according to size and quantity of wool produced as a result of skin folding. The A and B types are heavy types, with the A type possessing heavy neck and hide folds. The B type also carries heavy neck folds, but is relatively free of hide folds. The heavy folding of the skin as occurs with these types creates difficulty in shearing, may result in a lower quality fleece, and promotes parasite infestations. Limited production of A and B types occurs within the United States and the focus has been toward the C type, or Delaine Merino, which is a smooth or nearly smooth type. The Delaine Merino is of medium size with ewes weighing one hundred twenty-five to one hundred eight pounds and rams one hundred seventy-five to two hundred thirty-five pounds.

The Rambouillet breed originated in France in the late 1700s and was developed using Spanish Merino lines. However, the Rambouillet was selected for greater size over its Merino ancestors and breeding records trace to 1801 in France. The importation of the Rambouillet into the United States occurred around 1840 and it is estimated that 50% of the United States crossbred sheep population today were influenced by Rambouillet breeding. Today, the Rambouillet is the seventh most popular breed of sheep in the United States, prevalent in the western

states and preferred as a producer of fine wool. This breed is large bodied with the mature weight of ewes ranging from one hundred fifty to two hundred pounds and the weight of rams two hundred fifty to three hundred pounds. Rambouillets are fast growing, produce a high quality white fleece, and are adequately muscled.

■ **Fig. 10.1.5**   Rambouillet. (Courtesy The American Rambouillet Sheep Breeders Association.)

■ **Fig. 10.1.4**   Merino. (© John Carnemolla, 2009. Under license from Shutterstock, Inc.)

■ **Fig. 10.1.6**   Finnsheep. (Courtesy Elizabeth H. Kinne, Stillmeadow Finnsheep, DeRuyter, NY.)

The Finnsheep, also known as the Finnish Landrace, originated in Finland. First importations into the United States occurred in 1968 and although further importations since this time have been limited, the breed has gained popularity in the United States in recent years. The primary use of Finnsheep was for cross-breeding programs. The breed is renowned for multiple births, commonly producing three or more offspring per lambing. Lambs are characterized by their high survivability resulting in increased lamb crops in cross-bred flocks. However, Finnsheep display a reduced growth rate relative to other maternal breeds, reach maturity at an early age, and are of small mature size. Females average one hundred twenty to one hundred ninety pounds at maturity, whereas males average one hundred fifty to two hundred pounds. The breed is polled and wool is of medium quality and not as desired as that originating from the Merino or Rambouillet breeds, however, the fleece is high yielding.

Paternal or Meat Breeds

Paternal breeds are defined according to growth rates and carcass characteristics. Breeds are usually dark faced and further classified by the weight of the offspring at market, which includes heavy, medium, and light weight types. The Suffolk and Hampshire are popular meat breeds of the United States that produce heavy weight lambs for market.

The Suffolk breed originated in southern England from the breeding of Southdown and Norfolk Horned sheep. First importations in the United States occurred in 1888. Currently the most popular breed of sheep in United States registries, the Suffolk is white woolen and characterized by black face, ears, and legs that are free from wool. The breed is large bodied with mature ewes weighing one hundred eighty to two hundred fifty pounds and mature rams two hundred fifty to three hundred fifty pounds. Suffolk are polled, display prominent muscling, and produce rapidly growing, lean muscled offspring. Selected for carcass quality, Suffolk yields a lightweight quality fleece that contains black fibers, elements that detract from its market value.

The Hampshire originated in England and was imported into the United States in the mid 1800s, prior to the Civil War. These original imports were lost during the war and subsequent imports of 1880 served as the foundation for the breed in the United States. Today, the breed is the second most populous breed of sheep in the United States. Hampshires resemble the Suffolk in both size and marking similarities; however, the Hampshire tends to be smaller and has a dark brown face, ears, and legs. At maturity, ewes weigh a minimum of two hundred pounds, whereas the minimum mature weight of rams is two hundred seventy-five pounds. Wool of Hampshires extends onto the legs and head giving rise to a wool cap. As the wool is of similar

■ **Fig. 10.1.7**   Suffolk. (Courtesy Paul Kuber, The Ohio State University.)

■ **Fig. 10.1.8**   Hampshire. (Courtesy Paul Kuber, The Ohio State University.)

■ **Fig. 10.1.9** Dorset. (© Paul Cowan, 2009. Under license from Shutterstock, Inc.)

■ **Fig. 10.1.10** Columbia. (© Steve Shoup, 2009. Under license from Shutterstock, Inc.)

quality as the Suffolk and of little market value, the increased wool of Hampshires is not desired. Hampshires are efficient utilizers of forage and the rapid growth weight of lambs, usually gaining a pound or more per day till marketing, contributes to this breeds use in the meat industry.

Dual-Purpose Breeds

Breeds that produce both wool and meat of acceptable quality are termed dual-purpose breeds. The wool and carcass produced by these breeds are of lesser quality relative to the maternal and paternal breeds, respectively, but both products are adequately produced in one animal. The dual-purpose breeds include the Dorset, Polypay, and Columbia.

The Dorset originated in the counties of Dorset and Somerset, England. The breed was first imported into the western United States in 1860 and the eastern United States in 1885. The Dorset is now the third most popular breed of sheep in Unite States registration numbers. The breed has a white face, ears, and legs that lack a woolen covering. The breed may be horned or polled with the number of polled Dorsets currently exceeding the number of horned. The polled trait results from a mutation of the horned breed first identified and selected for in a flock from North Carolina in the 1900s. The ewes are known for their prolificacy, milk production, and their ability to breed out of season and reach mature weights of one hundred fifty to two hundred pounds. Rams weigh two hundred twenty-five to two hundred seventy-five pounds at maturity. The ability to breed out of season is the single most important quality that contributes to the breed's popularity. Most of the demand for out of season lambs lies in the eastern United States; hence, most Dorsets are raised on the east side of the

Mississippi River. The wool is of a medium grade and its carcass traits are average.

The Polypay breed was developed in 1968 at the United States Sheep Experiment Station at Dubois, Idaho and is a composite breed of Finnsheep, Rambouillet, Targhee, and Dorset breeding. The breed was developed to meet the goal of lifetime prolificacy, ability for ewes to lamb at least twice per year, production of rapidly growing lambs, and a high quality carcass. It was the intent of the breeders to develop a more productive sheep with the potential of stabilizing the declining industry. The name Polypay was meant to insinuate that this breed of sheep could produce more than two paying goods per year, fleece and two lamb crops. The Polypay is characterized by early maturity and is a medium sized sheep with a white face, ears, and legs. The wool covering extends onto the top of the head and down the legs of the sheep.

The Columbia breed was developed by the Bureau of Animal Industry at the King Ranch in Laramie, Wyoming in 1912. Developed from the crossing of Rambouillet ewes with Lincoln rams, it is now the eighth most populous breed of sheep in the United States. The goal during breed development was to replace cross-breeding practices on the range by providing a true breeding type that maximized wool and lamb production. With ewes weighing one hundred fifty to two hundred twenty-five pounds and rams two hundred twenty-five to three hundred pounds, these animals are well suited for life on the range in the northwestern regions and are adequately suited to pastures of the Midwest. The ewes are adequately fertile, good mothers, and produce acceptable market lambs. Columbians are polled and white with woolen face and legs.

■ **Fig. 10.1.11**   Katahdin. (© Sajko, 2009. Under license from Shutterstock, Inc.)

Hair Breeds

All sheep have a combination of woolen fibers and hair fibers. Primitive breeds of sheep are distinguished by their hair fibers that mask the undercoat of woolen fibers, which provide warmth. The majority of sheep breeds were selected and bred to maximize the presence of woolen fibers and minimize the hair fiber. However, in arid climates there is relatively little need for wool. Sheep in these parts of the world have been selected for an emphasis on hair fiber that naturally shed in response to the climate, which negates the need of shearing. Although hair sheep may resemble goats by the novice, they are indeed of sheep descent and are raised for their meat. Considered tropical sheep, hair sheep have exceptional fertility rates, viability, parasite resistance, and extended breeding seasons. However, these sheep have reduced growth rates, lighter mature weights, and reduced carcass merit. The two most globally utilized hair sheep breeds include the Barbados Blackbelly and Saint Croix, whereas the Katahdin is the most common hair sheep of the United States and is sixth in registrations. The

Katahdin traces its origins to Maine. Named after Mount Katahdin, the highest peak in Maine, the breed was developed for land management to graze power line easements as an alternative to spraying or mowing. Multiple breeds of sheep including the St. Croix, Tunis, Southdown, Hampshire, Suffolk, and Wiltshire Horn have influenced the Katahdin, which is known for its adaptability, exceptional maternal qualities, adequate carcass, and low maintenance in the absence of shearing.

Dairy Breeds

Just as sheep breeds were developed to meet the demands for meat and fiber, specialized breeds have been developed for the dairy industry as well. Dairy sheep breeds include East Friesian, Awassi, Assaf, Lacaune, Sarda, Manchega, and Chios. These breeds are exceptional milk producers that produce four hundred to eleven hundred pounds of milk per lactation; however, with the exception of the East Friesian and Lacaune, these breeds are not available in the United States. East Friesian is noted as the highest producing dairy breed, producing over one thousand pounds of milk in a two hundred to two hundred twenty day lactation. The breed was developed in the regions of Germany and the Netherlands and are best suited to small-scale operations. The Lacaune is available in only limited numbers within the United States. The breed originates from France and is noted for increased milk solids when compared to the East Friesian, making the breed an ideal selection for cheese production. The United States dairy industry is reliant on adapting currently available sheep breeds for milk production and includes the use of Dorset, Polypay, and Rideau Arcott breeds. The Rideau Arcott was developed in Canada and is predominantly influenced by Finnsheep, Suffolk, and East Friesian breeding. Originally created for improved lambing rates, the breed is noted to routinely give birth to triplets and lambs every eight months. When compared to the specialized dairy breeds of European and Mideastern countries, the available meat and wool sheep breeds used in United States dairying are inferior, producing only one hundred to two hundred pounds of milk per a ninety to one hundred day lactation. Crossbreeding improves milk yields and extends lactation length. Dorset-East Friesian crossbred sheep produce three hundred to six hundred fifty pounds of milk in a one hundred twenty day average lactation. Currently, 50% of the United States dairy sheep are listed as crossbreds.

United States Sheep Industry

At 5.2 million head, the current sheep inventory represents only 9% of the peak inventory reported during World War II. Market lambs are the predominant source of income and reflect 25% of total inventory, with production confined primarily to the western United States. Texas, California, Wyoming, Colorado, and South Dakota are the leading states in total sheep and lamb inventory and are representative of larger operations. Production in the eastern United States is characterized by smaller operations. Operations with less than twenty-five head account for 67% of sheep production systems, but only 10% of total sheep and lamb inventory. Sheep inventory includes estimates for breeding and non-breeding stock. Breeding stock represents the majority of the sheep inventory, but has declined in recent years reflecting withdrawal from the industry. Recent lamb crop estimates of 3.4 million head represent further decline in the industry as does shorn wool production. The status of the sheep industry is reliant on consumer demand. The majority of lamb consumed in the United States is concentrated in the Northeast and western states that support growing immigrant populations. Lamb represents less than 1% of dietary protein in the average Americans diet, equating to per capita consumption of 0.8 pounds per person annually. As the industry traditionally focuses on high valued cuts for market, it has not capitalized on consumer demand for high quality, inexpensive, convenient products. Nearly 50% of the lamb consumed originates from outside the United States, with 99% of imports coming from Australia and New Zealand. While these countries aggressively market the quality of sheep products produced, the United States market structure and pricing system rewards producers on market weight. Adoption of a value-based marketing system to improve product consistency and consumer confidence is needed for domestic lamb to be globally competitive.

Sheep Management Systems

Purebred-flock Management
Purebred-flock management involves the production of high quality breeding stock for other pure-

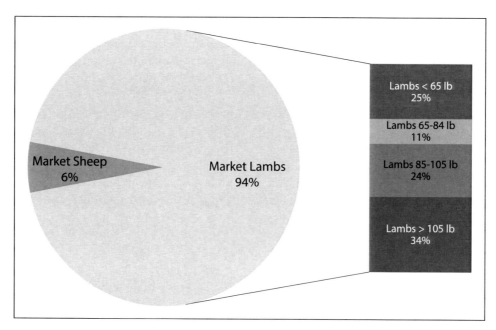

■ **Fig. 10.1.12** United States sheep inventory structure and size distribution of market lamb inventory. Lambs dominate the market industry. Considerable weight variation exists within the industry, with the majority of marketed lambs weighing more than 105 lb. (United States Department of Agriculture-Economic Research Service)

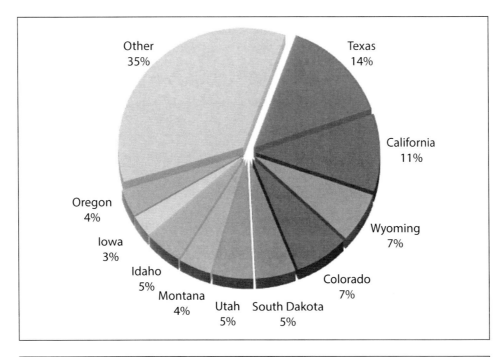

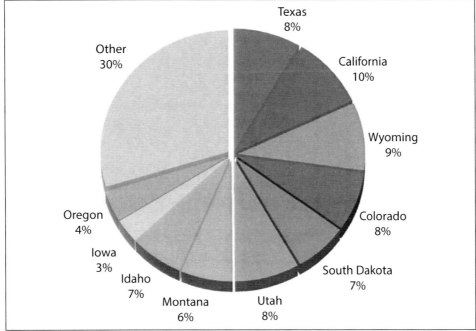

■ **Fig. 10.1.13**   Top ten states in sheep and lamb inventory (top) and wool production (bottom) during 2013. (United States Department of Agriculture-Economic Research Service)

bred operations, commercial, or specialty industries. The purebred industry is the predominant source of breeding animals used in crossbreeding programs of commercial operations. Usually a manager is assigned to supervise all aspects of production and detailed data is recorded on each animal. The purebred farm gives careful consideration to each mating so as to produce the highest quality and performing animals possible. This farm also is involved in regimented feeding and actual advertising of their stock.

■ **Fig. 10.1.14** The majority of sheep raised in the United States are kept on range and managed by shepherds. (© Hu Xiao Fang, 2009. Under license from Shutterstock, Inc.)

Purebred flocks are commonly kept on the farm to allow closer supervision; however, some are located on the range. Overall, this is a highly specialized segment of the sheep business in which only a few participate.

Commercial Production

Commercial operations include farm-flock and range systems. Farm-flock operations are used predominantly for the production of sheep for meat, whereas sheep used in meat and wool production are common to the range system of production. Farm-flock systems are representative of the small operations of the Midwest and eastern United States. This system of production uses confined, fenced pasture land for raising sheep of an average flock size of less than fifty head. Often sheep production is not the sole source of income, and sheep may be companion grazed. Farm-flock systems represent the greatest number of operations in the United States, but produce less sheep than range systems. In range systems, sheep are managed in a range-flock method of production. Sheep are not strictly confined but controlled through the use of shepherds. Because range systems graze larger sized flocks, averaging more than two hundred head and reaching as many as one thousand head of ewes and their lambs, greater land availability is required. Thus, range systems are located in the western United States and approximately 25% of the sheep raised graze public lands managed by the Forest Service of the United States Department of Agriculture and the Bureau of Land Management for a fee. Sheep are marketed through three primary channels: traditional, early harvest, or direct lamb marketing. Traditional lamb marketing accounts for 75% of marketed sheep. Lambs that began on farm-flock or range systems are transferred to pasture or feedlot operations at fifty to seventy pounds. The operations provide high quality feeding systems for finishing to an average weight of one hundred forty pounds. Feedlot operations provide an alternative to pasture finishing when vegetative growth is limited and insufficient to produce a quality carcass. The re-

■ **Fig. 10.1.15** The number of specialty dairy sheep flocks has increased in response to the rising demand for domestic sheep cheese. (*Left:* © Ashley Cooper/Corbis. *Right:* © anistidesign, 2009. Under license from Shutterstock, Inc.)

sulting finished lamb is sold for harvest through auction or contract arrangements between the producer and processor. Early harvest lamb marketing involves the direct sale and harvest of lambs that were placed on high quality pasture at an early age. The greatest demand for early harvest lambs are centered near urban areas. The last marketing channel, direct lamb marketing, involves the sale of live lambs from the farm to individuals.

Specialty Flocks

Prior to 1980, sheep dairying was nonexistent in the United States. However, in the countries surrounding the Mediterranean, sheep dairying has been a long-standing and important agricultural enterprise. Sheep milk is used in the production of some of the world's most expensive and gourmet cheeses, including feta, ricotta, Fiore Sardo, pecorino, Roquefort, and Brebicet. Sheep milk is especially well suited for cheese production. It has a high percentage of solid components averaging 18% and can produce nearly twice as much cheese than what would be rendered from the same quantity of cow milk. The United States imports a large majority of sheep cheese, averaging more than seventy million pounds annually, while domestic production is only 1.5 million pounds. To capitalize on the increasing domestic demand for sheep cheese, specialty dairy flocks have risen in popularity and it is estimated that over one hundred farms in the United States are milking ewes. The largest concentration of dairy flocks are located in northwestern Wisconsin with fifteen flocks registered for dairy production.

Voluntary surveys of dairy sheep enterprises in Wisconsin reveal that the majority of milk produced is marketed and sold through cooperatives and consumed locally. Annual production is estimated at nearly three hundred ninety pounds of milk per ewe with a value approaching $50 per hundred weight, twice the value of cow's milk. However, the sheep dairy industry is considered a high risk enterprise that involves knowledge of both sheep and dairy production and the availability of cooperatives for competitive marketing. As a new enterprise in the United States, there remains uncertainty of the markets future despite growing domestic demand for sheep dairy products.

Health Management

Causes of Morbidity

Approximately 80% of sheep operations have added replacement animals from an outside flock. The introduction of animals poses the greatest risk of introducing new diseases into a system. To minimize the transmission of disease, biosecurity measures including quarantine of the newly acquired animals should be strictly enforced. Isolation and confinement of the animal from the remainder of the flock should occur for a minimum of four weeks to allow for the incubation and identification of certain diseases. During quarantine, animals can be observed for signs of illness, tested for diseases, and vaccinated and treated for parasites. Despite the advantages of quarantine, it is estimated that less

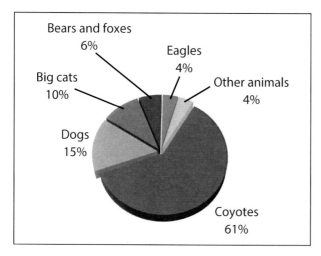

■ **Fig. 10.1.16** Over one third of death losses to the sheep industry are a result of predation, with coyotes being the number one predator. Flock protection is a necessary management practice in the industry and may include the use of guard dogs, donkeys, or llamas. (Bottom: © Robert Saroseik, 2009. Under license from Shutterstock, Inc.)

than one third of sheep operations practice such precautionary measures. Currently, of greatest economic concern to the sheep industry are the prevention and treatment of internal parasites, which are prevalent in 74% of sheep operations. The most common internal parasites affecting the industry are round worms. For some of these species that inhabit the gastrointestinal tract, sheep are the sole host. Adult parasites lay eggs within the gastrointestinal tract; these eggs are subsequently shed through the feces and contaminate the environment. Under appropriate conditions, the eggs will hatch and larvae develop. Sheep that graze the contaminated environment will ingest the larvae, which will fully mature to an adult capable of laying additional eggs within the gastrointestinal tract, thus repeating the cycle and spread of the parasite. Younger lambs,

older sheep, and animals with a compromised immune system are more susceptible to the effects of gastrointestinal parasites, which include reduced appetite, reduced growth, and even death with increased parasite load. Anthelmintics are commonly administered drugs to treat internal parasites, however, limited availability and extensive use leading to resistance has led producers to seek alternative means to treat internal parasites. Selection of breeds more resistant to infection and improved pasture management techniques are being adopted as new strategies in the treatment of internal parasites.

Cause of Mortality

Although disease has a significant impact on the sheep industry, predation is the single leading cause of sheep mortality and accounts for over one third of death losses documented in the industry. Predator control accounts for almost 12% of the total cost of sheep production. Lethal predator control measures include: shooting, trapping, snaring, and the use of poison. It is important that lethal control methods target the predator without risk to other animals. For this reason, lethal control collars are used to administer poison. The collars are worn around the neck of the sheep and administer a lethal dose of poison into the predator when the neck is bitten. An alternative to lethal control methods are non-lethal methods including the use of guard animals, electric fencing, and noise devices; however, these measures only reduce and do not eliminate predation. Predators include mountain lions, dogs, coyotes, bears, foxes, eagles, and bobcats. The coyote is responsible for the most damage, accounting for 61% of predator kills on sheep, followed by dogs at 15%.

Capacity of Production

The majority of sheep breeds are seasonally polyestrus, displaying estrous cycles averaging sixteen to seventeen days, but only in response to the decreasing ratio of light to dark within the photoperiod. In general, sheep are termed short day breeders and most fertile from September to November, while experiencing anestrus or reproductive inactivity during the Spring and Summer months. Because of the ewe's gestation period of one hundred forty-eight days, this places most lamb births

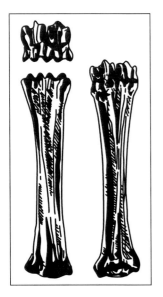

■ **Fig. 10.1.17**
Skeletal maturity is assessed by the presence of break joints (left) or spool joints (right) and is used to determine the physiological age of sheep and carcass quality.

within the spring months; however, length of breeding season is dependent on the breed and some breeds, including the Dorset, have extended breeding seasons and will lamb in the fall. Breeds originating in higher altitudes generally have shorter breeding seasons, and breeds of sheep originating from tropical regions are considered non-seasonal breeders as a consequence of temperate photoperiods. Conditions can be artificially manipulated to allow for multiple breedings within one year and can be accomplished through selecting breeds that are non-seasonal breeders, administering hormonal treatments to induce estrus, or using artificial light to simulate the most favored light-to-dark ratios. Rams also are seasonal breeders affected by the shortening of day. It is during this time that the rams produce the greatest volume of quality semen and display the greatest interest in the female. Unlike other food animal industries, artificial insemination is not routinely used in commercial flocks and is reserved for purebred operations. The anatomical structure of the ewes cervix limits penetration and resulting conception rates associated with cervical AI are less than 50% and even as low as 25%. Techniques used by the purebred industry involve a trained and skilled technician for laparoscopic insemination that places semen in the uterus. Although high conception rates are documented with this procedure, it is currently not an economically viable procedure for wide spread acceptance by the commercial industry. Because of the difficulty of artificial insemination techniques, most

matings occur by natural service. Two methods of natural service are hand mating and pasture mating. Hand mating requires the use of teaser rams to identify ewes in estrus. Once located, the ewes are removed from the flock and mated to a selected ram. If the ewes are to be pasture mated, they are sorted into groups according to the ram which they are intended to be mated. The brisket, or ram's chest, is marked with paint which is transferred onto the ewe's back during mating. This paint acts as a means of identifying when the ewe is bred and the color is changed every seventeen days to identify ewes rebred.

Sheep as a Food Commodity

Lambs are marketed in the United States according to age and weight. Sheep are marketed at any age and weights typically range from twenty to one hundred sixty pounds. Young lambs marketed prior to weaning are commonly referred to as milk-fed or hot house lambs. Hothouse lambs are produced primarily for east coast markets, namely in Boston and New York. These lambs are harvested between six to twelve weeks of age, twenty-five to fifty pounds, producing a desirable fat cover and resulting in extremely tender meat. These lambs are born and marketed from December to April and thus require their ewes to be bred out of season. The majority of lamb consumed in the United States is from lambs that were born the previous spring and marketed at or prior to one year of age. The average weight of these market lambs is one hundred forty pounds. When lambs exceed one year of age, but are less than twenty-four months, they are classified as yearling mutton. Mutton represents the meat of market sheep greater than twenty-four months of age. It should be noted that chronological age is not the sole determining factor in lamb classification, physiological age is considered as well. As sheep mature, the cartilage at the surface of the cannon bone on the fore shank undergoes ossification, or hardening, as a result of calcification of the joint during bone maturation. In lamb carcasses, the joints on both foreshanks are soft, red from increased vascularization, and porous. At this stage the joints are referred to as break joints that can easily be severed to remove the lower shank during harvest. In yearlings, there is the presence of one imperfect break joint and one spool

■ **Fig. 10.1.18** Lamb represents less than 1% of the protein consumed in the average Americans diet. (© Dale Berman, 2009. Under license from Shutterstock, Inc.)

joint, which represents the hardening of the joint from ossification. Carcasses classified as mutton are marked by the absence of break joints and the presence of two spool joints.

Nutritional Value of Lamb

Lamb is an excellent source of protein, with one serving containing 43% of the recommended daily allowance of protein. Lamb also contains essential amino acids and is rich in vitamin B12, zinc, niacin, iron, and riboflavin. Lamb is comparable in fat to other meats; however, the fat is not marbled within the meat, but rather is located to the exterior of the muscle and can easily be trimmed. Furthermore, only 36% of the fat is saturated, with the remaining being monounsaturated and polyun-saturated.

Sheep for Fiber and Conservation

Fiber

Sheep are normally sheered in the spring, in preparation for the upcoming summer months. In conven-

tional shearing methods, the sheep is placed on its dock and cradled between the shearer's legs. The shearer can manipulate the sheep by changing leg position. An electric powered shearer is used to remove the wool. When properly clipped, the wool can be removed as a complete fleece. The fleece removed from the legs is usually coarse and soiled and is separated from the quality fleece. It is necessary that all sheep are sheared, not just those involved in fine wool production, as the fleece can lead to heat retention and heat stroke in the animals. The shorn wool that has not been washed is termed grease wool.

After sheared, wool is placed in plastic bags and the bags are weighed and identified for marketing. Each fleece in a producer lot is sorted, graded, and baled based on grade, with the grade determining the value of the raw wool. The six hundred to eight hundred pound bales of wool will be shipped to manufacturers where it is cored and micron evaluated to determine the value of the bale. Wool is evaluated based on yield, diameter, length, and strength. The most desirable staple length is three to four inches, equivalent to one year growth on the animal. The wool must be strong to withstand processing of wool fibers and should not have breaks in the fibers, which may be caused by fever, illness, or parturition. The finer (less diameter) fleeces are more valuable and are white in color and clean. White, clean fleeces also are associated with increased yields. There are three systems for the grading of wool: micron range, spinning count, and American blood grade. The micron range measures the wool fiber diameter in microns. The spinning count determines the number of hanks (560 yards in length) that can be spun from one pound of wool top. This system is no longer practiced. The American blood grade system is based on the percentage of Merino wool that is found in the bloodlines of the sheep. Grades include: fine blood, which equates to 100% Merino; half blood; three-eighths blood; quarter blood; low quarter blood; common; braid; and hair (no value as wool). Today, fine blood refers to fine wools that are associated with Merino and Rambouillet breeds; half blood, three-eights blood, quarter blood, and low quarter blood are medium wools associated with the Cheviot, Columbia, Corriedale, Dorset, Hampshire, Polypay, Southdown, and Suffolk breeds; common and braid are coarse (long) wool breeds including

Fig. 10.1.19 Sheep are sheared by hand, which removes the wool as a complete fleece. The shearer will position the sheep so that it is resting on its dock and is cradled between the shearer's legs. Shearing is a necessary component of sheep production, especially during the summer months. (Both images: © Gail Johnson, 2009. Under license from Shutterstock, Inc.)

the Leicester, Cotswold, and Lincoln; lastly, the hair grade system refers to the hair breeds Katahdin, Barbados, and St. Croix.

The wool is subsequently scoured, or washed in 140° F water, to remove lanolin and other impurities that can represent 30% to 70% of the weight of the raw fleece. The lanolin is separated from the wash water, purified, and used in creams, soaps, and cosmetics. The scoured wool is dyed and carded, which straightens the wool fibers and prepares the fibers for spinning, the process of yarn formation. The yarn is weaved or knitted through traditional looms or weaving machines. The resulting wool fabric is considered a versatile, natural fabric. Wool is absorbent and regulates body temperature, supporting its use in military uniforms. Furthermore, it resists snagging, breaking, and is flame resistant. In 2002 following an oil spill in Southern Australia, conservationists used knitted wool pullovers to protect penguins during recovery and rehabilitation. The wool pullovers guarded against hypothermia and accidental digestion of the oil until the birds recovered for release.

Fig. 10.1.20 Sheep can be companion grazed with cattle. The more selective grazing habits of sheep and preference for sloped land allow greater pasture use when sheep and cattle are companion grazed. (© clearviewstock, 2009. Under license from Shutterstock, Inc.)

Conservation

As grazers, sheep are being evaluated as tools to manage the landscape of public and private lands. Sheep naturally consume a range of vegetation, including woody and broadleaf plants and grasses.

Sheep grazing habits have been used in silviculture, riparian management, wildlife habitat establishment, brush management, and noxious weed control. Sheep are considered a healthy alternative to the use of herbicides. In Washington and California, conifer growth has increased as sheep graze brush and weeds that would crowd saplings. These grazing habits also are used as a low-cost method of undergrowth control to minimize risks of wildfires. In California's Angeles National Forest, sheep are the primary means of controlling flammable underbrush. Similarly, noxious and invasive weeds such as leafy spurge, that would otherwise threaten and destroy native vegetation, are consumed readily by sheep. In Montana, sheep effectively control up to 90% of the leafy spurge infestation at a cost of $0.60 per acre compared to the $35 per acre costs of herbicides.

GOATS

"And one goat for a sin offering, to make an atonement for you."

—*Bible*

Goats: A Historical Perspective

The domestication of goats was concurrent with that of sheep, occurring ten thousand years ago in South West Asia, independently in Turkey and Iran. All domesticated goats trace to the Bezoar goat through DNA analysis. Archaeological evidence suggests that goats may have been preferred over sheep in higher elevations and more arid regions due to their adaptability to these climates. Not only were goats an important source of food, they were an important commodity to agrarian societies and were used to clear underbrush for establishment of farm land. Goats were brought to the Americas along with sheep during the voyages of Christopher Columbus. De Soto and other Spanish explorers brought goats as well, as a source of food. These early goats importations serve as the foundation for the feral goat populations of the southern United States. Traditionally, goats have been used in small numbers for meat and milk production. The exception is Angora goats used for fiber production. Following

the 1995 removal of subsidies for wool and mohair, the goat fiber industry began a steady state of decline. Meanwhile, the meat industry increased, as producers of the fiber industry switched to producing meat goats and demand for meat increased to support ethnic populations. Today, the meat industry is the largest segment, followed by the use of goats in dairying. Goats are used to lesser extent in fiber production, conservation, and biotechnology.

Breeds of Goats

The goat is a multi-purpose animal that has underwent less selection and breed development than other food animals, though distinct breeds have emerged reflecting the animals use. Considered an adaptable, intelligent animal, the goat quickly reverts to a feral state if unattended by humans. Goats are categorized according to their use for milk, meat, and fiber. In addition, dwarf breeds occur. Sizes of goats significantly vary according to breed, with the pygmy goat only reaching twenty pounds at maturity, while large breeds can weigh in excess of three hundred thirty pounds.

Meat Breeds

There are approximately eleven breeds of goat marketed in the United States for meat, with three of these breeds contributing to the majority of the industry: Boer, Kiko, and Spanish. Initial goats used

■ **Fig. 10.2.1**   Boer. (© blewisphotography, 2009. Under li-cense from Shutterstock, Inc.)

in the meat industry were dual-purpose animals, supplying meat and milk, due to limited demand of goat meat. In recent years, the demand for goat meat has led to the establishment of meat goat breeding herds distinct from dairy herds. Meat-type goats are adaptable and capable of producing adequate muscling from foraging. Most meat-type goats are horned; however, the horns of the male are usually greater in size than those of the female.

The South African Boer goat has had the greatest influence on the United States goat meat industry since its introduction in 1993. Indigenous to South Africa, the Boer originated from European, Indian, and Angora goat breeding, the breed was first characterized in the early 1900s and the first registry established in 1959. The name Boer was derived from the Dutch word for farmer. The Boer goat is noted for its adaptability, rapid growth rate, high rates of gain exceeding 0.4 pounds per day when finished in feedlots, improved carcass quality as a result of early maturity, and maternal abilities. Twinning is common in the breed and females are capable of three kiddings every two years. The Boer breed is commonly distinguishable by its red head with floppy ears and white body, though a variety of color variations have been documented. The mature buck can weigh between two hundred to three hundred forty pounds and does between two hundred to two hundred twenty-five pounds.

The Kiko goat was developed in New Zealand for the specific purpose of meat production. The breeds name implies flesh and the breed traces its ancestry to feral does of New Zealand crossbred with Anglo-Nubian, Toggenburg, and Saanen bucks. Selection was based on survivability and growth rate and led to the establishment of the breed in 1986. Limited imports to the United States occurred in 1990, which served as the foundation for breed establishment in this country. In the United States, the breed is noted for the ability to produce meat on pasture without supplemental feeding. The breed performs well in both arid and mountainous climates and has performed well in the Southern United States where Boer goats have not been as successful.

The Spanish goat developed from feral populations in the southern United States. The breed is considered hardy and adaptable, able to survive under more harsh climatic conditions. Originating from feral goats, the breed is generally smaller in size with a smaller udder. Spanish goats were originally used to clear brush. Limitations on importations of animals into the United States led producers to rely on previously established goat breeds within the country to build and sustain the goat meat industry. And as the demand for goat meat increased, the breed was selected for improved carcass qualities. The breed is often crossbred with the Boer goat for commercial meat goat production.

■ **Fig. 10.2.2** Kiko. (Courtesy Mia Nelson.)

■ **Fig. 10.2.3** Nubian. (© saiva_l, 2009. Under license from Shutterstock, Inc.)

The myotonic goat is another breed of meat goat with herds primarily in Texas and Tennessee. Also known as Tennessee Fainting, Wood Leg, or Stiff Leg goats the breed is commonly marketed as pets. The origin of the breed is not fully known. The goats are characterized by their behavior when approached unexpectedly. Due to a recessive, spontaneous mutation the goats become immobilized when startled and may topple. The response appears as though the goat is fainting, but in fact it is the result of the inability to follow muscle contraction with relaxation.

■ **Fig. 10.2.4** Angora (top) and Cashmere (bottom). (*Top:* © rscreativeworks, 2009. Under license from Shutterstock, Inc. *Bottom:* © Falk Kienas, 2009. Under license from Shutterstock, Inc.)

Dairy Breeds

Unlike the meat industry, which is moving toward commercialization, the dairy industry retains its hold through small farms with an average herd size of thirteen and distribution of product through rural communities. Goat milk is a niche market, with the fastest growing segment being cheese production. The American Dairy Goat Association recognizes six breeds of dairy goats including Alpine, La Mancha, Nubian, Oberhasli, Saanen, and the Toggenburg.

The Nubian is the most popular dairy goat breed of the United States. Developed from the British goats and improved by the influences of goats imported from Asia, the breed is not noted as a high producing dairy breed; however, it produces milk with the greatest butterfat content of 4.6% or greater. The second most popular breed is the Alpine goat, with an average annual milk production of 2,083 pounds per doe and milk fat content of 3.5%. Second and third in annual milk production are the Saanen and Toggenburg breeds, with averages of 1,921 and 1,852 pounds of milk per doe, respectively. Milk from the Saanen and Toggenburg breeds is intermediate in milk fat content, 3.3 to 3.5%. The La Mancha, characterized by their gopher or elf ear, and Oberhasli produce 1,687 and 1,472 pounds of milk per doe, respectively, with a milk fat content of 3.7%. In general, the Swiss breeds (Alpine, Saanen, Toggenburg, and Oberhasli) are less tolerant of arid climates requiring shade for optimal performance.

Fiber Breeds

Fiber breeds of goats are kept for their high quality production of hair. The two most popular breeds of fiber goats in the United States are the Angora and the Cashmere. Angora are further classified by the type of hair produced. Type C produces a ringlet type of hair, whereas type B Angoras are known to produce a flat lock. Mohair was an important commodity in the nineteenth century and to increase saleable mohair, the original Angora was crossbred with goats of larger stature to increase the size of the breed. The modern Angora is a result of this crossbreeding as no attempts were made to maintain purity of the original lines. Angoras are predominantly white, but may be colored, and are usually raised under range conditions. Both sexes have horns, although occasionally polled specimens will surface.

■ **Fig. 10.2.5**   Pygmy (top) and Nigerian (bottom). (*Top:* Courtesy Dori Lowell. *Bottom:* © Susie Prentice, 2009. Under license from Shutterstock, Inc.)

The Angoras have been selected for a large size as this provides more surface area for mohair production. The Cashmere industry is relatively new in the United States and its future is uncertain. Cashmere is the fine undercoat fibers produced by all breeds of goat, except the Angora. The Cashmere breed has been developed to produce a significant amount of this undercoat for market. Cashmere goats are sheared once annually and produce an average 2.5 pounds of fleece, of which only 20% is recovered as Cashmere.

Dwarf Goats

Dwarf goat breeds trace their origins to Africa and are considered multi-purpose breeds, with compact, well muscled bodies and proportionately short legs. In the United States, two dwarf breeds have gained popularity: Pygmy and Nigerian Dwarf. Pygmy goats are heavily muscled goats noted for their milking abilities. The maximum height for females is 22.4 inches, whereas males may reach 23.6 inches. Unlike other dairy goats, Pygmy goats are non-seasonal breeders. Both sexes of the Pygmy breed have horns and usually range in colors from white, gray, and black accompanied by dark points and lighter coat areas around the nose, forehead, eyes, and ears. Nigerian Dwarf goats reach a maximal height of 23.6 inches. As with Pygmy goats, they are non-seasonal breeders and exceptional milk producers for their size capable of producing five hundred pounds of milk over a three hundred five day lactation. Nigerian dwarf goats are distinguished from Pygmy goats by their conformation and are bred to have a body structure similar to that of other dairy goat breeds. Although dairy breeds, dwarf goats are popular as pets for humans and horses due to their miniature size and their docile behavior.

The United States Goat Industry

The structure of the United States goat industry is complex, consisting of small farm size and direct live market sales to consumers. Products are marketed as speciality items and many are not readily available in mainstream markets. Currently, the USDA does not track goat product consumption and the value of the industry is uncertain. The average herd size varies across the industry, whereas the average herd size of meat goats in the United States is twenty goats, the dairy industry reports and average of fourteen goats. In Texas, Angora goat herds exceed one hundred head on average. Texas claims the greatest number of goats, an estimated eight hundred seventy thousand raised for meat, seventy-five thousand Angora goats for fiber, and twenty thousand goats for milk. Tennessee is the second leading state in meat goats, while Arizona is second in Angora goat inventories. The dairy industry, however, is concentrated in Wisconsin and California, with an estimated forty-six thousand and thirty-eight thousand head respectively. Between 1997 and 2012, Angora goat numbers fell substantially. However, total goat numbers increased 9% as a consequence of increases in the number of meat

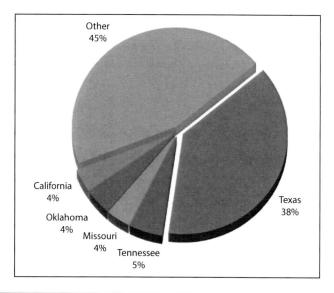

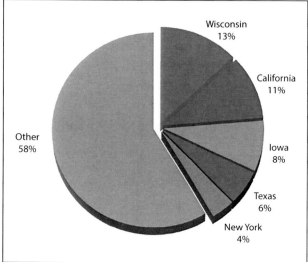

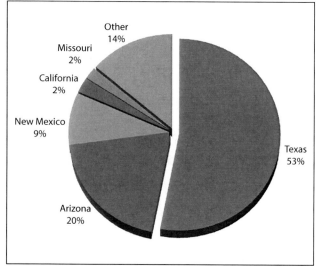

■ **Fig. 10.2.6** Top five states in goats used (clockwise) for meat, mohair, and milk during 2013. (United States Department of Agriculture-Economic Research Service)

and milk goats, which increased 1.9 and 2.4 fold, respectively. The United States goat industry is expected to remain strong in part due to changing population demographics. Ethnic diversity, gender, and increased interests in small-scale farming have had substantial influences on the industry. Ethnic diversity is a primary factor that has contributed to the growing goat meat industry. The most influential ethnic populations are Hispanic, Muslim, Caribbean, and Chinese. The number of women owned goat farms increased 34% between 1997 and 2002 with the greatest increases occurring in Delaware, Tennessee, and Kentucky. Ownership of small ruminants in both small scale dairying and fiber production has been a female dominated production prac-

tice in recent years as women seek opportunities to participate in agricultural industries. Additional potential for sustained growth arises from observations in the average ages of goat operators. Goat farming is conducive to smaller farming practices and inviting to individuals returning to rural communities and visiting farming as a hobby. Over 43% of goat operators are over the age of fifty-five and less than 30% are under the age of forty-four.

Goat Management Systems

Goat enterprises consist primarily of purebred and commercial operations. The goal of the purebred operation is to specialize in providing a specific breed, while maintaining the breeds quality and integrity. In the meat goat industry, limited purebred operations exist in the United States, impart due to the meat goat breeds being developed in Africa and New Zealand and few imports made into the United States. In the dairy industry, increased importance placed on registered and pedigreed animals has increased involvement in purebred operations. The need for registered dairy goats originates from the emphasis placed on the show animal that is ideal in conformation and milk production. Commercial operations are increasing in number in the meat industry, though limited in the dairy industry. Less than thirty-five commercial goat dairies are reported. In commercial operations, either non-registered purebred animals are maintained or crossbreeding strategies are used to improve the characteristics of marketed animals. Most meat goat operations have relied on the infusion of Boer goat genetics to improve carcass quality. The dairy industry will utilize meat-type bucks to improve the marketability of dairy offspring for the meat markets.

Capacity of Production

Most dairy goat breeds are seasonally polyestrus, displaying estrous cycle regularity averaging twenty-one days when the period of light decreases within the photoperiod. Exceptions include the dwarf goats, which are non-seasonal. Non-seasonal estrous cycles are common amongst fiber and meat breeds as well. Kidding occurs following a one hundred fifty day gestation period. Multiple births are dependent on the breed. Boer goats are known to routinely twin, whereas triplets and quadruplets are common to the Nigerian dwarf breed. Length of lactation is breed dependent as well. Boer goats average lactations of four months in length, whereas the lactation of the dairy breeds is commonly ten months. For the dairy goat, peak milk production is coincident with the third or fourth lactation and the doe can produce up to eight pounds of milk per day during this time. Male horned goats posses musk glands, located on the crown of the head overlapping the horn bud. These glands are responsible for the characteristic odor of male goats during the breeding season. Producers may deodorize their bucks by removal of the musk glands; however,

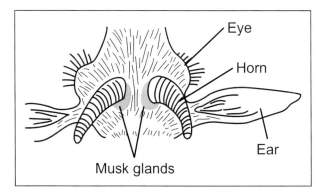

■ **Fig. 10.2.7** Male goats naturally mark their territory through musk glands, which are located on the crown of their head, overlapping the horn bud. Removal of the musk glands, known as deodorizing, is performed during the debudding process by making two additional burns with the debudding iron in the area of the musk glands.

when left intact the odor emitted from the glands stimulate estrus in the doe and improves conception. If the doe is being bred to produce milk, the bucks should be kept separate from these does as the odor can taint the milk.

Goat as a Food Commodity

Goats are covered under the Federal Meat Inspection Act of 1906, and any goat commercially slaughtered for sale must be inspected. Federally inspected goat harvest increased seven fold between 1984 and 2010 to seven hundred seventy-nine thousand head. However, live goats direct marketed to consumers and freshly harvested goats fall outside the statistics for federally inspected harvests, and therefore, the number of harvested animals is much greater than current federal inspection estimates. New Jersey and Texas dominate in terms of federally inspected harvests, representing 30% and 18%, respectively, of the total industry. The demand for goat meat is not for individual cuts of meat, but for whole carcass or portions of the carcass including quarters. These larger cuts lend themselves to the traditional cooking methods of slow roasting and stewing. Kids, goats less than one year of age, are harvested between three to five months of age weighing twenty-five to fifty pounds and the meat is

■ **Fig. 10.2.8** In the early twentieth century goat milk was consumed predominantly by the lower class as goats were viewed as the poor man's cow. In recent years, there has been a resurgence in the use of goat's milk to manufacture cheese sought by the artisanal food movement. (Library of Congress.)

marketed as chevon. Meat from kids marketed within the first week of life is termed cabrito and is highly sought by select ethnic populations. Older, or mature, goat meat that sources from goats of the meat, dairy, and fiber industries is used primarily in processed foods. Currently there is no quality grading system in effect for marketed goat meat.

As limited commercial dairy goat enterprises are in operation, the majority of fluid goat milk marketed in the United States originates from California. Increased fat content makes goat milk ideal in the cheese industry and in fact, goat cheese is the most rapidly expanding market of the dairy goat industry. In 2005, there were a reported one hundred goat cheese producers in the United States, most products being marketed through farmers markets, local retailers, and direct sales. Goat's milk can be used to make any type of cheese traditionally made from cow's milk. Popular cheeses made from goat's milk include chevre, feta, and gjetost.

Nutritional Value of Goat Products

As a result of the age of harvest, there is minimal fat covering of the goat carcass, contributing to goat meat as one of the leanest of all domesticated meats. When compared to beef, lamb, and pork it has the lowest amount of total fat, saturated fat, calories, and cholesterol per serving. Goats' milk offers numerous advantages to the consumer when compared to cows' milk. These include: increased concentrations of vitamin A and niacin, greater concentrations of short- and medium-chain fatty acids that are more easily digested by the young and elderly, and reduced allergenic response, and thus it is more tolerated by people with allergies.

■ **Fig. 10.2.9** Goats are browsers and will consume twigs and young shoots of trees and shrubs. (© ason, 2009. Under license from Shutterstock, Inc.)

■ **Fig. 10.2.10** Considered stronger than steel, spider web silk is sought after by industries that require a durable, yet lightweight, fiber. Prior to the development of transgenic goats that secrete the spider silk protein in their milk, spider silk fiber was unavailable commercially due to the inability to farm spiders for the product. (© Arthur Eugene Preston, 2009. Under license from Shutterstock, Inc.)

Goats for Fiber, Conservation, and Biotechnology

Fiber

Mohair is the cleaned, long, outer hair coat of the Angora goat. Mohair may be divided into three types: kid, young, and adult. The mohair from younger animals is finer and more valuable. After the fourth shearing, Angora goats are traditionally sold for meat as the quality of fiber is decreased. Mohair is smoother than wool, but lacks the felting properties. It is commonly used in upholstery fabrics due to its strength, durability, ability to accept dyes, and flame resistance. Cashmere fiber is obtained predominantly from the Cashmere breed of goat, collected once annually. This soft undercoat begins to grow in late July and is complete in December. The cashmere must be collected before the end of its growth or it will be shed naturally. The cashmere is separated from the regular goat hair fiber by combing or using a commercial dehairer. Sixty percent of the world's supply of cashmere is produced in China and the remainder in Turkey, Afghanistan, Iraq, Iran, Kashmere, Australia and New Zealand. It is a new industry for the United States appearing in the last thirty years. Cashmere is fine in texture, soft, light, and resilient. It is considered one the worlds most luxurious fibers; historically documented as the fiber of kings, cashmere was reported to line the Arc of the Covenant in biblical accounts.

Conservation

Similar to the sheep industry, goats may be used for ground cover management to clear brush for fire control, remove invasive weeds and noxious plants, and clear roadways or passageways. Goats are browsers and prefer broadleaf plants, twigs, and young shoots of trees and shrubs, plants not utilized by cattle. The selection of browse by goats and the grazing habits of cattle for mature grasses allow the two species to be companion grazed alongside in the same pasture. Although estimates suggest relatively few operations practice companion grazing, this practice may be used to improve pasture quality and promote plant diversity for sustainable rangeland management, while providing the benefit of dual-income. In the southern United

States goats are being used for the management of the invasive Asian vine Kudzu and when compared to sheep, goats are able to more effectively manage and clear brush in significantly less time than sheep.

Goats are being used as pack animals as well, capable of navigating terrain that is less accessible to larger animals including horses, donkeys, and llamas. Goats are easily trained as pack animals when less than one year of age and can sustain 25% of their body weight in pack over five to fifteen miles of mountainous terrain per day. While allowed in National Forests and lands supervised by the Bureau of Land Management, goats are not allowed in National Parks.

Biotechnology

The use of transgenic goats in biotechnology has received considerable attention in the literature. The most wide-spread use is the production of pharmaceuticals that can be harvested from the milk of transgenic lactating females. Products from these animals, however, cannot be used for human consumption. One application of transgenic goats is the production of spider silk. Spider silk, one of the strongest materials in the world, can be used for the manufacture of products ranging from bullet proof clothing to the manufacture of artificial tendons, ligaments, or limbs.

CHAPTER 11

Pigs

"I like pigs. Dogs look up to us. Cats look down on us. Pigs treat us as equals."
—Winston Churchill (1874–1965)

This chapter introduces the primary breeds of the United States pig industry, which historically were classified as lard, bacon, or meat type. The structure of the industry and the physical facilities involved in pig production are emphasized. Basic knowledge in the proper management of pigs toward their well-being, health, and performance is presented. Lastly, the nutritional benefits of pork and its contribution to the human diet are described.

Pigs: A Historical Perspective

According to fossil records, wild pigs are estimated to date forty-five million years ago. However, the domestication of pigs did not occur until 9000 B.C. in Eastern Turkey, 4900 B.C. in China, and as late as 1500 B.C. in Europe. Dispersal of pigs through human migration was limited and wild pigs were actually domesticated independently by various human settlements. Many early domesticates of pigs became extinct, and the domestic pigs of today are considered descendants of the European wild boar.

The pig was first brought West in 1493 by Columbus and supplied a food source during his second voyage to the New World. However, it is Hernando DeSoto that is accredited as being the father of the American pork industry. In 1539, DeSoto brought thirteen hogs to what is now Tampa Bay, Florida. During his three years of exploration, the initial population increased to over seven hundred. Pigs readily revert to a feral state, acquiring many characteristics of their wild ancestors, and it is DeSoto's original swine herd that served as the founder population of the feral razorbacks that originated in the southern United States.

Free-ranging pig populations within the United States trace their origins to early domesticated pigs that escaped confinement and became feral over generations and/or European wild boars that were imported into the United States for gaming. Hybrid offspring of feral pigs and European wild pigs have largely replaced wild boar stock and the term feral pig more accurately describes the current free-ranging pig population throughout the United States. Feral pigs have been identified in forty-five states, with the greatest populations confined to the south eastern United States from Texas to Florida. The population is estimated at over five million animals with annual damages of $1.5 billion in forest, field, and crop destruction and control costs. Population expansion into urban areas, destruction of wildlife habitat and agriculture commodities, and potential for disease transmission to farmed livestock and pets necessitate methods to control and/or eliminate feral pigs. Current control and eradication methods involve trapping and shooting. Animals can be marketed for meat for human consumption.

Pigs were brought to the Jamestown Colony in 1607 by Sir Walter Raleigh. As colonies dispersed throughout the east coast, they took pigs with them. Without confinement semi-feral pigs became a nuisance over time and in Manhattan a wall was constructed to control the roaming animals. Some suggest this wall served as the origin of Wall Street.

■ **Fig. 11.1**   Domesticated pigs are descendants of the wild boar. (© Eric Isselee, 2009. Under license from Shutterstock, Inc.)

■ **Fig. 11.2**   Pig breeds of the United States were originally classified as lard or bacon type. As the market for lard declined post War World II, breeders began to focus on the production of pigs that deemphasized lard and increased lean muscle mass to produce meat type pigs. Lard-type pigs are compact and fatten rapidly on grain. The Mangalitsa is a Hungarian breed of lard type hog with a carcass that is 50% fat. The breed is marketed for speciality hams and bellies in limited numbers, primarily in major metropolitan areas. The long, thickened appearance of the coat has earned the breed the name wooly pig. (© meischke, 2014. Used under license of Shutterstock, Inc.)

Records were maintained as early as 1633 documenting the number and weight of marketed swine. These animals were caught by hounds and marketed and shipped as salted pork. There were no efforts to manage product consistency. The majority of these free roaming animals foraged from fields and forests, with some producers fattening their stock on surplus grain, milk, and excess table waste. These practices began the development of swine as an important agricultural animal. In early history, the pig's primary contribution was lard, rather than meat. The lard was a highly valued commodity and a trend toward the development of breeds of pigs for their ability to generate excessive body fat began. Corn feeding also helped in fattening. Surplus corn grown in Tennessee, Ohio, and Kentucky was used in the swine industry and these states became leading swine production centers during the nineteenth century. Cincinnati, Ohio flourished from the industry due to its centralized location and distribution of pork by boat along the Ohio and Mississippi rivers. It was here that the first pork-packing center was constructed in 1850 and the city became known as Porkopolis. By the 1860s, expansion of farming west and the development of railroads relocated the center of pork packing activity to Chicago, Illinois. Chicago remained the primary marketing center until post World War II when marketing decentralized and packing plants were built near areas of production. World War II also marked the transition toward the development of pigs that maximized lean cuts (ham, loin, picnic, and Boston butt) and minimized lard, which was now considered waste.

Breeds of Pigs

There are over four hundred breeds of swine reported worldwide; however, only eight of these breeds have gained popularity in the United States. With the exception of Berkshire, Landrace, and Yorkshire other initial breeds in America were developed in the United States in the mid to late nineteenth century. Initial breeds were developed in response to the surplus grain and consumer demand for fat. But consumer demands changed, as did feed availability, and profitability and swine breed development followed. Throughout history, swine have been raised for three different purposes, giving rise to lard, bacon, and meat type breeds. During colonial times, the pig was raised for its lard and producers sought a pig which could deposit extreme amounts of fat. This type persisted into the late nineteenth century. Short stature, early maturity, compact build, and small litter sizes were characteristic of this type, which gained the name cob roller. In 1915, the desired qualities reversed entirely. Breeders demanded a pig of increased body length, which allowed for

■ **Fig. 11.3** Yorkshire. (Courtesy of the National Swine Registry.)

■ **Fig. 11.4** Landrace. (Courtesy of the National Swine Registry.)

maximum yield of bacon meat. Bacon type swine were the converse of lard type with their increased length and lean carcass; however, lameness and slow maturity contributed to their failure. Thus, in 1925 the American swine breeders sought to produce animals that were an intermediate between the lard and bacon type animals, the meat hog. The meat type pig combines length of body, muscling, and the ability to quickly reach market weight without excess fat. Today meat type animals are the goal of producers who emphasize leanness and rapid growth first and foremost and are usually not tied to a particular breed of swine. As such, 90% of the industry relies on crossbred animals. However, eight major breeds of pig exist within the United States swine industry and play an important role in production.

Maternal Breeds
Maternal breeds are typically white in color and excel in litter size, fertility, and milk production. In recent years, improvements in growth rate and carcass quality have been realized. Yorkshires, Landrace, and Chester White are recognized as maternal breeds.

The Yorkshire breed is the leading breed in United State's registries. Yorkshires trace their origins to northern England and in England the Yorkshire breed is known as the Large White. Its first importation into the United States occurred in 1830 when the breed was brought to Ohio. The Yorkshire is always white in color, but may possess black freckles over the body. The ears of this breed are erect and the face is slightly dished. In addition to milk production and large litter sizes, the Yorkshire is known for its durability and soundness.

■ **Fig. 11.5** Chester White. (Image courtesy of Stephen Moeller, The Ohio State University.)

Yorkshires also excel in lean meat production, reduced backfat, and muscular stature.

The Landrace breed was developed in Denmark by crossbreeding native pigs with imported Yorkshire. Denmark held a monopoly on the Landrace breed, granting their export to the United States in 1934, but only permitting their use in crossbreeding. In 1950, Denmark allowed surplus Landrace to be used for pure breeding. American Landrace was subsequently developed from Denmark, Norwegian, and Swedish Landrace. The breed is mainly white although some small black spots are common. The Landrace has ears that droop forward and are known for their body length, muscling in the ham and loin, and their ability to farrow and raise large litters.

The Chester White breed derives from its county of origin, Chester, Pennsylvania. The breed originated from the cross breeding of various white breed imports from England in the early nineteenth century and was influenced by Yorkshire breeding. The original crossbreeds had reached such uniformity by the mid nineteenth century that they were designated their own breed. Like other maternal breeds, this breed is white and a limited degree of freckling is acceptable. The Chester White are characterized by medium sized drooping ears. The breed is noted for their prolificacy, early maturity, and adaptability.

Paternal Breeds

Paternal breeds of swine are typically colored and excel in leanness, muscling, and increased growth rates. Such breeds include the Duroc, Hampshire, Spotted, and Berkshire.

The Duroc breed was developed in New York and New Jersey in 1812. Interestingly, the breed was named in honor of the founding owner's champion trotting stallion, Duroc. Although originating in New York and New Jersey, most of the breed's development occurred in Ohio, Nebraska, Indiana, Iowa, Kentucky, and Illinois. All Duroc pigs are solid red in color and can range from golden hues to dark mahoganies. The breed has drooping ears and is known for its rapid growth and maturity, heavy muscling in the ham and loin area, and good finishing ability.

The true origin of the Hampshire breed is uncertain. The Hampshire color pattern is similar to the English Saddleback and suggests breed influence.

Pigs native to Hampshire England served as the foundation for original imports into the United States in the 1820s to 1830s. The offspring from these early imports were sent to Kentucky, which led to further development of the breed, most notably, in Boone County, Kentucky. The breed was originally known as The Thin Rind and the name later changed to Hampshire to reflect founder origin. Hampshires are reportedly the original breed selected for use by Smithfield Hams in the late 1800s. Hampshire swine are distinguished by their black body with white belt, that encompasses the shoulder area and two front legs, and erect ears. The breed is noted for leanness, carcass quality, reduced backfat, and increased loin eye area. They are considered adaptable to outdoor rearing systems. Today, the Hampshire is the third most populous breed of swine.

The Spotted breed was developed predominantly in Indiana during the mid nineteenth century. The Poland China breed developed in Ohio and the English Gloucester Old Spot breed are the foundations of the Spotted breed. The breed is noted for their rapid growth rates, improved feed efficiencies, and desirable meat qualities. Breed characteristics are similar to Poland China and interestingly, infusion of Poland China into the Spotted breed has been documented into the mid twentieth century. The breed is always spotted in large black and white patterns, with both colors presenting equally.

The Berkshire present in the United States trace their lineage to the founding Berkshire animals of Berkshire and Wiltshire, England. The Berkshire has been an established breed for over two centuries

■ **Fig. 11.6** Duroc. (Courtesy of the National Swine Registry.)

■ **Fig. 11.7** Hampshire. (Courtesy of the National Swine Registry.)

and were first imported into the United States in 1823. The original imports were absorbed into current swine breeds of the United States. Subsequent imports in 1875 were maintained pure and served as the foundation of today's United States Berkshire herds. In fact, the current registry requires that all registrants lineage trace to these imported lines. The Berkshire has erect ears and is black in color with six points of white, including the lower legs, muzzle, and tail. Berkshires produce small litters sizes, have reduced growth rates and feed efficiency, and increased fat deposition relative to other popular breeds; however, the meat quality is considered exceptional with darker well marbled pork that is consistently tender and palatable. The darker muscling and excellent quality has led to an increased demand for Berkshire pork. Certified Berkshire pork, which ensures the traceability of Berkshire pork claims by following the product from conception to processing, allows producers to market their product at 10$ per hundred weight premiums. Both domestic and international demand drives production. Japan is a leading importer of United States pork valued at 1.9 billion dollars in 2012. A primary market is for 100% Certified Berkshire pork, where the product is called kurobuta or black pork.

The Poland China breed was produced from foundation stock of Warren and Butler County, Ohio. Breeding of the foundation stock with Russian and Big China breeds led to the development of the Warren County Hog. This breed was further improved by breeding with Berkshires and Irish Grazers and in 1872, the breed was established as the Poland China. The Poland China is markedly similar in appearance to that of the Berkshire; however, the Poland China's ears are flopped. The Poland China are exceptional feeders which allow them to be the heaviest of all swine, no matter what age. The breed is recognized for its increased rate of gain, high percent lean, and prolific litter sizes.

Purebred operations function primarily to provide foundation breeding that are rarely used in commercial production systems. Prior to 1950, most commercial swine producers of market animals generated breeding stock from their own herds; however, an increasing trend in the industry is the purchase of breeding males and females from specialized breeding suppliers. These suppliers generate and supply purebreds, crossbreds, inbreds, or hybrids that have been developed to emphasize performance and quality.

■ **Fig. 11.9** Berkshire. (Image courtesy of Stephen Moeller, The Ohio State University.)

■ **Fig. 11.8** Spotted. (Image courtesy of Stephen Moeller, The Ohio State University.)

■ **Fig. 11.10** Poland China. (Image courtesy of Stephen Moeller, The Ohio State University.)

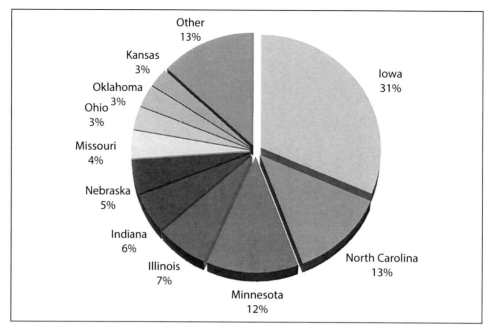

■ **Fig. 11.11** Top ten states in total hog inventory, includes market and breeding hogs, during 2013. (United States Department of Agriculture-National Agricultural Statistics Service)

United States Pig Industry

Swine account for 12% of the economic value of the food animal industry with an annual income of $22 billion. The United States produces 9% of the world's total pork with only 6.2% of the world's pigs. Global demand for pork has allowed the United States pork industry to grow beyond the constraints of domestic demand to become the leading exporter of pork and pork products. Swine are produced in all fifty states. In most cases, the geographical location of dominant swine producing areas parallels the production of corn and other small grains. Proximity to grain production is driven by the fact that feed represents 70% of production costs in swine production. Thus, 70% of swine are produced in Iowa, Indiana, Illinois, Minnesota, Nebraska, Kansas, Missouri, and Ohio. These states collectively occupy a region of the United States referred to as the Corn Belt. Interestingly, Iowa has dominated the swine industry since 1880. The second ranked state in swine production, North Carolina, is not a grain producing area. Here, pigs farrowed are transported to the Corn Belt region to be finished. Prior to the 1997, North Carolina was the most rapidly expanding swine producing state. The rise of swine production in North Carolina was

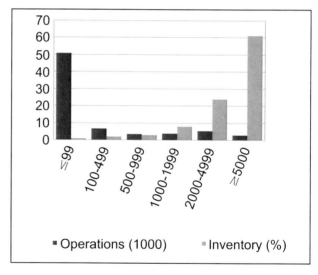

■ **Fig. 11.12** The number of swine operations has decreased nearly 90% in the last three decades, met with an increase in the size of operations. (United States Department of Agriculture-National Agricultural Statistics Service).

driven by declining profits from the tobacco and cotton industries as producers sought new avenues of income. As large scale farms were initially exempt from zoning regulations, the swine industry expanded without regulation. Expansion continued until 1997, when through the Clean Water Act a moratorium was enacted to limit construction and expansion of swine operations. The structure of the swine industry in North Carolina emerged as a vertically integrated system of production. Vertical integration is defined by a limited number of producers controlling the majority of swine produced and all aspects of animal production from conception to market. The trend of vertical integration dominates the current United States swine industry. The top three swine producers own 25% of sows in production. These corporations own all segments of production including seedstock production, breeding and farrowing, nursery, finishing, and product processing. These segments of production are maintained at different physical locations to minimize the spread of disease. The benefits of vertical integration are a reduction in production costs and greater consistency when producing the market product. Vertical integration, however, results in increasing numbers of pigs owned by fewer numbers of people. In 1950, 2.2 million swine producers were in operation. This number has decreased approximately 90% in thirty years. Although the number of operations has declined, the size of these operations has increased. For many swine farmers, the term owner has been replaced with contract grower, whereby producers agree to raise pigs under the specifications set forth by a contractor. The impact of this trend on the future of the swine industry and small scale producers remains uncertain. As smaller producers view the rise of corporate owned vertically integrated production systems as a threat, proposed legislation has attempted to limit the practices of these corporations and the changing face of the swine industry remains a controversy.

Pig Management Systems

Purebred and Seedstock Production

The first step in developing quality market animals is the establishment of proven lines of maternal and paternal genetics for the inheritance of desirable traits. The swine produced in this stage are selected for their superior genetics and sold to breeding facilities in hopes of passing on these exceptional traits to their offspring. In some cases, the specimens developed may be of outstanding quality and kept by the seedstock producer to collect and distribute the semen nationally for artificial insemination purposes. Purebred, crossbred, hybrid, and inbred seedstock are used in this stage of the industry.

Breeding and Farrowing

The animals produced from the seedstock stage are used in the breeding phase. Gilts that have not reached sexual maturity may be kept in close proximity to boars to aid in an earlier onset of estrus. Once estrus is reached, the gilts and sows are inseminated using a pen-mating system, hand mating, or artificial insemination. Artificial insemination is the dominant method and is practiced by over 90% of the industry. The successfully inseminated females are either housed in group pens or in individual gestation stalls. There are several options for group housing that differ in the number of animals allowed per group space, flooring, and feeding systems. Group housing requires more space and can accommodate only 60–70% of the animals housed in comparison to gestation stalls. Gestation stalls are suggested to reduce physical injury that is incurred by social aggression and reduce early embryonic losses due to physical trauma. However, gestation stalls are currently a focus of welfare debates as these stalls completely confine pregnant females for the duration of their one hundred fourteen day gestation.

Gestation stalls are currently banned or in the process of elimination in several states, with Florida as the first state to ban gestation stalls in 2002. In most states with established or planned bans, pork production is a minor agricultural commodity, the exception being Ohio. A growing number of food companies have also enacted policies to source product from operations that do not use gestation stalls. Following suit, prominent United States meat packers have announced phase-out plans for gestation stalls.

Sows are transitioned from gestation housing to a farrowing facility prior to the anticipated farrowing date. Sows are individually housed in crates designed to provide the litter access to milk, yet minimize loss of the pigs due to crushing by the sow. It is in this stage that the newborn pigs are subject to procedures such as castration, tail docking, clipping of needle teeth, and iron injections. The dam and

■ **Fig. 11.13** Gestation stalls (left) are at the center of controversy concerning the humane practices of pregnant swine. With planned phase-outs of gestation stall use, methods of group housing (right) are being implemented. (Image courtesy of Stephen Moeller, The Ohio State University.)

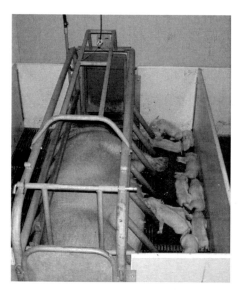

■ **Fig. 11.14** Farrowing crates (left) house the female and litter from birth until weaning and provide protection for both the growing litter and human handlers Alternatively, farrowing systems may involve pasture based farrowing huts (right) that individually house the sow and litter in outdoor operations. (*Left:* Image courtesy of Stephen Moeller, The Ohio State University. *Right:* © Mark William Richardson, 2009. Under license from Shutterstock, Inc.)

litter are maintained in farrowing crates until the pigs are weaned at fourteen to forty-two days. Determination of weaning age is dependent on factors including: disease transmission, rate of pig weight gain, prediction of performance during nursery and finishing stages, subsequent reproductive performance of the sow, and space and inventory considerations. Once weaned, the pigs are transferred to nursery facilities and the sows are reinseminated at estrus.

Nursery (Feeder) Pigs

The weaned pigs are grouped according to age, size, and sex and placed into nursery pens. The floors of these pens may be slatted to allow excretory waste to pass through and retain a clean environment. The pigs are fed from a common feed unit a complex, protein rich diet that is highly palatable. It is especially important during this stage to regulate ventilation and temperature to maintain health and ensure optimal growth rate. Once the pigs achieve a weight between thirty-five and fifty pounds, they are moved to the final stage of swine production.

Growing and Finishing

At the end of the nursery phase, pigs are moved to a larger housing facility where they are kept and fed until a reasonable market weight between two hundred thirty and two hundred ninety pounds is reached. The pigs are maintained in groups according to age, size and sex due to the fact that barrows and gilts require different nutritional management for optimal growth. It is especially important during this stage to maintain proper temperature control, particularly during the summer months. Because pigs lack sweat glands and are unable to perspire, they are especially susceptible to heat stress and can easily become over heated during this stage. Two popular methods exist for cooling the finishing facilities. One system requires the use of large fans that are placed at one end of the barn, opposite to vents that are exposed to the outside. This system draws fresh air through the vents and pulls the expired air into the outside environment. The other method is a fan-less system and necessitates more vents to compensate for the reduced airflow. Often times an entire side of the housing facility is composed of the large vents. During extremely hot periods the swine may be further cooled with watering systems.

■ **Fig. 11.15** Group housing conditions of the grow-finish stage of production. (© Dario Sabljak, 2009. Under license from Shutterstock, Inc.)

In the 1990s, 65% of swine operations practiced farrow-to-finish production. Today, less than 20% of pigs marketed are from farrow-to-finish operations. Instead, most producers specialize in a single stage of management: farrow-to-wean, wean-to-feeder, or feeder-to-finish.

Facilities

There are different facility options for swine management systems and include confinement barns, hoop barns, or pastured systems. Confinement barns have the greatest investment costs, but reduced labor and improved disease control. Individual and group housing is accommodated with confinement barns, which can be used for all stages of production. Hoop structures are a lower-cost option to confinement barns, but require availability of bedding and methods for its disposal. Pigs may be more difficult to observe in hoop structures and disease risk is greater compared to confinement systems. Hoop structures are commonly used for gestation and grow-finish stages of production. Pastured systems are also an option for all stages of production. Pastured systems are more commonly used within niche marketing schemes. Pigs are allowed to forage to partially fulfill nutrient requirements. However, disease and predation are of concern and the additional input of land management must be considered.

Health Management

Swine diseases are difficult and costly to treat and often have lasting effects on the surviving animals. Even completely recovered, animals affected by diseases may never perform efficiently. Pigs are susceptible to several viral diseases and transmission through a herd can occur rapidly once the disease is introduced. Biosecurity is a major factor when developing a swine operation, and adherence to strict biosecurity measures is necessary to maintain the health of the herd. It is a physical impossibility to eliminate all disease causing pathogens; therefore the goal of biosecurity is to prevent new strains of pathogens, those to which the herd has not been exposed, from entering the facility. An effective biosecurity

■ **Fig. 11.16** In today's modern farrowing operations, average litter sizes of approximately twelve pigs are targeted. (© WizData,inc., 2009. Under license from Shutterstock, Inc.)

program can result in improved growth and reproductive rates along with a decrease in preweaning mortality. Each of these factors contributes to the overall productivity of the swine operation. Effective biosecurity measures include the physical location of the operation, a minimum distance of 1.5 miles from other swine producing facilities, and proper personnel training that incorporates a work system hierarchy from the highest to the lowest herd health status. In farrow-to-finish operations movement occurs from youngest to oldest pigs. Newborn pigs and weaned pigs have the weakest immune systems and are susceptible to a greater number of pathogens. The daily chores should start with these younger animals that are considered of greater health status risk. Breeding sows, have the strongest immune systems and are likely to have previously come in contact with the pathogens of the farrowing and nursery facilities and should not be affected by any cross-contamination resulting from the previous tasks. However, it is important to properly sanitize before working with the pigs and in between moving from one stage of production to another. Greatest biosecurity operations require employees to shower before entering the facilities and before leaving the premises. In addition, disinfecting solutions for boots are commonly placed at the entrances of individual room for use before entering. Lastly, facilities are cleaned and disinfected after each rotation of pigs into and out of the building.

Capacity of Production

In most animal industries, the major aim of reproductive management is to successfully breed the female and ensure the female carries the offspring to term. However, in the swine industry, additional emphasis is placed on conceiving litters of increased number. The swine industry as a whole has made astounding advancements during the last century. For example, the average litter size in 1930 was six pigs born per litter. Today, that number is twelve pigs born per litter with ten pigs or more weaned. Another great achievement has occurred with the average litter per sow ratio. In 1930, every sow produced an average of 1.3 litters per year; however, that number has increased to a national average of nearly 2.4 litters per year. With only these two improvements taken into consideration, the sow today is able to produce twenty more pigs per year than in the past. The potential for further improvements exist. The relatively short gestation of one hundred fourteen days allows for each female to produce three litters in a little over one year, ideally averaging 2.7 litters per year. The current benchmark for the industry is for each reproductive female to wean thirty pigs annually, increasing litters per year to 2.5 and pigs weaned to twelve. Providing a micro-environment of proper temperature and free from drafts is important to reducing mortality of newborn pigs. Practices including cross-fostering are also implemented to improve weaning rates. Cross-fostering equalizes litter number and size by moving pigs between sows in the immediate postnatal period. Cross-fostering is ideally done within the first twenty-four hours after birth and before social order of the litter is established. Cross-fostering balances the chances of survival and health as pigs are given equal opportunity to nurse. Poor reproductive or maternal abilities are a primary cause to cull, or remove, sows from the production herd. Culling rates routinely average 30% for each gestation period.

Pigs as a Food Commodity

In 2005, the National Pork Board adopted a new standard for all swine producers known as Symbol III. Symbol III illustrates a phenotypically desirable specimen that excels in all aspects of production and serves as a target for the pork industry in terms of performance, profitability and nutritional value.

Symbol III standards set for the male pig are in terms of the barrow. This is because nearly all market pigs are castrated. Not only does castration control aggressive tendencies of the males, it also alleviates the production of hormones created in the testes that influence the taste and smell of pork. If left uncastrated, androstenone and skatole, the hormones associated with male reproduction, can accumulate in fat tissue. Once cooked, the compounds may be detected and are considered undesirable. Only male pigs used for the purpose of breeding are left intact. When castration is not practiced on males intended for market, intact males are harvested at a younger age to avoid the reduction in meat quality. Ideally, the perfect barrow should reach a market weight of two hundred seventy pounds in one hundred fifty-six days, with a loin area greater than 6.5 square inches. The gilt should reach an equivalent weight in one hundred sixty-four days with a loin area of 7.1 square inches. Both barrows and gilts are expected to have feed efficiencies of 2.4 pounds of feed per pound of gain. Symbol III also denotes a pig with increased leanness at a backfat thickness less than .7 inches, and a gene pool free from undesirable qualities.

In the selection for leanness, a detrimental genetic condition known as porcine stress syndrome, or PSS, has emerged. Porcine stress syndrome is a blanket term that incorporates autosomal recessive defects that are characterized by sudden deaths resulting from acute stress. Breeds with heavy muscling are more likely to be a carrier of PSS. When carriers endure what is perceived to be a stressful situation, increased body temperature, decreased blood pH, and uniform muscle contraction

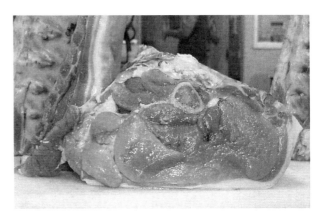

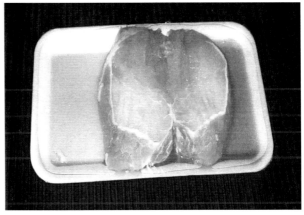

■ **Fig. 11.18** Porcine stress syndrome results in pale, soft, exudative meat that is undesirable to the consumer. Note the loss of water from the surface of the meat. (Courtesy of Henry Zerby, The Ohio State University.)

occur. Symptoms of the onset include muscle tremors, facial twitches, rapid respiration, and red blotched skin. Death associated with PSS is rapid and occurs within fifteen to twenty minutes, but affects less than 1% of carriers. Post-mortem indication of PSS is rapid onset of rigor mortis only minutes after the death, resulting in unpalatable meat known as PSE (pale, soft, and exudative). The pork is undesirable since it contains a higher percentage of water and is thus paler and softer than normal pork. The increased water percentage has unfavorable affects when cooking the meat, as it easily evaporates from the pork, leaving it dry and tough. Thus, PSE pork is of low value and sells for significantly less, costing the pork industry millions of dollars each year. A study conducted by the Pork Quality Solutions Advisory Group suggested that 3.34% of all pork marketed falls under the PSE category.

■ **Fig. 11.19** World wide, pork is the most consumed meat. (© Tatiana Belova, 2009. Under license from Shutterstock, Inc.)

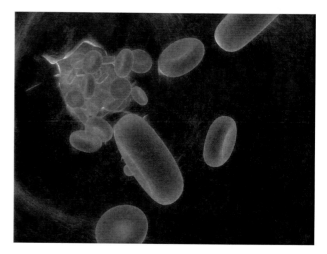

■ **Fig. 11.20** Atherosclerosis accounts for over five hundred thousand deaths annually in the United States. An accumulation of hardened lipid in the artery narrows the arterial lining and ultimately deprives the heart of blood and oxygen. Swine fed high fat diets develop atherosclerosis, similar to the condition of humans. Research is underway to tease out the roles of genetics and lifestyle habits in the development of atherosclerosis using the pig model. (© Sebastin Kaulitzki, 2009. Under license from Shutterstock, Inc.)

Pork is relatively uniform in quality and tenderness and although inspection is mandatory by the United States Department of Agriculture, grading is voluntary and only reflects whether the pork is of acceptable or less than acceptable (utility) quality. Acceptable product is further distinguished by yield of four main cuts: ham, loin, picnic shoulder, and Boston butt. Grade 1 is the top grade and is indicative of a carcass yield of 60.4% or greater. Anything below grade 4, a carcass yield of 54.4% or less, is also considered as utility grade.

Nutritional Benefits of Pork

Worldwide, pork is the most popular meat. In the United States, the average person consumes slightly less than fifty pounds of pork per year. However, this consumption has reached a plateau in past years, which has spurred pork producers to bring about a reemergence of their product in the food market. The pork industry continues its efforts to market pork as a health food and to produce a product that is leaner and less in calories. Pork contains many essential nutrients needed to build and sustain a healthy body. These include the essential vitamins: thiamin, riboflavin, niacin, vitamin B6, vitamin B12, and pantothenic acid. Pork is also an excellent source of the minerals phosphorous, magnesium, iron, and zinc. Pork is one of the primary sources to obtain thiamin and rivals milk as the best source of riboflavin. Furthermore, pork is naturally low in sodium and a rich source of quality protein. A single one hundred gram serving of lean pork can contain half the required daily protein, 15% of iron, 30% of zinc, and 70% of vitamin B12. The fat found in pork consists of saturated and unsaturated fatty acids. Research has been conducted to promote the deposition of conjugated linoleic acid (CLA) in pork, an important fatty acid that is suggested to provide heart health benefits and play a role in the prevention of certain cancers.

Pigs as Model Organisms

Pigs are believed to be the ideal donor of organs for transplant into humans. Swine organs including heart valves and skin are currently used in medical practice. Pig skin, similar to human skin, is tightly attached to subcutaneous tissues and is relatively hairless with similar cutaneous blood supply characteristics, although it is thicker and less vascular. Porcine skin graphs are used in hospitals for wound healing, to protect open lesions from infection, and aid in the regeneration of the patient's own skin cells. Porcine heart valves have been routinely used for over thirty years in human medicine and are considered durable for ten to fifteen years once transplanted.

The pig is also an important research model in the understanding and development of treatments for atherosclerosis, diabetes, and neural disorders. Atherosclerosis is a type of heart disease in which plaque forms and hardens the arterial walls of the heart. This is a markedly prevalent disease that affects millions of Americans. The swine heart shares anatomical and physiological similarities with the human heart, including hemodynamics or blood flow, and is susceptible to atherosclerosis following prolonged high fat feeding. Using the swine heart as a model, research is underway to determine the role of genetics and lifestyle in the prevention and treatment of atherosclerosis. In 2001, using a cytotoxic drug that destroys the insulin producing cells of the porcine pancreas; a diabetic condition was induced in swine. Because of the similarities between swine and humans, proposed treatments for diabetes can be tested, monitored, and controlled within these pigs. In addition, studies have been conducted involving the xenotransplantation of normal, functioning pancreatic cells of pigs into other animals displaying a diabetic condition. In preliminary studies, the diabetic conditions of monkeys were reversed. Some scientists believe that this breakthrough has potential within the human population for eradication of the diabetic condition if barriers to xenotransplantation can be overcome. Lastly, the pig is proving to be an important research model for the study of neural disorders such as Parkinsons and Huntingtons disease. In studies, porcine fetal neuronal cells were transplanted into the brains of human patients suffering from these diseases. Amazingly, some neuronal function was restored in these patients. The increasing use of pigs in biomedical research will continue to shed light on diseases that afflict the human population.

Horses

"There's nothing so good for the inside of a man as the outside of the horse."
— *Henry John Temple (1784–1865)*

This chapter introduces the historic role of the horse and its current place in society. The horse's anatomical evolution is highlighted along with the primary breeds and the purposes they serve. The distinction between classifications of coat color and the genetics responsible for these phenotypic variations is illustrated. Health and reproductive management concerns specific to the horse are described. Lastly, the impact of the horse industry in the United States and its contribution to the economy is discussed, as well as the use of horses in equitation.

Horses: A Historical Perspective

There are seven species of *Equus* in existence today: the asses or donkeys of North America, Plains zebra, Grevy's zebra, Mountain zebra, onagers or Asian Wild Asses, Przewalski's horse that is considered the oldest living species of horse, and the true horse or *Equus caballus*. The modern horse originated in North America. The earliest known ancestor was eohippus, a small, primitive, four toed horse that was estimated to stand fourteen inches in height. The earliest remains of eohippus trace to fifty-four million years ago in North America. Orohippus appeared two million years after eohippus. Orohippus shared similarities to eohippus with the greatest differences observed in the skeletal anatomy of the fore and hindfeet. Eohippus and Orohippus co-existed with fossilized remains found in Washington and Oregon, though Orohippus remains are less prevalent. Both eohippus and Orohippus persisted into the mid-Eocene epoch, followed by Mesohippus approxi-

mately thirty-five million years ago. Also known as middle horse, Mesohippus was intermediate in appearance to the primitive forest-dwelling horses that preceded and the more modern appearing horses that followed. The first modern appearing horse was Mercyhippus. Appearing twenty million years ago, Mercyhippus was the first grazing horse with longer limbs and a lengthened face, however it retained the three toes of its ancestor. The first prototype of todays one-toed horse was Pliohippus. Although Pliohippus actually had three toes, two toes were regressed giving the external appearance of a one-toed, hoofed animal. Pliohippus was also the ancestor of Dinohippus, the closest known relative of *Equus*, which first appeared two million years ago. Several other genera appeared and disappeared in the evolution of the horse from eohippus to *Equus* coinciding with the Earth's temperature and climatic changes as the North American swamps were replaced by grassy plains. Collectively, horses that prevailed underwent a reduction in the number of toes, an increase in the size of cheek teeth, a lengthening of the neck and face, a repositioning of the eyes, and an increase in body size, whereas other lines went extinct.

Predecessors of the horse spread from North America to other parts of the world through existing land bridges. As the horse flourished throughout the world, it became extinct in North America. China and south west Asia were among the earliest places of domestication, a process that began around 4000 B.C. As with many animals, the domestication of the horse began as a means to secure a stable food supply. Subsequent use as a draft animal did not occur until 2000 B.C. and evidence of riding did not appear until 1500 B.C. With riding came the increasing use of horses in battle. Following the fall of the Roman Empire, the use of horses in sport began to emerge and remains a popular use today.

■ Fig. 12.1 The domestication of the horse has been linked to the rise and fall of many civilizations. The greatest warrior in Grecian history, Alexander the Great, rode his horse Bucephalus in nearly every battle until it succumbed to battle wounds in 326 B.C. (© Yiannis Papadimitriou, 2009. Under license from Shutterstock, Inc.)

The reintroduction of horses to the Americas is attributed to the voyage of Columbus in 1493. Cortez and other Spanish conquistadors, such as DeSoto, were the first to reintroduce the horse to North America in the early 1500s. The mustang herds in the American West are feral horse populations viewed as descendents of strays from the early Spanish expeditions. Spanish missions were the first to educate the Native Americans on the ways and value of the horse. Native Americans across North America acquired and spread the horse, which led to the development of the rich Native American horse culture. Over the course of colonization, various breeds were imported to the continent by settlers and by the late 1800s the horse was an integral component of life. Used in agriculture and of central economic importance, horse numbers approached eight million by the mid 1800s. Horse numbers increased to twenty-five million by the 1920s as the horse was used for draft in agriculture, fire protection, and transportation. As horses were shipped to Europe for battle during World War I and with increasing technology, such as the combustion engine, the American lifestyle began to change and horse numbers soon declined. In 1960, only three million

horses remained in the United States. However, the horse has since seen resurgence in use through recreation, sport, and show, and in 2007 9.2 million horses were recorded in the United States.

Selected Horse Breeds

Horse breeds were developed for specific desirable characteristics and to fulfill certain societal needs and are classified as draft horses, light horses, and ponies. The classification of the horse breeds is dependent on the height and weight of the horse. In the horse industry, the hand is considered the standard unit of measurement with one hand equaling four inches. Draft breeds, also termed cold bloods, were developed for heavy work and are a large powerful horse standing between 15.2 and 18 hands tall at the withers. These heavy muscled breeds weigh sixteen hundred to twenty-two hundred pounds. Light horses, also termed hot bloods or warm bloods according to origin, are 14.2 to 17.2 hands tall at the withers. Light horses are classified as stock, hunter, or saddle type and are used primarily for riding, driving, showing, racing, or utility as a farm or

■ **Fig. 12.2** The Percheron, Belgian, and Clydesdale (clockwise) are the most common breeds of draft horses in the United States. (*Top:* © SF photo, 2009. Under license from Shutterstock, Inc. *Bottom left:* © Cathleen Clapper, 2009. Under license from Shutterstock, Inc. *Bottom right:* © Margo Harrison, 2009. Under license from Shutterstock, Inc.)

ranch horse. Weights for the light horse breeds range between nine hundred and fourteen hundred pounds. Ponies are classified as horses that are under 14.2 hands and weigh five hundred to nine hundred pounds. Ponies may be further classified as draft, heavy harness, or saddle type.

Draft Horse Breeds

Draft horses are said to descend from the horses of knights in armor during the Middle Ages. The breeds of draft horses in the United States trace their origins to northern Europe and are named according to place of origin. At the beginning of the twentieth century, draft horse breeds accounted for 75–85% of the total horses in the United States and were a source of power in farming, hauling freight, and transportation. The development of the combustion engine and its use in farming and transportation led to the decline of the draft breeds. Today, these

breeds are most commonly used in show and competition, although farming cultures including the Amish continue their use in farming. The draft breeds are heavy muscled horses that are considered calm and patient. There are five breeds common to the United States: Percheron, Belgian, Clydesdale, Shire, and Suffolk.

Light Horse Breeds

The Arabian is claimed to be the first breed of horse and the original ancestor of all other horse breeds in existence today. The origins of the breed trace to modern day Iran, Iraq, Syria, and Turkey and while the history of the Arabian remains controversial, most agree that the breed has had the greatest influence worldwide, used in the refinement and overall improvement of all other horse breeds. Selection for improvement of the breed has been occurring for over two thousand years; however, it was not until the twentieth century that the breed acquired popularity in the United States. Originally registered by the Jockey Club, this organization began to solely register Thoroughbreds in 1943. Subsequently, the Arabian Horse Registry was established in 1949 and the International Arabian Horse Association was formed to promote Arabians in the United States. Several types of Arabians have been established worldwide and are named according to the country of origin. Recognized for its intelligence, durability, and stamina, the Arabian stands 14.1 to 15.1 hands at the withers. The breed is small to medium in size and is known for its dished head, high arching tail carriage, and refined stature characterized by a long, arched neck set high on the shoulder. Solid colors are preferred by breeders and include bay, brown, chestnut, grey, and black.

The Thoroughbred provided foundation stock to many light breeds of the United States including the Quarter Horse, the Standardbred, the Morgan, and the Saddlebred; however, its ancestry traces to Arabian lines. In fact, all modern Thoroughbreds trace to three stallions imported into England in the 17th and 18th centuries: the Byerly Turk, the Darley Arabian, and the Godolphin Arabian. Furthermore, approximately 90% of all Thoroughbreds trace to the foundation sire Eclipse born in 1764 and a descendant of a Darley Arabian. The Thoroughbred was developed with the intention of establishing a breed that would excel in speed sustained over distance. First recordings of Thoroughbreds occurred in England in 1791. United Stated registrations began in 1873, originally by the American Stud book that was later purchased by the Jockey Club, which maintains Thoroughbred registrations today. While the Thoroughbred remains the fastest horse in 1¼ miles, the breed also is popular in polo, hunting, jumping, and pleasure riding. The ideal Thoroughbred is defined by speed and not appearance. Traditionally, the breed is tall at 15.1 to 16.2 hands, long, and slender with exceptional muscling in the rear quarters. Color is not important, but is recorded with registration for identification purposes.

Considered one of the most versatile horses recognized for its beauty, style, gait, and temperament, the American Saddlebred was developed in Kentucky as a utility breed. For this reason, the Saddlebred was developed for riding comfort, rather than speed. Several breeds have been recorded in the development of the Saddlebred. The Morgan and Canadian Pacer contributed the compact body and stamina; whereas, the natural gait was acquired from the Narragansett Pacer, a breed from Massachusetts that is no longer in existence. Current breeding and use has been for show purposes and the breed can be three- or five-gaited. The basic three-gait includes the walk, trot, and canter. Additional training is required for the five-gaits, which include the walk, trot, canter, rack, and running walk. The breed is expected to perform each gait distinctly, without hesitation. In addition to show, the breed is used as a pleasure or harness horse.

■ **Fig. 12.3** Arabian (© Karen Givens, 2009. Under license from Shutterstock, Inc.)

■ **Fig. 12.4** Thoroughbred (© Robert Ranson, 2009. Under license from Shutterstock, Inc.)

American Saddlebreds stand fifteen to sixteen hands in height and can be bay, brown, black, chestnut, grey, or roan; however, dark colors are preferred.

The Appaloosa is characterized by its distinctive spotted color pattern; the mottled appearance of the skin of the lips, nose, and genitals; and the black and white stripes of the hooves. Spotted horses trace their origins to over twenty thousand years ago in France, however, the Appaloosa is a breed developed in the United States by the Nez Perce Native American tribe, located in Oregon, Washington, and Idaho. The Nez Perce carefully selected their breeding stock, developing a sturdy horse that could travel distances over mountainous terrain. The breed was developed in the region known as the Palouse River area, from which the breed's name is derived. The Appaloosa was nearly erased from American culture when the Nez Perce surrendered to the United States Calvary. The establishment of the Appaloosa Horse Club in 1938 is accredited with the preservation of the breed. Color patterns of the Appaloosa vary and include: snowflake, leopard, white blanket, spotted blanket, marble, and frost. However, all horses classified under this breed must have a fully or partially spotted coat, mottled skin, striped hoofs, and white sclera around the eye. The

■ **Fig. 12.6** Appaloosa (© Zuzule, 2009. Under license from Shutterstock, Inc.)

■ **Fig. 12.5** Saddlebred (American Saddlebred Horse Association.)

■ **Fig. 12.7** Morgan (American Morgan Horse Association.)

■ **Fig. 12.8** Quarter Horse (© Stephanie Coffman, 2009. Under license from Shutterstock, Inc.)

■ **Fig. 12.9** Standardbred (© Kenneth Graff, 2009. Under license from Shutterstock, Inc.)

Appaloosa is known for its gentle disposition and can be found in virtually every riding discipline.

The Morgan breed originated in the New England states and is named after the foundation sire Justin Morgan. Ancestry remains uncertain, however, the refined head and arching tail carriage suggest Arabian lineage. A popular breed of the mid nineteenth century, the Morgan was noted for its all-purpose ability in work and transportation. Mechanization replaced the Morgan on the farm and for transportation and current use is primarily for pleasure riding. The Morgan is a compact horse, usually ranging from 14.1 to 15.2 hands. The breed is versatile and known for endurance and docility. All solid color patterns are acceptable except white.

The Quarter Horse is claimed to be one of the oldest breeds established within the United States. Breed development traces to colonial America in the 1600s, where colonists raced English stock in quarter-mile races. In an attempt to develop a faster and more sturdy breed, English horses were bred to the Chickasaw horse, descendants of the Spanish Barb horses brought to the United Stated by early Spanish explorers. The resulting colonial horse was a compact, heavily-muscled breed recognized for its speed over short distances. Further improvements to the breed would occur through colonial importation of the grandson of the Godolphin Arabian, one of the three recognized foundation sires of the Thoroughbred. The use of the Quarter Horse in racing lost favor to Thoroughbred racing. However, further breed development would continue in the southwest, during the westward expansion of pioneers. Post Civil War the Quarter horse was adopted for ranch and farming and recognized as an ideal

horse for working cattle. Today, the Quarter Horse is the most populous breed in the United States and also claims the largest breed registry in the world. Opinions for ideal conformation differs amongst breeders. Whereas the Quarter horse's use as a stock horse that is trained and utilized for work involving cattle and other livestock favors a stockier, more muscular animal; others demand a horse of speed for racing. The uses of the breed currently include racing, pleasure, show, and rodeo events such as cutting, roping, barrel racing, and reining. The breed stands at 15.2 to 16.1 hands tall and color is of no importance except to the owner; however, animals with markings that suggest Appaloosa or other spotted horse breed influences are not accepted for registration.

The Standardbred originated in United States and is a breed of harness racing and riding. In the late 1800s the ability to trot or pace a mile in the standard time of 2.30 or 2.25 minutes, respectively, was a requirement for registration and the basis of the breed's name. The modern Standardbred has been selected for increased speed and the requirements for time are no longer maintained. The trot is defined as a two-beat diagonal gait where the opposite front and rear feet leave and return to the surface at the same time. The pace is defined as a two-beat lateral gait where the front and rear feet on the same side of the body leave and return to the surface at the same time. Few Standardbreds are capable of both gaits and the majority favor one gait as influenced by genetics and training. Standardbreds trace their ancestry to the Narragansett Pacer. The size of the breed is attributed to the introduction of Thoroughbred blood, tracing to the sire Messenger,

■ **Fig. 12.10** The trot. Note how the opposite front and rear feet of the horse leave the ground at the same time during the trot. (© Perry Correll, 2009. Under license from Shutterstock, Inc.)

and the trotting abilities originate from Canadian stock. Standardbreds range in height from 14.2 to 16.2 hands and are predominantly bay in color, though solid colors including chestnut, brown, black, and grey also occur.

The Tennessee Walking horse was developed in Tennessee and owes its conformation to the influences of the Thoroughbred, Morgan, Standardbred, Saddlebred, Narragansett and Canadian Pacer. The breed is most noted for its natural overstride when performing the running walk, whereby the horse places the back hoof ahead of the print of the front hoof at a distance of fifty inches or more. It also is known for the flat walk. Tennessee Walking horses were originally developed as a general-purpose breed for riding, driving, and farm work. It is more commonly used today for pleasure riding and show. Standing at fifteen to sixteen hands tall, the breed displays all solid colors and white marking are common. Black and dark colors are most popular, whereas gray and roan and not desirable.

Ponies

Pony breeds are known for their sturdy build and hardiness. Pony breeds of the United States include Shetland, Hackney, Welsh, Chincoteague, and Pony of the Americas (POA). The Shetland pony originated in the British Isles and is the smallest of the pony breeds, standing at a maximum of 11.2 hands in height. It is a gentle breed and a popular pony with children. American Shetlands developed in the United States are more refined than the British counterparts originally imported. The Hackney pony originated in England and may reach fourteen hands in height. It is routinely used as a trotting and carriage horse. The Welsh pony of Wales also is known for its trotting ability as well as jumping. The history of the Chincoteague pony remains unknown. The ponies are native to Assateague Island, Virginia and are reported to have inhabited the island for over four hundred years. The breed stands twelve to thirteen hands in height and are considered feral with the official breed registry established in 1994. The POA were established in the United States in the mid 1950s. Developed from the Appaloosa and Shetland they are a newer breed considered a good riding pony for children. The height may reach 13.2 hands and the breed comes in a variety of coat patterns due to the Appaloosa influence.

Coat Colors

While the color of a horse is not related to performance, color classification is an important tool in horse identification. Furthermore, certain breed registries display a preference for coat color and despite advancements in DNA testing in validation of breed lineage, coat color continues to be used as an indicator of parentage. Nearly all breeds require a description of coat color for registration, thus, it is important to understand and differentiate between coat colors.

Basic Coat Colors and Modifications

There are five basic coat colors recognized in horses: black, brown, bay, chestnut, and white. Horses that are considered black are those which are wholly black, including the hair around the muzzle, eyes, and flanks. True black horses are uncommon and have dark brown eyes, black skin, and a black coat that lacks any permanent reddish or brown sections. However, black horses can possess white markings in which the skin underneath the markings is pink. Foals are not born with a black coat but gradually develop the black coat as they mature. Some black horses will fade to have a smoky appearance, and during the summer months a black

horse can be bleached out by the sun's rays but will return to the true black coloring during the winter.

Brown horses may be confused as black, but are distinguished by brown or tan hairs about the muzzle and flank. The mane, tail, and legs are always black. Horses of dark brown are sometimes termed seal brown.

Bay horses are characterized by a reddish brown body ranging from a light tan-brown to dark shades of mahogany. In order to be classified as a bay, all horses must have black points, which include the mane, tail, tips of the ears, and lower legs. These points, usually on the lower leg, can be covered by white markings and do not alter the horse's classification as a bay. The skin is darkly pigmented except where white markings occur where the skin is pink. Bay is one of the most common coat colors among many horse breeds.

A chestnut horse varies from a light yellowish brown (light chestnut), to a bright and saturated copper (sorrel), to a dark chocolate (liver chestnut). The mane and tail of a chestnut horse are usually the same shade as the body; however, when they are a lighter shade of cream, the mane and tail are referred to as flaxen. Darker shades of red or brown are common as well and may appear black, how-ever, black points are never found as the chestnut is classified by the absence of black hairs. The absence of black hairs on the legs provides further distinction between chestnut and black horses.

White is the rarest coat color in horses. True white horses are born white and remain so throughout their lifetime. White horses are not albino and have completely white hair, pink skin, and usually brown eyes or, less common, blue eyes. The majority of horses that appear white are actually gray horses that become progressively white with age.

Each of the basic color schemes can be modified to yield the coat colors of gray, roan, dilutions, and pinto-paint. Gray coloring is achieved through a combination of white and colored hairs or simply white hairs on darkly pigmented skin. The foal coat of a gray horse is initially solid in color with the addition of increasing white hairs with each successive coat. Gray horses can be born any of the basic colors, lightening into permanent gray coloring as they mature. Variations occur and can result in flea bitten gray (white coat with red or brown flecks throughout the coat), dapple gray (darker gray coat with light rings of white hair), and rose gray (gray coat with a reddish or pink undertone). With regards to skin pigmentation, even if the horse is completely

■ **Fig. 12.12**   Although black is a less common coat color, some breeds of horses, including Friesians, are predominantly black. (© Alexia Khruscheva, 2009. Under license from Shutterstock, Inc.)

white, it is still considered to be gray if it has black skin. The skin is most easily observable around the eyes and the muzzle.

Roan is a color pattern that results in white hairs becoming intermixed within the basic body color. Often confused with gray, roan horses are born with the same proportion of white hairs that will remain with aging, gray however is progressive. Roan horses generally have solid colored head, neck, and legs. The roan pattern may occur with any background color. White hairs with bay background are considered red roan, white hairs with chestnut are considered to be a strawberry roan, while white hairs with a black background coat results in a blue roan.

Dilutions occur when the intensity of the base color is lessened as an effect of each individual hair. Bay dilutes to buckskin, which are horses tan or golden in color with black points. Duns are very similar to buckskins in that they both have yellow, golden, or sandy body coloring accompanied by black points. However, duns usually have unique and distinguishing primitive markings including a dark dorsal strip along the top of the back, a shoulder stripe and sometimes zebra-like striping along the lower legs. As in buckskins, the dun coloring is produced by dilution of bay. Palominos are produced with the dilution of chestnut and produce a horse of golden color that lacks the black points that the buckskin and dun possess. Palominos also have a lighter, flaxen, tail and mane. Double dilution of the chestnut produces the cremello coat color and dou-

■ **Fig. 12.13**   One of the most common coat colors among horse breeds is bay, which is characterized by black points including those of the lower legs, mane, and tail. (© Karen Givens, 2009. Under license from Shutterstock, Inc.)

■ **Fig. 12.14**   Dapple gray mare and bay foal. Note the white marking on the feet of the bay foal. (© Abramova Kseniya, 2009. Under license from Shutterstock, Inc.)

ble dilution of a bay produces a perlino coat color, also a black is diluted by the dun gene to grulla.

Pinto or paint refers to a spotted body with either color on white, or white on color. The spotting is independent of the markings commonly viewed on the head and lower legs of many horses of varying coat colors. The white markings are present at birth and overlie pink skin. Tobiano and overo are color patterns commonly displayed in paints and pintos. Tobiano horses are characterized by white that crosses the top line. The head is colored, whereas the legs are white (may occur below the knee or hock) and one or both flanks are colored. Conversely, in overo horses white originates from the belly, but seldom crosses the top line. White spotting is less defined in comparison to the Tobiano and appears splotchy. One or more legs are colored except for white markings.

Coat Color Modifying Genes

The pigmentation of skin and hair is determined by melanin, produced by specialized cells known as melanocytes. In horses there are two types of melanin, one responsible for black/brown pigmentation and the second red/yellow pigmentation. The amount and distribution of melanin is controlled by the presence or absence of alleles that define the common coat colors of the various breeds. Consequently, all horses carry the gene for black or chestnut; however, other genes will modify or mask the expression of the black or chestnut allele. For example, the basic colors of bay and brown are modifications of black. It should be noted that the basic color of white is a consequence of a lack of melanocytes and not a result of modifier genes. The genetic underpinnings of coat color in horses have been a long held interest of breeders. Although considerable progress has been made, color inheritance is complex and a complete understanding of the genetic interplay between alleles responsible continues to evolve.

Extension Gene
The control of black, bay and chestnut resides with the extension gene *E*. This gene controls the extent

■ **Fig. 12.15** Buckskin, Dun, and Palomino (clockwise). Note the presence of the dorsal stripe on the Dun and the absence of black points on the Palomino. (*Top:* © Chris Hill, 2009. Under license from Shutterstock, Inc. *Bottom left:* © Stephanie Coffman 2009. Under license from Shutterstock, Inc. *Bottom right:* © mariait, 2009. Under license from Shutterstock, Inc.)

that black hair is expressed on the body of the horse. The gene extends the amount of melanin responsible for black/brown pigment and reciprocally reduces the melanin responsible for red/yellow pigment. The dominant E allele (*EE, Ee*) results in a black or bay horse, whereas *ee*, is expressed as chestnut, a red horse that lacks any black hairs. The *E* gene is epistatic to the A locus and animals with the dominant *E* will have black hair expressed somewhere on the body, however, the A locus will affect the distribution of black hair.

Agouti Gene

Gene *A* is the Agouti gene and is the determining factor that controls the distribution of black hairs. It is this gene which establishes whether a horse is bay or black. The Agouti coloring is a dominant-recessive trait. Any horse with gene *A* and a gene for black coloring (the presence of at least one gene *E*) will have black points while the rest of the body remains a shade of red-brown. All bays are consistent with this gene pattern. However, if two recessive Agouti genes are present, *aa*, in combination with at least one gene *E*, then the horse will display black hair over the entirety of its body. Thus, the genotypes *AA* or *Aa* restricts black to the points, whereas, the recessive genotype aa results in uniform distribution of black. The horse that does not possess the dominant gene for black hair does not express the Agouti trait phenotypically (a chestnut with the genotype *ee*); however, the chestnut can pass one of its unexpressed Agouti genes on to its offspring.

Gray

The progressive accumulation of white hairs that contributes to the gray coat is due to the presence of the dominant gene *G* at the gray locus. Gray is epistatic to all other coat color genes and the *GG* or *Gg* genotype will produce a gray horse regardless of coat color at birth. Gray horses must have at least one gray parent.

White

True white in horses is due to a single dominant allele, *W*, which does not allow a horse to form pigment in its coat and skin. Gene *W* is epistatic to all other coat colors; therefore, a horse will still carry the typical genes for coat colors, such as gene *E* or *G,* but will be unable to produce the melanin necessary to express these colors if a dominant gene *W* is present. All white horses are of the genotype *Ww* as *WW* is known as a homozygous lethal geno-

type that results in embryonic or early postnatal death. A homozygous recessive combination, *ww,* allows the horse to be fully pigmented according to the other color genes.

Roan

Roans are considered to contain 50% white hair at birth due to the gene *Rn*. The presence of gene *Rn* allows white hairs to uniformly penetrate the basic coat colors of black, chestnut or bay. Similar to gene *W,* embryonic lethality is associated with the homozygous dominant genotype *RnRn* and all roan horses are heterozygous.

Tobiano and Overo

The Tobiano pattern is due to the dominant allele *To*. Any horse with the genotyp*e ToTo* or *Toto* will be Tobiano. The amount of white spotting may be determined by modifier genes at additional loci. Interestingly, the Tobiano gene is absent in the Quarter Horse, Thoroughbred, Standardbred, and Arabian. The overo pattern is controlled by the Overo allele, *O*. The genotype for overo horses is *Oo,* homozygous dominant horses (LWO) are associated with a lethal defect marked by the absence of the colon. Homozygous recessive genotypes (*oo*) results in the absence of the overo pattern.

Dilution Genes

Dilutions of basic coat colors of horses have resulted in further variations of possible coat colors. Dilution of coat color is attributed to four distinct genes at different loci. These modifier genes ultimately lighten

■ **Fig. 12.16** Tobiano. (© Eric Stacy Bates, 2009. Under license from Shutterstock, Inc.)

the base color and are termed diluter genes. Gene *C* is named for cream effects. This gene is expressed in an incomplete-dominance pattern. If a horse is genotype *CC*, its color will be expressed without dilution. However, the genotypic combinations of *CC^{cr}* and *C^{cr}C^{cr}* result in a dilution of the base color. If a chestnut horse is heterozygous, *CC^{cr}*, its red color will be diluted to a yellow, resulting in a palomino; whereas a bay horse will be diluted to buckskin, retaining its black points. Black hairs are generally not affected by the heterozygous form of this gene. A horse of any base color which carries the homozygous combination of *C^{cr}C^{cr}* will exhibit a greater degree of dilution. All hair colors will be washed to a very pale cream with pink skin and blue eyes resulting in the cremello coloring when the base coat color is chestnut and perlino when the base coat color is bay. The dilution gene *D* is responsible for dun coloring. Unlike gene *C*, this gene is expressed in a dominant-recessive pattern. Hence, the homozygous dominant *DD* does not dilute colorings to the extreme as seen in gene *C*. Any horse carrying the dominant allele *D* will express the coloring specific to the dun. The homozygous recessive, *dd*, has no effect on the original base coloring. A third dilution gene, *Z*, is responsible for silver dapple coloring. This gene affects only black pigment and is dominant. The presence of this gene dilutes black coat color to brown. The rare pale coat color, mottled skin, and amber eyes of the Tennessee Walking horse is due to dilution by the dominant champagne gene (*Ch*). Champagne dilution results in a pale brown horse that carries the basic black coat color, yellow horse with brown points with the basic bay color, and a gold horse with yellow points when the horse carries the basic chestnut color.

Facial and Leg Markings

Base coat colors may be further modified by the presence of white on the face and/or legs. Face and leg markings are commonly used in the identification of horses. Facial and/or leg markings are present at birth and do not change over the life of the horse. The most common facial markings include blaze, bald face, strip, star, and snip. Blaze is an elongated marking of uniform width that extends the length of the face. A broad blaze that radiates around the eyes and extends down to the upper lip is known as bald face. Strip, or stripe, is a narrow elon-

gated marking from the forehead to nostrils. Star is used to identify any marking on the forehead whereas snip refers to any white marking between the nostrils. In addition, combinations of a star and snip; or a star, strip, and snip are used as facial marking identifiers as well. Leg markings are identified relative to their coverage along the length of the leg. Leg markings that extend to the hock on the hind leg or knee on the foreleg are identified as stocking. Sock leg markings may extend halfway along the length of the cannon bone, also known as boot, or may extend just above the fetlock. A marking that extends above the foot and includes the entire length of the pastern is referred to as pastern; whereas half or partial pastern describes a leg marking that includes only half the pastern. The lowest of the leg markings is the coronet, which extends just above the hoof. The presence and extent of white facial and leg markings are the result of interactions across three genes. Depending on the breed, markings may be of aesthetic and economic importance.

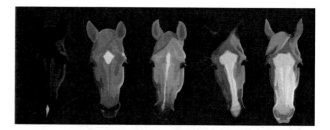

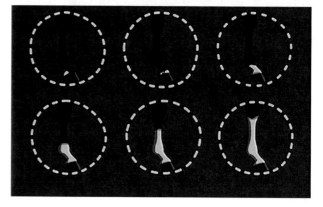

■ **Fig. 12.17** Facial (top) and leg (bottom) markings of the horse. Common facial markings include: (left-right) snip, star, strip, blaze, and bald face. Leg markings include: (clockwise) half-pastern, coronet, pastern, stocking, sock-boot, and sock fetlock. (Shutterstock)

Color	Gene	Alleles	Genotypes and Corresponding Phenotypes	
White	White (W)	W	*WW:*	Lethal
		w	*Ww:*	Born white. Horse lacks pigment in skin and hair (white)
			ww:	Fully pigmented
Gray	Gray (G)	G	*GG:*	Born nongray color but progressively grays as the horse matures. Pigment is present in skin. (gray)
		g	*Gg:*	Same as GG
			gg:	Does not grey with age. Remains original color
Chestnut	Extension (E)	E	*EE:*	Has the ability to form black pigment in skin and hair.
		e	*Ee:*	Same as EE
			ee:	Cannot form black pigment in hair but can in skin
Bay Black	Agouti (A)	A	*AA:*	Directs black hair to only the points (bay when present with gene E)
		a	*Aa:*	Same as AA
			aa:	Directs black hair formation over entirety of the body (black when present with gene E)
Palomino Buckskin Cremello Perlino	Cream (C)	C	*CC:*	Fully pigmented. No dilution occurs
		C^{cr}	*CC^{cr}:*	Red pigment is diluted to yellow. If black pigment is present, it is unaffected (Chestnut → palomino; Bay → buckskin)
			C^{cr}C^{cr}:	Both red and black pigments are diluted to cream. Skin and eye colors also are diluted
Dun	Dun (D)	D	*DD:*	Dark points are unaffected. Body color is diluted and contain additional dark points including a dorsal strip, shoulder stripe and zebra stripping on the lower legs
		d	*Dd:*	Same as DD
			dd:	No dilution occurs
Tobiano	Tobiano (TO)	TO	*TOTO:*	Tobiano spotting pattern
		to	*TOto:*	Same as TOto
			toto:	No tobiano pattern present
Overo	Overo (O)	O	*OO:*	Lethal
		o	*Oo:*	Overo spotting pattern
			oo:	No overo pattern present
Roan	Roan (RN)	RN	*RNRN:*	White hairs are present along with any other body color (roan)
		rn	*RNrn:*	Same as RNRN
			rnrn:	Lethal

United States Horse Industry

The United States horse industry has a direct economic impact of $38.8 billion dollars and this value escalates to $102 billion dollars when indirect and induced spending is included. Approximately two million people are reported to own horses, whereas 4.6 million Americans are involved in the equine industry as service providers, employees, and volunteers. These estimates do not include spectators of equine related sporting events. Currently, leading horse states include Texas, California, and Florida.

Although classified as a livestock species by legislation, equine use extends beyond agricultural in-

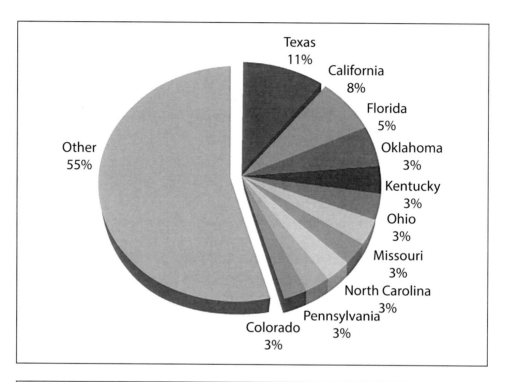

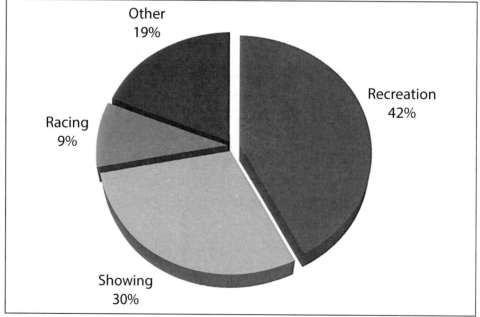

■ **Fig. 12.18**   Horse distribution by state (top) and use (bottom). (American Horse Council, 2005)

terests. Originally consumed for meat, a commercial market for horse production as a food commodity has never been established in the United States. Since the early 1900s state legislation has restricted the consumption of horse meat in the United States, though peak consumption occurred post World War II. The taboo of eating horse meat is not shared worldwide and several European and Asian countries still permit the marketing of horse meat for human consumption. The United States horse industry is diverse and the reasons for keeping horses can be classified into four categories: showing, recreation, racing, and other activities. These other activities can include rodeo, polo, ranch, or police use and breeding. Of these four categories, recreation is the largest segment. Horses are found in all fifty states and are owned by people in all income brackets. Approximately 34% of horse owners have an annual income of less than $50 thousand and 28% have an income that exceeds $100 thousand. The equine industry meets the needs of both rural and urban communities, though the activities and use of horses differ. In rural communities, breeding and training are dominant activities, whereas urban sectors contribute to the use of horses in racing, show, and sale.

Health Management

Horses are subject to a variety of diseases and conditions. Diseases are broadly categorized as infectious when the etiology is parasitic, bacterial, or viral; or non-infectious when the underlying cause is environmental, nutritional, or genetic. Some of the more common diseases include: strangles, colic, and laminitis. Strangles is a highly contagious disease that results from a streptococcus bacterium infection. Although horses of any age may be affected, young horses are more susceptible. The disease is often diagnosed by the presence of abscesses and identification of *Streptococcus equi*. The introduction of new horses can prolong the disease occurrence within a herd. Ruptured abscesses easily contaminate the environment and the disease remains highly infectious for months. Colic is a broad term that refers to the clinical diagnosis of abdominal pain and is the leading cause of death in domesticated horses. Multiple etiologies contribute to the occurrence of colic and include: inflammation of the small intestine, parasitic infection and decreased

intestinal blood flow, impaction of the digesta (may include feed, sand, or dirt and occurs in the colon), and gaseous distension of the large bowel. A reduction in appetite, decreased fecal output, pawing at the ground, pacing, increased sweating, head tossing, flehmen response, repeated lying down and rising, and abdominal distension are all symptoms of colic. Horses in severe discomfort may attempt to roll, which can lead to twisting of the intestines. Treatment may involve walking to relieve intestinal pressure, mineral oil as a laxative to encourage fecal output, or in severe cases, surgery. As parasites are a leading cause of colic, a parasite control program is an important preventative measure against colic. Additionally, maintaining a regular feeding schedule, avoiding sudden changes in diet quality or quantity, and providing access to clean water also reduce the incidence of colic. Lastly, laminitis represents inflammation of the lamina of the inner hoof wall. Often confused with founder, which represents displacement of the coffin bone of the foot, laminitis can occur independently of founder. As with colic, there are multiple etiologies of laminitis and in fact, colic can lead to laminitis in instances where gut microflora are disrupted. Disruption of gut microflora can occur with excess soluble carbohydrates from grain or grass, which leads to the over production of lactic acid. Lactic acid damages the intestinal wall and leads to the presence of toxins in the blood, known as endotoxemia. In addition, dystocia, repetitive stress to the leg, exhaustion, and infection have been noted as causes underlying laminitis. Repetitive stress to the leg results in decreased

■ **Fig. 12.19** Horses with colic may attempt to roll in an effort to relieve abdominal distension. (© Otmar Smit, 2009. Under license from Shutterstock, Inc.)

blood flow to the lamina. The condition progresses rapidly and is often observed as a shift in weight onto the heels to alleviate pressure on the toes and increased restlessness and agitation. If not diagnosed within the first few days, the sole of the hoof may bulge downward leading to lameness followed by founder. Without treatment, the coffin bone may rotate downward to perforate the sole. Treatment involves support of the sole, anti-inflammatory medications, hot- and cold-water soaks to encourage circulation to the lamina, and deep bedded stalls to cushion the hoof. With prompt and proper diagnosis and treatment the horse often returns to its normal level of activity.

Capacity of Production

Mares are seasonally polyestrous, displaying estrus several times annually, but only in response to increasing photoperiod. The mare begins to cycle irregularly in early spring with regularity in cycles corresponding to peak breeding season of June. As day length shortens in September and October, a return to irregularity in cycling occurs. The occurrences of irregular cycling during the early and later months of the year are associated with reduced fertility and are known as breeding transition months. The mare's estrous cycle can be manipulated by the use of artificial lighting by gradually extending lighting to imitate the natural lengthening of the photoperiod that occurs in the spring and summer.

Artificial lighting allows the mare to ovulate at any wanted time during the year; however, preparation must begin sixty to ninety days before the first ovulation is desired. Manipulation of a mare's estrus is most often used in timing foaling to occur in the first months of the year. This is the ideal timeframe for some breeds as the first of January is the universal birthday for registries, including the Jockey Club. Thus, horses born early in the year have had additional time to grow and develop compared to horses born in the later months of the year. While advantageous to producing a larger foal, reproductive efficiency is sacrificed. For mares cycling regularly, estrus occurs every eighteen to twenty days and lasts three to eight days with ovulation occurring twenty-four to forty-eight hours before the end of estrus. Successful conception can be detected by rectal palpations, ultrasonography, or blood testing for the presence of pregnant mare serum gonadotropin, also known as equine chorionic gonadotropin. The average gestation period of the horse is three hundred forty days, corresponding with spring foaling and an advantageous environment for rapid growth and development of the foal due in part to the warm weather and availability of new pasture. The mare will often return to estrus within six to twelve days postpartum. This first estrus, known as foal heat, is associated with the uterus returning to a non-pregnant state. Mares should be ideally bred during the second estrus, which is likely to occur approximately thirty days postpartum. Some mares may not dis-

■ **Fig. 12.20** Horses with laminitis may shift from a typical standing position (left) to a stretching stance (right) to reduce weight bearing on the hooves. (Both images: © Eric Isselée, 2009. Under license from Shutterstock, Inc.)

play signs of estrus during this period and are said to display a silent heat. Failure to recognize the mare in estrus contributes to national conception rates of 50% to 60% in the equine industry when artificial insemination is practiced.

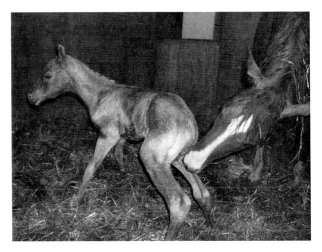

■ **Fig. 12.21** Newborn foals should stand within two to four hours after birth. (© Melissa Dockstader, 2009. Under license from Shutterstock, Inc.)

Today artificial insemination is used widely in the horse industry. However, acceptance of this reproductive technology lagged behind the other animal industries, owing to the role of horses as an animal of sport and pleasure and not a food commodity. Slow adoption of reproductive technologies by the equine industry was due in part to resistance by some breed registries that worried artifical insemination would weaken the breed. Even today, the Jockey Club still does not permit horses generated through artificial insemination to be registered as Thoroughbreds. Conversely, the American Quarter Horse registry recognizes animals resulting from artificial insemination with fresh, transported or frozen semen. Initial restrictions by many breed registries have progressively decreased over the past ten to twenty years and artificial insemination is the predominant breeding method for the industry today.

Equitation

The riding and managing of horses is the art of equitation, which there are two primary styles: English

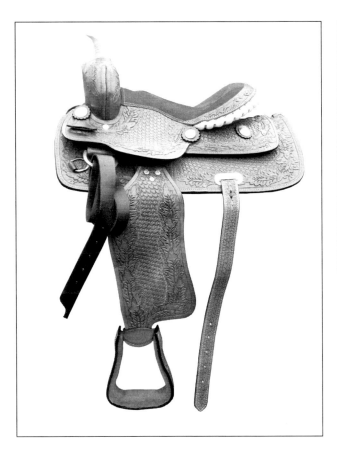

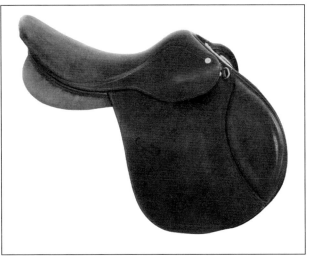

■ **Fig. 12.22** Western (left) and English (right) saddles. (*Left:* © Margo Harrison, 2009. Under license from Shutterstock, Inc. *Right:* © marekuliasz, 2009. Under license from Shutterstock, Inc.)

and Western. The two styles differ in the methods of sitting in the saddle. Although there are many similarities in the two riding styles, discrete differences begin with the type of saddle. Western riding combines elements of ranching and the use of horses by cowboys of the West. The saddles used in Western style are large and relatively heavy and have a deeper seat. They are designed to distribute the weight of the rider more evenly over a larger area of the horse, allow comfort in long distance and rugged travel, and minimize the likelihood of the rider being unseated.

Additionally, Western seats are characterized by a prominent pommel and cantle. Horns are also present for securing the lariat, or rope. In contrast, English saddles are distinguished by a flat seat and the absence of prominent pommel and cantle. The use of English seats requires greater training to ensure a secure seat. As the rider sits closer to the horse, he or she is able to communicate more readily with the horse through movement in the seat and leg positioning. English style allows the horse freedom of movement and is the style used in most Olympic equestrian events, the exception being reining, which is a Western discipline. With regards to riding, English riders are often required to use both hands on the reins whereas Western riders traditionally use one hand. Certain gaits of the horse are also reflected by riding style. The trot and canter are gaits of English style riding, whereas jog and lope are used to describe the same gaits in Western style riding.

CHAPTER 13

Poultry

"Poultry is for cookery what canvas is for painting . . . It is served to us boiled, roast, hot or cold, whole or in portions, with or without sauce, and always with equal success."
— *Jean-Anthelme Brillat-Savarin (1755–1826)*

This chapter introduces the poultry industry, including the dominant species that contribute and the primary foundation breeds that are instrumental in the highly efficient strains used in today's commercial operations. The stages of poultry production for layer and broiler industries are a focus, along with health and reproductive management of flocks. Lastly, the nutritional contribution of poultry meat and eggs to the human diet is defined.

Poultry: A Historical Perspective

Origins of the modern domesticated chicken are debated among archeologists and geneticists. Recent mitochondrial DNA analysis provides evidence that a species of wild fowl native to Thailand serves as the maternal ancestor of today's domesticated chicken. The Red Jungle fowl, scientifically known as *Gallus gallus,* still exists undomesticated in the jungles of Thailand and in other parts of Asia. The mitochondrial DNA evidence challenges the previous archaeological estimates of domestication, which dated to 2000 B.C. in India. It is now recognized that the chicken was domesticated at least twice, in different geographical regions and at different times.

The domestication of the chicken in India was co-incident with the origins of cockfighting, an ancient and popular sport that was a favored pastime for many civilizations. Cockfighting is considered not only the catalyst behind the fowls domestication, but the birds distribution throughout the world. The sport remains a popular pastime to the majority of the world's population and is often referred to as the world's largest spectator sport. In the United States, the sport is banned in all fifty states with Louisiana the last state to impose the ban in 2008. Poultry were also alluring to individuals intrigued by the rarity of exotic species. During past centuries, it was common to maintain small flocks aboard ships during long voyages at sea. The exotic poultry were frequently transported to and from varying ports because of the traveler's interest; the poultry's relatively small, compact size and ease of maintenance; and provision of food during times of food shortages. Christopher Columbus was the first to introduce chickens to the West during his second voyage in 1493. The formation of the Jamestown Colony in 1607 marked the introduction of the species to the North American continent. During this time, it was a common practice for individual families to keep a small flock of chickens that served as a source of both meat and eggs with extra produce being sold or traded. This small-flocking system persisted into the middle of the twentieth century.

The first layer breed, Single Comb White Leghorns, was imported into the United States in 1828. This breed would ultimately become the most important egg-producing breed in the nation. To follow, the first incubator was patented in 1844 and eventually allowed the long-distance shipment of chicks throughout the nation. This development marked the beginning of the commercial hatchery industry, which was later boosted by the United States Postal Service's decision to allow shipment of chicks by mail, which continues today. In 1918, the United States Department of Agriculture introduced the first federal-state poultry grading standards and programs to implement such grading. The 1920s marked a century that began to produce poultry on an industrial

■ **Fig. 13.1** The Red Jungle fowl continues to occupy regions of Thailand and is the ancestor of domesticated chickens. (© Robert J. Beyers II, 2009. Under license from Shutterstock, Inc.)

■ **Fig. 13.2** Cockfighting was a revered pastime of ancient civilizations of the Indus valley, currently occupied by Pakistan. The sport has been banned in all fifty states following enactment of the ban most recently in Louisiana during 2008. (© Hugo Mases, 2009. Under license from Shutterstock, Inc.)

scale. The first classes, standards, and grades of eggs were first announced by the United States Department of Agriculture in 1923, which also marked the development of the electrically heated incubator by Ira Petersime. Subsequent development of the brooder, a device that controls heat and light requirements of newly hatched chicks, would result in nearly all chicks being hatched in incubators by the 1960s. The growing industry relied on transportation from the production facilities to areas of consumption and often utilized railroad systems. An outbreak of fowl plague in the New York City railroad yard sparked poultry inspection programs, which lead to the development of the United States Department of Agriculture Poultry Inspection Service in 1926. Legal standards and provisions for uniform marketing were brought into effect in 1934.

In the 1920s chicken meat was considered a by-product of egg production and prior to 1940, chicken meat was reserved for special occasions. As a measure of prosperity, Roosevelt declared a chicken in every pot as a campaign promise toward reduced unemployment in the late 1930s. Chicken meat, however, did not find its destination as an everyday food until mechanization of poultry processing. A mechanical dresser that quickly delivered ready to cook product

from live chickens spurred the meat-producing sector of the poultry industry. New developing technologies in the 1960s pertaining to improved diets, equipment, genetics, flock health, and processing reduced the cost of production and prompted the movement of the industry across the Midwest. Shortly after, the poultry industry became vertically integrated as feed producers began raising their own flocks. Today, nearly all of the poultry industry is vertically integrated as the majority of producers are responsible for all stages of production, from the hatchery to processing. It is interesting to note that Colonel Sanders' franchise of Kentucky Fried Chicken, which was initiated in 1956, helped create the commonality of everyday chicken consumption. The introduction of buffalo chicken wings in 1964, the egg McMuffin™ in 1973, and chicken McNuggets™ in 1983 promoted further increases in poultry product consumption. In 1981 the combined cash receipts from the poultry industry bested those of the swine industry, and in 1997 the poultry receipts surpassed those from dairying. Per capita consumption of poultry meat and products is still increasing today, due in large part of the industry's abilities to capitalize on cost effective methods of production and their capability to incorporate and market new and convenient poultry products.

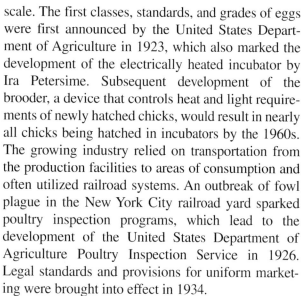

4165. The Dinner Party.

■ **Fig. 13.3** Poultry initially were raised in small outdoor flocks that provided a source of readily available eggs and meat. This system of poultry production persisted until the mid 1900s. A resurgence in backyard poultry production has occurred in urban areas in recent years. (Library of Congress.)

Breeds, Varieties, and Strains of Poultry

The term poultry is applicable to all types of fowl including chickens, turkeys, geese, ducks, pigeons, peacocks or peafowl, chukars, partridges, and guineas. Poultry are classified into classes, breeds, varieties, and strains. A class is the broadest form of categorization and refers to a group of birds that has been developed in the same, expansive geographical location. Such classes include American, Asiatic, English, Continental and Mediterranean. A breed is a group of related fowl that when bred, produce consistent traits that are identifiers of its lineage. These traits are usu-

ally physical in nature and include the body shape and type, skin color, and number of toes. Currently, the American Poultry Association recognizes fifty-three large breeds of chicken, eight breeds of turkey, seventeen breeds of duck, and eleven breeds of geese. Varieties are subdivisions of a breed and are based on specific characteristics such as plumage color and comb classification. For example Leghorn varieties include single or rose comb and plummage colors of white, buff, light or dark brown, silver, Columbian, golden, red, and black tailed red. Strain is the most closely related classification of poultry. A strain indicates a family or breeding population that is more

■ **Fig. 13.4** White Leghorn rooster. (*Left:* © Elena Butinova, 2009. Under license from Shutterstock, Inc.)

States today and are used solely for the production of eggs. Leghorns and their strains are characterized by yellow skin color, which is especially apparent on the legs and beak. The pigmentation is a consequence of dietary carotenoids and is associated with stage of production. As stage of production increases the carotenoids stored in the skin are transferred into the egg yolks and the skin becomes bleached. Increased yellow pigmentation of the skin is associated with birds that have laid relatively few eggs, whereas, decreased pigmentation of the skin is indicative of birds that have laid many eggs over the productive life. To obtain desired yolk color, carotenoids such as the xanthophylls, are added to their diets of laying hens. Hens allowed access to pasture will naturally obtain the carotenoids through foraging.

alike than animals within a breed or variety. The term is not associated with purebred lines and is often used to describe the products resulting from a highly systematic inbreeding program. The commercial chicken industry of today is primarily based upon the selection of quality and high performing strains that are bred to excel in essential factors important to each industry. For layer, or egg producing systems, desired traits include the number of eggs and their size, eggshell quality, and fertility. For broiler, or meat, production these traits are white plumage, egg production, fertility, hatchability, growth rate, and carcass quality. Strain development is also important in the turkey industry for the production of uniform, broad breasted birds for roasting. Ducks, geese, and other poultry industries rely primarily on pure breeds of birds when marketing their products.

Unlike other livestock industries, no individual registries exist for recognizing specific poultry breeds. Instead, the American Poultry Association functions as a comprehensive breed registry and maintains a standard for over three hundred recognized breeds and varieties of chickens, bantams, ducks, geese, and turkeys in its American Standard of Perfection publication. This publication illustrates the size, shape, color, and various other physical features that describe each standard breed.

Layer Breeds

Commercial egg-producing flocks in the United States trace to the Single Comb White Leghorn, which was imported into the United Stated in 1823. These are the most popular breed of chicken in the United

> Egg shell colors range from various shades of white/cream, blue/green, and red/brown. Egg shell color is controlled through the bird's genetics and influenced by age. Color is deposited within the uterus during formation of the shell. A breakdown product of heme, biliverdin, is responsible for the blue/green shades of egg shells. Protoporphyrin, a product of heme biosynthesis underscores the red/brown colors of egg shells. The degree of transfer of these pigments to the egg shell determines the spectrum of colors observed in chicken eggs. In the United States, Aracauna, Ameracauna, Easter Eggers, and Olive Eggers are marketed for their blue/green egg shell color variations.

The Rhode Island Red is a dual purpose breed developed in New England in the mid nineteenth century. Although relatively limited numbers of purebred Rhode Island Reds are used in commercial production, they serve as the foundation breed for the brown egg laying strains used today. The breed is suggested to trace its origins to the red Malay breed of Asia. Because of their dual purpose use, producing quality eggs and meat, this breed is recommended for small outdoor flocks. With feed production costs of the Rhode Island Red being greater than the White Leghorn, egg costs for brown eggs exceed the costs for white eggs.

Broiler Breeds

The broiler industry of the United States has relied on the development of chickens that attain market

■ **Fig. 13.5** Rhode Island Red rooster (left) and hen (right). (*Left:* © Ant Clausen, 2009. Under license from Shutterstock, Inc. *Right:* © Brasiliao, 2009. Under license from Shutterstock, Inc.)

weight of four pounds by six to seven weeks of age and traces its origins to the Plymouth Rock and Cornish breeds. The Plymouth Rock was developed in the United States during the mid nineteenth century and is used predominantly as the maternal parent line. The broiler industry of the 1920s began with the use of the Barred Plymouth Rock, currently strains developed from the White Plymouth Rock dominate the specialized broiler industry. In general, white varieties of poultry are preferred due to the absence of dark pin feathers, feathers just emerging through the skin. These feathers are not easily removed during the processing of the carcass and if noticed, can detract from the value of the carcass. For this reason, white pin feathers are preferred because of their likelihood of remaining undetected. The paternal parent line is represented by the Cornish breed. Developed in England for muscling and conformation, fertility is reduced in the Cornish breed. Similar to the layer industry, the meat industry is dominated by the development of strains and purebred chickens are no longer commercially used by the industry.

Turkey Industry Breeds

TheAmerican PoultryAssociation lists eight breeds of turkey: Bronze, Beltsville Small White, Narragansett, Black, Slate, Bourbon Red, White Holland, and Royal Palm. Turkeys are native to North America and were domesticated within the last five centuries. With the exception of the White Holland, recognized turkey breeds were developed in the United States. The Bronze is found as an improved or unimproved type, with improved Bronze turkeys reflecting intensive se-

■ **Fig. 13.6** Barred Plymouth Rock. (© Amy Kerkemeyer, 2014. Used under license of Shutterstock, Inc.)

lection for increased breast muscle yield. Of commercial importance is the Broad-Breasted White, a strain developed in the 1950s from the crossbreeding of the Bronze and White Holland breeds. The Broad-Breasted White turkeys were developed for increased breast muscle with white feathering. Intensive selection for increased breast muscle has resulted in the unintended consequence of artificial insemination required for mating due to physical constraints imposed by the breast muscle size.

Duck and Geese Breeds

Most domestic ducks originate from the Mallard duck, the exception being the Muscovy that traces to Brazil. Ducks are considered one of the most

■ **Fig. 13.7** Royal Palm, Bronze, and Broad-Breasted White turkeys (clockwise). (*Top:* © 2009. Under license of Shutterstock, Inc. *Left:* © Joy Brown, 2009. Under license from Shutterstock, Inc. *Right:* © Babusi Octavian Florentin, 2009. Under license from Shutterstock, Inc.)

versatile of poultry, able to adapt to a wide range of climatic conditions and unsusceptible to diseases of other poultry including Marek's disease. Duck breeds are classified as meat or egg producers. The breeds of ducks most used for meat production include the White Pekin, Aylesbury, and Muscovy. The White Pekin breed originated in China and was introduced to the United States in 1870s. It is considered the most popular breed of duck raised in the United States because of its excellent market carcass achieved at seven to eight weeks of age. White Pekins do not usually possess maternal qualities and will rarely raise a brood. Today, this breed is highly recognizable to the general public due to its long running use in the Aflac insurance commercials. Muscovy is another recognizable breed of duck due to the red crest above the beak and around the eyes. The most productive of the egg-laying breeds of

duck are the Khaki Campbell and Indian Runner, first and second in production, respectively. The Khaki Campbell originated in England and although uncommon, production rates of three hundred sixty-five eggs per duck per year have been reported. The Indian Runner duck originated in present day Indonesia and was developed into an egg producing breed in western Europe.

Domestic geese were domesticated from different species of waterfowl in different geographical locations. The Greylag goose is the wild ancestor of American and European domestic geese, while Asian and African breeds descend from the swan goose. Domestication traces to five thousand years ago. There are eleven breeds of geese recognized by the American Poultry Association: Toulouse, Embden, African, Sebastopol, Pilgrim, American Buff, Saddleback Pomeranian, Chinese, Tufted

Roman, Canada, and Egyptian. The Embden, Toulouse, and African are the most popular geese breeds in the United States. The Embden was one of the first breeds imported into the United States and originated from Germany. It is characterized by rapid growth and early maturity. The Toulouse originates from France and is one of the largest of the breeds. Its large appearance is further heightened by its loose feathering. The African is distinguished by the protuberance on its head. It is noted for its rapid growth rate and early maturity, however, its commercial use is limited by the presence of dark pin feathers. In addition to their use as a food commodity, geese are used for eggs, show, and as guard animals.

Ratites

Ostriches and emus are large, flightless birds that originate from Africa and Australia, respectively, and were domesticated within the last two centuries. Commercial production in the United States is lim-

ited, with emus having the greatest impact. Lack of knowledge of ratite nutrition, disease prevention, and optimal environmental conditions as well as failure of widespread adoption of their meat in the diet has contributed to their limited use. Original uses of the birds included processing for the hide, feathers, and oil; the meat was considered a by-product. Collectively, the birds provide 95% usable product. Marketing ratite meat as a meat of similar quality to lean beef, with reduced calories, has promoted the ratite industry, but at this time it is considered a speciality product.

United States Poultry Industry

The per capita consumption of broiler meat has increased steadily since the 1960s, contributing to a threefold increase that equates to an average annual consumption of nearly sixty pounds per person. Per capita consumption of turkey is approximately four-

■ **Fig. 13.8** Pekin, Indian Runner, and Muscovy ducks (clockwise). (*Top:* © Alistair Scott, 2009. Under license from Shutterstock, Inc. *Bottom left:* © Istomina, 2009. Under license from Shutterstock, Inc. *Bottom right:* © Sally Wallis, 2009. Under license from Shutterstock, Inc.)

■ **Fig. 13.9** Embden, Toulouse, and African geese (clockwise). (*Top left:* © Andy An, 2009. Under license from Shutterstock, Inc. *Top right:* © lynnlin, 2009. Under license from Shutterstock, Inc. Bottom: © 2009. Under license from Shutterstock, Inc.)

teen pounds, whereas annual consumption of duck is less than 0.4 pounds. Per capita egg consumption was the greatest in the United States during 1945 with an average of four hundred three eggs eaten per person per year. Per capita egg consumption has been variable with a general trend toward decreased consumption. The decline in egg consumption is attributed to concerns over cholesterol intake and a lack of ready to eat packaged products. Per capita egg consumption today is around two hundred fifty eggs, with over one third of the eggs produced being consumed by the Hispanic population.

The value of the poultry industry, estimated from the sales of broilers, eggs, and turkeys, is $44.1 billion; $30.7 billion attributed to broilers and $8.5 billion attributed to eggs. The value of turkeys is $4.8 billion. The poultry industries in the United States have evolved from multiple small flocks owned by multiple individuals, to a limited number of large, consolidated production systems. The poultry industry was the first animal industry to adopt vertically integrative production and marketing practices and served as the catalyst for similar practices in other animal industries. In fact, vertical integration traces to early events of the broiler in-

dustry. In the mid 1900s when the broiler industry was expanding, profits from the industry began to decrease. At this time, each stage of poultry production was owned by different individuals. Separate ownership occurred for the hatcheries, production units, processors, even feed suppliers. For producers to regain lost profit margin, they began to increase flock size. Initially, producers turned to the feed supplier to finance expansion by providing credit to purchase hatchery birds. The debt would be repaid with the sale of the flock, but there were risks associated with financing farms. In response, companies began to acquire their own flocks and integrate the different stages of production into one system. Through integration and control of each stage of production, the efficiency of production was improved through uniform genetics, nutrition, breeding, and management. Today, commercial broiler production is 100% vertically integrated. Similar trends follow in the layer and turkey industries. As discussed with other livestock industries, broiler operations have decreased in number, but increased in the number of birds kept per operation. The top three broiler companies own over 50% of the market. The majority of the broiler industry is located in

■ **Fig. 13.10** Emu (left) and ostrich (right). (*Left:* © Arvind Balaraman, 2009. Under license from Shutterstock, Inc. *Right:* © Natalia Yudenich, 2009. Under license from Shutterstock, Inc.)

the southern and southeastern United States because of the climate, which reduces housing costs, labor costs, and provides nearby populations that act as markets for the meat products. The top producing broiler states include Georgia, Arkansas, Alabama, North Carolina, and Mississippi. The egg industry averages three hundred fifty million laying birds and produces ninety-five billion eggs annually. In the 1980s egg production increased due to an increase in number of eggs laid annually by hens, with an approximate increase of one egg per hen per year. The top producing egg states include Iowa, Ohio, Pennsylvania, Indiana, and, California.

The turkey industry has experienced consolidation akin to that of the broiler, but growth has occurred to a lesser degree. Nearly two hundred seven million turkeys are produced annually in the United States, which corresponds to over seven billion pounds of turkey meat. An average of 70% of this meat is marketed as processed products as opposed to sold whole. The turkey industry is not geographically concentrated. Minnesota and North Carolina are the leaders in turkey production and are followed by Arkansas, Missouri, and Virginia. Turkey production has increased over the years due in large part to the increasing human population, exports, and the marketing of turkey outside of the traditional holiday fare.

In 2012, there were over five thousand duck operations reporting sales. Total sales exceeded twenty-three million birds sold with nearly two million ducks slaughtered. Geese operations with sales totaled one thousand seven hundred forty-three, with approximately two-hundred thousand birds sold. Ducks are marketed as broiler, roaster, or mature ducks and geese are marketed as young or mature. Marketing is based on age and weight, with mature birds representing layer and breeder birds past production and the meat used in processed products. Ducks were originally farmed on Long Island, New York, but the industry relocated under rising real estate prices that compromised profitability of production. Currently, the duck industry is focused in Texas, Indiana, and Wisconsin; whereas the geese industry is prominant in Texas and California.

The ratite industry has contracted in recent years and has been crippled by overproduction relative to demand. Effective 2002, meat from ostrich and emu became subject to mandatory United States Department of Agriculture inspection. Federal inspection increased market opportunities; however, these opportunities have not been realized. Sold as a specialty item in restaurants with limited availability retail, the price of ostrich and emu exceeds that of beef, pork, chicken, and turkey. Currently, the market for meat is not sustainable and the industry lacks the infrastructure needed to expand. Between 2007 and 2012, the number of emu farms reporting sales decreased by one half, and the number of ostrich farms decreased by two-thirds. There are only two hundred seventy-two emu farms and forty-two ostrich farms reporting sales. The combined total number of emus and ostriches sold in 2012 was sixty-two hundred birds. Texas reports the greatest number of farms with thrty-five emu and seven ostrich farms in operation.

Poultry Management Systems

Specialized production segments within the layer and broiler industries include: breeding flocks, hatcheries, grow-out operations, and market egg producing operations. Poultry production begins with primary breeder flocks, which represent the genetic stock for broiler and layer industries. Birds generated from primary breeder flocks serve as the parent stock in multiplier flocks. Multiplier flocks,

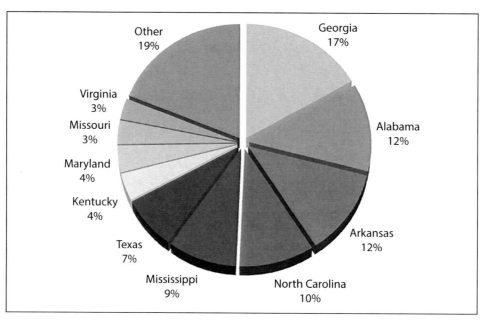

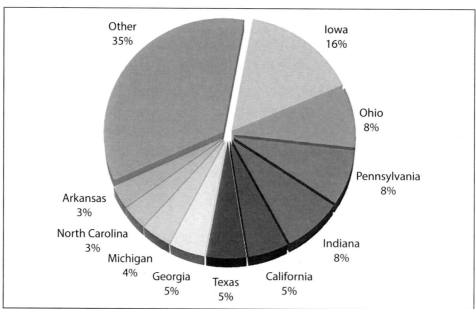

■ **Fig. 13.11** Broiler (top) and layer (bottom) distribution by state during 2013.
(United States Department of Agriculture-National Agricultural Statistics Service.)

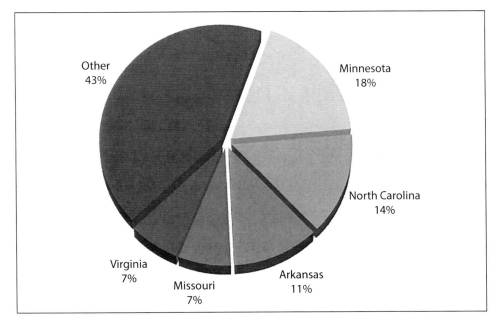

■ **Fig. 13.12** Turkey distribution by state during 2012. (United States Department of Agriculture-National Agricultural Statistics Service.)

in turn, produce eggs that will hatch for production birds in the broiler and layer markets. Broiler breeders are susceptible to obesity, which severely decreases breeding performance. As such, breeder flocks for the broiler industry must be restrict fed. Restrict feeding includes limiting daily feed intake or every other day feeding. As the appetite and growth rates of males are greater, males may be raised separate from females until breeding age. Eggs produced by flocks are incubated and hatched at the hatchery, hatched chicks enter grow-out operations and provide birds for the breeder flocks, broiler, or layer segments.

Incubation and Hatching

Incubation is the act of bringing an egg to hatch. With many poultry, artificial methods of incubation support the hatch of a fertilized egg independent of the hen. Poultry are oviparous and development of the young occurs outside the uterus. Structural development occurs soon after fertilization and continues through incubation. Before the egg is laid cells begin to acquire specialized function that is central to embryonic development. Within the first twenty-four hours of incubation, the nervous, digestive, and visual systems of the chick embryo begin to appear. By the end of day two, the heart begins to beat. At day nine, the embryo acquires a birdlike appearance, coincident with the initiation of beak development.

The protein rich albumin is completely depleted and the yolk is the main energy source for the embryo once day sixteen is reached. Day nineteen signals the remaining yolk and sac to enter the body of the embryo through the umbilicus. The internalized yolk will allow early survival until a post-hatch food source is identified. The embryo officially becomes a chick on day twenty and hatching follows on day twenty-one.

Poultry differ in incubation periods and it is crucial when raising poultry to be knowledgeable about

Reproduction in Poultry

Species	Incubation (days)	Eggs/Year
Chicken (layer type)	21	240
Chicken (broiler type)	21	170
Turkey	28	105
Goose	28–32	15–60
Duck (Pekin)	28	110–175
Emu	57–62	25
Ostrich	42	50

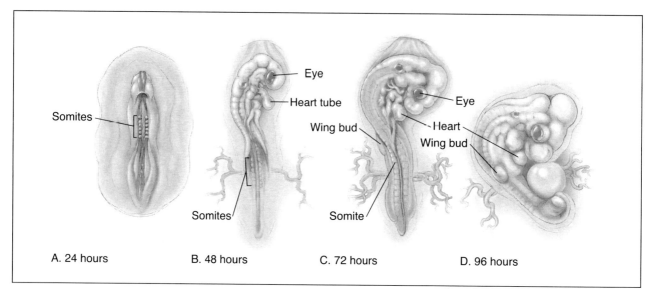

Somites

Eye

Heart tube

Somites

Wing bud

Eye

Heart

Wing bud

Somite

A. 24 hours   B. 48 hours   C. 72 hours   D. 96 hours

■ **Fig. 13.13**   The chick embryo grows and develops rapidly and within three weeks of incubation emerges as a fully formed chick. During the first day of embryonic development the nervous, digestive, and visual systems begin to develop. By day three, discernible features of the chicken appear, including the wing buds. The somites will differentiate into skeletal muscle and skin with further development. (© Kendall Hunt Publishing Company.)

the individual breeds in order to raise successful broods. A central aspect to successful incubation is maintaining a uniform and correct environment. Forced-air, or fan, incubators are the most commonly utilized incubation system in the commercial industry. The incubators control temperature and humidity, rotate the eggs, and facilitate an adequate supply of fresh air. The sizes of the incubators can vary and range from one hundred egg capacity to one hundred thousand eggs or more. Provided adequate techniques are used, hatchability of artificially incubated eggs may exceed 90% for chickens, although hatchability estimates of 65–70% may be more common for geese and ducks.

Temperature is arguably the most critical requirement for successful incubation and hatching. When an egg is laid the drop in external temperature relative to the internal temperature of the hen arrests embryonic development. The egg may be held at a temperature of 65° F for seven days before beginning incubation, which will reinitiate embryonic development. While this may reduce hatchability, it allows for eggs that were laid at different times to hatch within a narrow window of time. For chickens, 99 to 103° F is an optimal beginning temperature range for chickens. At the end of the fourth day of incubation, the temperature should be lowered by

0.25 to 0.5° F. Three days before hatching, the temperature should be lowered a second time by 1.0 to 1.5° F. This last, and fairly dramatic, change in temperature is due to the chick's transition from embryonic respiration to aerobic respiration. Aerobic respiration produces increased heat, raising the overall temperature of the incubator. Greater than optimal temperatures may accelerate embryonic development, increasing the risk of mortality and malformation of chicks. Lower than optimal temperatures retards the embryonic process and also increases the risk of mortality and malformation.

The optimal humidity maintained by incubators can range anywhere from 55% to 70% depending on the stage of incubation. The later phases require more humidity for the upcoming hatching of the chicks. Humidity is important in order to avoid evaporation of water from the egg. Rotation of the eggs also is a critical aspect of incubation. If not frequently rotated, the embryo may become fixed to the shell membrane and prevent further development. The timing and control of egg rotation is mechanized on all modern, commercial incubators. Maximum hatchability is achieved when the egg is rotated five or more times daily. However, during the three days preceding hatching, the egg should not be turned. At this late stage, the embryos are

■ **Fig. 13.14** Artificial incubators maintain optimal temperature and humidity for successful hatching. (© Gila R. Todd, 2009. Under license from Shutterstock, Inc.)

moving into the ideal position for hatching and are not benefited by egg rotation.

Proper ventilation is necessary to maintain appropriate oxygen and carbon dioxide concentrations. The oxygen content should be 21%, while the carbon dioxide accumulation should never exceed 0.5%. During the incubation stage, the eggs are candled at least once. Candling uses a 75-watt light to detect fertility and embryonic mortality and development. Eggs that are infertile or contain demised embryos are removed from the incubator. It is common for chicken operations to candle on the third or fourth day of incubation, while turkey and waterfowl are commonly candled between the seventh and tenth day.

Broiler Production

The objective in managing broilers is to provide a clean, comfortable environment with appropriate feed and water access, which will promote efficient growth rates. Broilers are typically raised in long and narrow housing facilities. The water dispensers and feed systems are fully or semi-automated and allow for the administration of vaccines and medications through the feed and/or water. Broiler farms in the United States average a capacity of sixty-four thousand birds housed at a density of 0.5 to one square foot per bird. Adequate floor space is important to avoid adversely affecting growth and efficiency of production. Too little floor space can result in difficulty obtaining the feed and water supply

and instigates competition between the fowl. Too much space can contribute to boredom. Both situations increase the incidence of aggressive behaviors including feather pecking and cannibalism.

Prior to receiving young birds, the housing facilities are cleaned, disinfected, and dry litter placed into the facility. The major purpose of litter is to absorb moisture from fecal excretions. The entirety of the floor should be covered with a minimum of two inches of litter. Commonly used substances include wood shavings, sawdust, woodchips, peat moss, ground corncobs, and recycled paper. To minimize disease transmission, biosecure measures including an all-in-all-out system in which only birds of the same age are introduced to the facility at the same time, strict sanitation and disinfection protocols, prompt removal of deceased or health compromised birds, and restriction of human traffic, should be adhered to. The broiler chicken generally achieves market weight around six to seven weeks of age, thus broiler farms raise and market multiple flocks per year. Although mechanized catchers are available for collecting birds ready for processing, collection is routinely done by hand and remains one of the few aspects of poultry production that is not completely replaced by mechanization.

Layer Production

Similar to the broiler industry, the laying industry is concerned with raising chicks in an environment that is clean, comfortable, and allows appropriate access to feed and water. However, the laying industry does not stress rapid growth rates like that of the broiler segment. Instead, the laying industry focuses on quantity and quality of eggs produced. There is a direct relationship between size of the bird at onset of lay and size of the egg, with smaller birds yielding smaller sized eggs. Producers aim to increase the weight of the bird or delay the onset of sexual maturity until adequate bird weight is achieved. A delay in sexual maturity is often achieved by reduced lighting programs that provide less than eight hours of light daily or gradually reduce light offered during the first eighteen weeks of age. After which time, the lighting is increased to greater than fourteen hours per day for rapid onset of sexual maturation. Lighting affects reproductive capacity of poultry through three mechanisms: increased light recognition by the eye, increased release of hormones including gonadotropin-releasing hormone

■ **Fig. 13.15**   Broiler housing facilities. (*Left:* © terekhov igor, 2009. Under license from Shutterstock, Inc. *Right:* © Mosista Pambudi, 2009. Under license from Shutterstock, Inc.)

■ **Fig. 13.16**   Caged layer housing facility. (© Enrico Jose, 2009. Under license from Shutterstock, Inc.)

and subsequent releases of follicle stimulating hormone, and maturation of the ovary. By twenty weeks of age, females should be moved from the rearing facilities into the laying house for acclimation to the surroundings prior to beginning the laying process.

Production Systems

There are different models for the production of poultry products and include: caged, confinement cage-free, and range or pasture raised. The caged system is the dominant layer system and permits greatest control of light, temperature, and ventilation. In traditional caged systems, known as battery

cages, birds, generally three per cage, are maintained in a stacked system composed of four to six tiers with wire flooring. The floors of these cages are slightly inclined to allow the egg to roll into a collection area at the front of the cage where they are collected. All of the cages have access to a common feed trough that is either manually or automatically filled. The manure from these systems is usually recovered by collection pits beneath the enclosures that are typically emptied by mechanical scrapers that push the manure to a central collection point. It is important in the stacked arrangements to prevent manure from dropping from high tiers to the cages below. Limited space that restricts birds from engaging in routine behaviors such as wing stretching has raised welfare concerns in the use of battery cages. In response, enriched colony cages are suggested as replacements. In enriched colony cages, space per bird is nearly doubled and nesting and perching resources are available. Confinement cage-free defines a system of production whereby layer birds are maintained in an indoor area that permits interaction between birds throughout the building. Aviary style systems offer perches and platforms positioned at different heights for an added dimension of interaction. Feed and water systems are automated. Eggs are gathered by using roll-away nesting equipment that automatically transports all eggs into a common holding area. All commercial broilers are produced in confinement cage-free systems, but are confined to a single level of the building. In range or pastured flocks, birds are

■ **Fig. 13.17**   Concerns and perceptions of confinement systems in poultry production has prompted the development of free-range poultry. (© Tony Campbell, 2009. Under license from Shutterstock, Inc.)

■ **Fig. 13.18**   Avian Influenza, H5N1, has been a major topic of global concern in the recent years. Millions of birds have died from the infection in Asia, Africa, and Europe. A barrier currently exists between the bird species and the human species which has helped prevent the spread of the disease to the human population. However, transmission across species does occur and human transmission has occurred. (© Jeremy Richards, 2009. Under license from Shutterstock, Inc.)

raised on the open ground and are often maintained within fenced areas for protection and management. Definitions for range and pastured systems are not interchangeable, range is defined by allowing continual access to the outdoors where pastured flocks require the additional management of pasture to partially fulfill the birds nutritional needs. Portable coops, known as tractors, may be used for housing. These portable coops facilitate ease of pasture rotation. There are increases in labor and land required for these systems of production. Disease and predation are both a concern in range and pastured systems. Production is decreased and mortality is increased in laying hens when compared to caged systems. Pastured systems are a primary system for finishing geese, which are raised indoors for the initial weeks of life before being transferred outdoors to finish on pasture and supplemental grain. Ducks, on the contrary, are raised primarily indoors. As consumer interest and demand for products raised under specific management strategies increase, commercial poultry production has responded by offering products across each management system.

Health Management

Temperature, moisture, ventilation, and insulation are important components in all stages of poultry production. Sudden fluctuations in housing temperature are detrimental for production. Especially in the non-incubation stages, temperature is influenced by many factors including the natural climate of the

housing location, solar radiation, wind speed, and heat production from the birds themselves. All of these factors must be considered when designing an appropriate temperature strategy to retain heat in the colder climates and dissipate heat in more arid climates.

The moisture content of the housing system can compromise production and increase risk of disease, with incidences worsened by sub-optimal temperatures. Spillage from watering systems, vapor from the surrounding air, and vapor expelled by the birds and their excreta contributes to increased moisture. However, both moisture and temperature can be controlled by proper ventilation. An effective ventilation system provides the birds with fresh air, removes the excess moisture, and plays an essential role in maintaining proper temperature. The most common ventilation systems rely on the use of exhaust fans coupled with air intake devices such as slotted areas found in the ceiling or near the top of the housing facility.

The poultry industry is especially concerned with the management and control of diseases due to the highly concentrated proximity in which birds are maintained. Just as in the swine industry, biosecurity measures are routinely followed in an effort to minimize disease transmission from outside sources and to reduce the spread of diseases between groups of birds. It is customary for those who work in poultry production to change clothes before entering housing facilities and after leaving the premises. Workers who actually interact with the flock should wear sanitized, protective clothing including boots and headgear. All poultry crates, egg cases, feed bags, and vehicles entering the premises should be disinfected. Dead or sick birds should undergo laboratory testing and those birds which have died should be properly disposed of by burial or incineration. Some poultry diseases have been eradicated from the United States while others can be controlled through the use of vaccines.

Capacity of Production

Capacity for egg production varies by species; however, the makeup and internal structures of the egg of all species are similar. The shell is the outer most covering of the egg and is composed primarily of calcium carbonate. The surface is perforated by six thousand to eight thousand microscopic pores and allows the exchange of gases through the shell. Immediately beneath the shell are the outer and inner shell membranes. The membranes help prevent bacteria from entering and water from evaporating too quickly from the egg. The pulling away of the outer membrane from the inner membrane during cooling creates the air cell found at the broad end of the egg. Such cooling occurs during the laying process when the egg leaves the internal temperature of the hen, around 106° F, and is deposited into the much cooler environment of the laying house.

The shell membranes encapsulate the albumin, or the white of the egg, which provides a source of protein for embryonic development. The yolk is the primary food source for the chick in the immediate post hatch period and is the sole site of fat deposition within the egg. The yolk also contains the germinal disk, which is the site of initial embryonic formation as fertilization occurs at that site. Surrounding the yolk is the vitelline membrane. This membrane re-

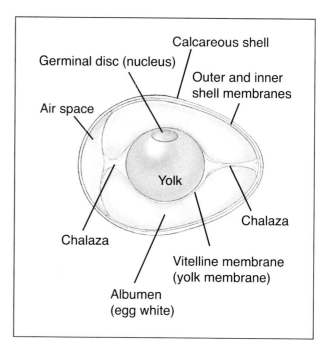

■ **Fig. 13.19** General structures of the egg. (© Kendall Hunt Publishing Company.)

tains the contents of the yolk. The yolk is held in the center of the egg by cordlike protein structures called chalazae. Prominent chalazae are an indication of the freshness of an egg.

Poultry as a Food Commodity

Egg Production

The United States Department of Agriculture provides five standards of classifying and grading eggs: AA quality, A quality, B quality, dirty, and check. Shell eggs available to the consumer for purchase include AA and A quality grades. Eggs that are B quality are broken-out and used to make liquid, dried, or frozen egg products. The size of the air cell, the shell quality, the white of the egg, and the yolk are all determinants of grading. With regards to interior egg quality, smaller air cells are an indication of the freshness of an egg as, over time, moisture will evaporate from the interior of the egg resulting in a larger air cell. The albumin should be firmer in fresher eggs with a reduction in firmness associated with prolonged storage. The yolk outline should be distinct in eggs of greater quality. These specifications for quality can be determined by candling the intact egg. Exterior quality involves an as-

Classification of Eggs for Commerical Distribution

Size standard	Minimum weight per dozen (oz.)
Jumbo	30
Extra Large	27
Large	24
Medium	21
Small	18
Peewee	15

sessment of the exterior of the egg and considers the intactness, cleanliness, and ideal shape of the egg. Eggs are dually classified according to weight per dozen and include: jumbo, extra large, large, medium, small, and peewee. Each of the quality classifications are found in each of the weight classes of eggs.

Meat Production

Grades of A, B, and C are assigned to poultry carcass by the United States Department of Agriculture and are generally dependent on the quality of processing and includes freedom from pin feathers, absence of tears or cuts, nonexistence of disjointed and broken bones, a lack of bruising and other blemishes, and the absence of freezing defects. Grading is also reliant on the confirmation of the bird, its muscling, and its fat covering. Grade A poultry has no deformities and normal flesh and fat distribution. The carcass of a grade A bird is free of feathers, incised flesh, broken bones, and discoloration. Grade B poultry meat permits some of these imperfections, while Grade C meat accepts the greatest degree of imperfection. Although not marketed on the grocery shelf, poultry meat sold in stores consists of Grade A and Grade B products. Grade C meat is reserved for processing into other chicken foods and is generally not sold directly to the consumer.

Nutritional Benefits of Poultry and Eggs

Chicken is an important source of daily requirements of protein, niacin, vitamin B6, vitamin B12, vitamin D, iron and zinc. If prepared in a health conscious way, chicken is one of the lowest-fat meats offered. The breast of the chicken is lower in fat content than other cuts including the thighs; how-

ever, a significant portion of the fat found in chicken is unsaturated

Eggs are often referred to as a complete food source because of the variety of essential components found within the egg. Of course this makes sense as the egg's purpose is to nourish and sustain a living organism and thus incorporates all necessary components essential to life. These include protein, carbohydrates, fatty acids, and a variety of vitamins and minerals. Egg protein is beneficial to humans as it contains all twenty of the amino acids found within proteins. The egg contains most vitamins, with the exception of vitamin C, and is a particularly good source of vitamin A, vitamin D, vitamin E, and the B vitamins. The mineral content of eggs is varied and contains iodine, phosphorous, zinc, selenium, calcium, and iron. Feeding practices of the hen can further improve nutrient profiles of the egg and commonly focus on enrichment of vitamin E and omega-3 fatty acids. Feeding hens diets with increased amounts of either nutrient leads to increased amounts of the nutrient within the yolk. As previously discussed, evidence supporting the health promoting benefits of diets increased in omega-3 fatty acids make enriched eggs an attractive food choice for the fat conscious consumer.

With its relatively complete nutrition, the egg can be an important part of a healthy diet, but over the years consumption has fluctuated in response to cholesterol content concerns. Cholesterol content of eggs is reduced when feeding hens a vegetarian diet, the resulting eggs are marketed as having a reduced cholesterol content of one hundred eighty-five milligrams compared to the two hundred to two hundred twenty milligrams of cholesterol in traditional eggs. Furthermore, cholesterol of the human body is a consequence of both diet and synthesis within the body and is highly variable between individuals. While a diet high in cholesterol can lead to high cholesterol for some individuals, it does not justify restriction of eggs in the diet for most.

Poultry for the Biomedical Community

Since the late 1800s chickens have been used in nutritional studies to validate the need for certain nutrients and continue to be a valuable animal model for these studies. Chickens are the model of choice

■ **Fig. 13.20** In the United States, poultry (including both chicken and turkey) is the most widely consumed meat. Popularity in part is driven by the industry meeting consumer demand for inexpensive, consistent, and easily prepared products. (© cappi thompson, 2009. Under license from Shutterstock, Inc.)

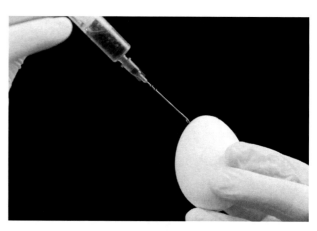

■ **Fig. 13.21** The chicken egg has proven to be a valuable resource in biomedical research. (© Ljupco Smokoviski, 2009. Under license from Shutterstock, Inc.)

for the study of embryonic development. Early embryonic development is similar across species, in the chick development within an egg removes maternal influences and the embryo can be easily manipulated. Another application of the chick embryo is in toxicology. Chickens are more sensitive to various toxicants than many other birds, which is important toward assessing the role of environmental toxins on developmental defects. Technology also has generated modified eggs, which produce compounds that can be harvested from the yolk. The most common compounds produced are antibodies. Exposure of the hen to an antigen (virus, bacteria, etc) results in the production of antibodies to neutralize the antigen. These antibodies circulate through the hen's body and are ultimately concentrated in the egg yolk. The yolk is extracted, and the antibodies are purified for use.

Llamas and Alpacas

*"Tina, you fat lard, come get some DINNER! …
Tina, eat. Food. Eat the FOOD!"*
—*Napolean Dynamite, 2004*

Relatively new animal production systems within the United States, the llama and alpaca industries are gaining ground providing animals primarily for companion purposes. This chapter introduces the history, production, use, and unique behaviors associated with the New World camelid species.

Lamoids: A Historical Perspective

Llamas and alpacas are members of the Camelidae family, which consists of Old World camels of Africa and Asia and New World camels of South America, also referred to as lamoids. Camelids evolved in North America an estimated eleven million years ago and approximately two million years ago, the family migrated from North America to Asia and South America. Between ten and twelve thousand years ago, camelids disappeared from the North American landscape. Animals that migrated were shaped by their new environments. There remain two types of Old World camels, the Arabian and Bactrian and four New World camels, the domesticated llama and alpaca, and the wild counterparts the guanaco and vicuna. Domestication of New World camels occurred four to six thousand years ago by the Incas of South America. The llama's main purpose was to function as a pack animal; however, it served as a source of meat, milk, fiber, and fuel and was used as a sacrificial offering to the gods. The alpaca, which is smaller in size than the llama, was bred for its fiber. Approximately five hundred years ago the role of South American camelids shifted as the Spaniards invaded and began to colonize the region. With colonization, the Spaniards began to eliminate llamas and alpacas and replace these species with their own domestic species including cattle and sheep. However, cattle and sheep were not suited for the altitudes of the Andes Mountains. As such, the South American camelids became well adapted to altitudes above thirteen thousand feet. The wild predecessor of today's domestic llama and alpaca remains a controversy. While it is argued that the llama was domesticated from the guanaco and the alpaca from the vicuna, evidence exists supporting the guanaco as the wild ancestor of both domestic species.

Llamas were imported into the United States in the late nineteenth century by zoos and private collectors who wished to display the exotic species. Alpaca importation was limited because of an 1843 Peruvian ban prohibiting the exportation of the animals from that country. In 1930, the United States enacted its own ban on the importation of all hoofed livestock from South America in reaction to the hoof-and-mouth pandemic. It was during this time that the Andean countries joined together to restrict the exportation of both llamas and alpacas. The only legal exports of llamas, which occurred from this time until the ban was lifted in the 1980s, were from a small Canadian herd. For these reasons, the ancestry of most llamas in the United States dates back to those specimens that were imported from South America prior to 1930, those which originated from the Canadian herd, or are the result of a limited amount of llamas imported illegally from Mexico. The United States Department of Agriculture is still wary of the transmission of hoof-and-mouth disease from other nations. For this reason, further importation of animals is costly and timely as the animals are required to undergo an extended quarantine period. Some quality breeding stock is imported annually;

■ **Fig. 14.1** The Camelidae family consists of the Old World camels of Africa and Asia, which includes Bactrian camels (background) and the New World camels of South America, which includes llamas (foreground). (© Eric Limon, 2009. Under license from Shutterstock, Inc.)

■ **Fig. 14.2** The guanaco, a wild ancestor of the llama, is well adapted to the high altitudes of the Andes Mountains. (© David Thyberg, 2009. Under license from Shutterstock, Inc.)

however, these limited numbers have not had a significant impact in expanding the United States gene pool.

Until recently, there was limited information regarding the biology, behavior, reproduction, nutrition, and diseases of South American camelids. Increased use of the animals in non-native countries has prompted studies to gain a more in-depth understanding of these animals. In their native environments, South American camelids are a preferred livestock species and may represent up to 30% of the livestock utilized by pastoral communities. In non-native environments, the animals are used primarily as companion or guard animals, and sources of fiber. Meat and milk are not used within the United States.

South American Camelid Species

Llama

Llamas are characterized by an ability to survive harsh, dry climates at high altitudes. This is supported by their dense fiber coat, large foot pads relative to their weight that ensures stable footing, and red blood cell physiology that ensures adequate oxygen supply. They are considered stronger than horses as they can support a pack weight of 100 pounds,

though 40 to 60 pound pack weights are more common. Llamas are medium sized mammals. The adult llama can weigh anywhere from two hundred forty to four hundred fifty pounds and stand approximately four feet at the withers and anywhere from five to six feet at the poll. The average lifespan of a llama is twenty to twenty-five years, with the female achieving mature size at age two and males achieving mature size at age three. Llamas can present in a variety of different coat colors that include solid coloring of white, black, red, beige, and brown, a combination of these, or in spotted and roan patterns. The fiber of the llama is present on the neck, back, and sides of the torso; while the head, abdomen, and legs are covered with hair. The fiber gives the llama protection against the cold weather while the shorter covering of hair allows heat to dissipate in the warmer months. The fiber is three to eight inches long, oil-free, and can be collected and spun into marketable goods. However, the fiber from llamas is inferior to the fiber of alpacas, and beyond pet ownership, the uses of the llama in the United States are primarily for packing in wilderness outfitting or guard animals.

The historical use and physical adaptations to high altitudes support the use of llamas as pack animals. In Bolivia, the llama remains a valued draft animal in Andean transport. In the United States

■ **Fig. 14.3** Llama (© Peter Baxter, 2009. Under license from Shutterstock, Inc.)

■ **Fig. 14.4** Llamas natural aggression toward coyotes and dogs make them a suitable choice as a guard animal in sheep flock protection. Advantages of guard llamas in comparison to guard dogs include increased working life expectancy and reduced management costs. (© David Gaylor, 2009. Under license from Shutterstock, Inc.)

there has been a gradual acceptance of the llama for packing in wilderness outfitting in some national forests. Relative to the horse, the llama has a low environmental impact when used in trekking. The soft foot pads in lieu of hooves allow the llama to maintain footing with minimal disturbance to forest ground. The llama has a lower feed requirement per body weight, forages over less distance, and produces less waste when compared to the horse. However, the llama is susceptible to disease and parasitic infection common to other ruminant animals. As such, there is concern of the transmission of diseases from pack llamas to wild ruminants.

As guard animals, llamas are naturally aggressive and effective against wolves, coyotes, and dogs. The primary animals guarded are sheep, and significant reductions in sheep mortalities occur when guard llamas are employed. When compared to guard dogs, the initial investment cost of a guard llama is greater. However, guard llamas offer the advantages of less investment time required for bonding and adjustment to the sheep flock, reduced annual expenses that include reduced feed costs as llamas can be maintained on the equivalent pasture of the flock, and increased longevity with llamas having a guard lifespan of ten to fifteen years. Over

50% of guard dogs are replaced within three years. The most commonly reported problem in guard llamas is aggressiveness and attempts to breed ewes, which was reported to occur in 25% of intact male llamas. These problems are reduced when castrated males or females are used. Furthermore, not all llamas successfully guard. Gelded llamas are the most effective. Guarding occurs through attack and relocation of the flock in the presence of a predator.

Alpaca

There are two types of alpacas: the Huacaya and the Suri. The Huacaya is the dominant type within the United States. The crimpy-fleece of the Huacaya grows perpendicular to the body, whereas the fleece of the Suri grows parallel to the body and gives rise to a fleece of locks or ringlets. Alpacas are smaller in size than llamas, with weight ranges from one hundred to one hundred eighty-five pounds and height of three feet at the withers and four and half feet at the poll. Alpacas have been bred for their fiber since the reign of the Incas, and fiber remains the primary use of alpacas in their native and non-native territories. The strength of alpaca fiber is three fold greater than merino wool. Second to the vicuna, alpaca fiber is considered the finest fiber in the world, but alpaca

■ **Fig. 14.5**   Suri (left) and Huacaya (right) alpaca types (*Left:* © Darren Begley, 2009. Under license from Shutterstock, Inc. *Right:* © James R T Bossert, 2009. Under license from Shutterstock, Inc.)

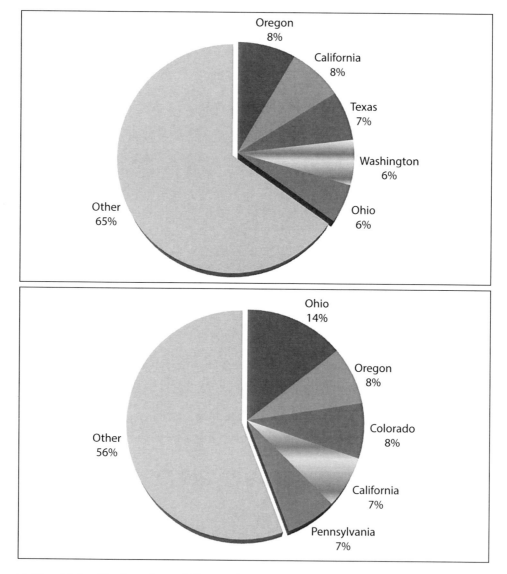

■ **Fig. 14.6**   Total registered llamas (top) and alpacas (bottom). (Internation Llama Registry and Alpaca Registry Inc.)

fleece vary considerable in quality due to differences in fiber medullation, diameter or fineness, color, and length. Whereas wool from sheep has a solid core, the fiber from alpacas is tubular with a hollow, or medullated, core. Meduated fiber is undesirable in an alpaca fleece, and the degree of medullation decreases with fiber diameter. The fiber diameter, as measured in microns representing micrometers, ranges from 20 to 36 microns. Thirty-six microns represents the most coarse of the fibers that are more medullated, resulting in weaker fibers that do not readily accept dye. Fibers from the more desirable fleece of the Suri are less meduated and finer when compared to Huacaya. Fiber diameter is highly heritable, as such progeny performance testing is used in the industry to provide predictions for the selection of mating pairs toward improvement of fleece quality. There are sixteen natural fleece colors classified in the United States. These colors include the primary colors: white, fawn, brown, black, beige, and gray. Additional colors occur through variations of the primary colors. For example, gray has six variations that range from light silver gray to dark rose gray. Fiber length ranges from three to nine inches. If the climate permits, it is desirable to shear alpacas only once every eighteen to twenty-four months to allow for increased fiber length. However, in warmer climates, yearly shearing is necessary to prevent overheating. Adult alpacas produce an average of four pounds of fiber per year, with greater product yield in comparison to shorn wool. Clean fleece yields are 85 to 95% accounting for losses in dirt and grease, with the grease content being less than lanolin of wool.

Worldwide, fiber production is the primary industry for alpacas. Alpaca fiber represents 0.1% of animal fiber on the world market, which equates to approximately four thousand tons marketed annually. Peru is the primary supplier of alpaca fiber. In the United States, the fiber market is a secondary market. While the fleece holds a market value of $24 to $80 per pound, the greatest profit in the industry is from the sale of breeding females.

The Llama and Alpaca Industry

Recent registrations place the llama population in the United States to be nearly one hundred eighty thousand animals and includes pure-bred llamas, and llama crosses with alpacas, guanacos, and vicunas.

Pure bred alpaca registrations are nearly two hundred thousand animals. The total number of alpacas registered more than tripled between 2000 and 2008 and reflected the annual increase in alpaca registrations that had persisted since the mid 1980s. Since 2008, this trend has reversed and annual registries have since declined to one-half 2008 annual registrations numbers. Of alpacas registered, 75% are Huacaya and 17% are Suri type. The remaining registered alpacas are of unknown type. The majority of llamas and alpacas are located along the West Coast; however, Ohio is the most significant, non-West Coast state leading the nation in alpaca registrations with 14% of the registered population. The United States represents only 6% of world alpaca numbers, which are estimated at 3.5 million head. Nearly 90% of alpacas are found within Peru, with the remainder in Chile and Bolivia.

In the United States, llamas and alpacas provide limited sources of income with the greatest profits earned through live sales. The price range for llamas and alpacas is variable and dependent on age, sex, pedigree, and origin. Imported animals demand a greater price due to increased genetic potential. The market value of llamas range from $200 to over $5000 and is greatest for miniature llamas. The market value of alpacas is further influenced by fleece color and quality. In 2008, the average price paid per registered breeding female was $25,000. Gelded alpacas for companion purposes are sold at a much lower price averaging $1000.

Defining Characteristics and Behaviors

Both the llama and alpaca are well adapted to high altitudes. The adaptation is a consequence of altered erythrocyte physiology. The blood of camelids contains more red blood cells per unit volume than other mammals due to a flattened, elliptical shape. In addition, the hemoglobin reacts more readily with oxygen and has a greater oxygen carrying capacity, and the red blood cell life span averages two hundred thirty-five days compared to the average one hundred day life span observed in other mammals. These unique features facilitate the absorption of oxygen from the relatively oxygen depleted environment encountered at higher altitudes.

Llamas and alpacas are herd animals and most content living with other members of their species. If the animal is to live alone, it will seek companion-

■ **Fig. 14.7** Llamas in Machu Picchu, an Incan site located eight thousand feet above sea level. The unique ability of the red blood cells of llama and alpacas to absorb oxygen in oxygen depleted environments allows these animals to survive at high altitudes. (© steve estvanik, 2009. Under license from Shutterstock, Inc.)

ship from other animals. The herd mentality is suggested to contribute to the llamas' success as a guard animal. Llamas and alpacas respond well to training and are generally receptive to handling. By nature, the animals are intelligent, gentle and curious about their surroundings. However, llamas are territorial and use both body language and vocalization to communicate. A humming sound is the first sign of discomfort and can be followed by a shrill alarm call. Perhaps one of the most commonly known, and often misperceived, facts with regards to llamas is their ability to spit. Although llamas are capable of spitting, it is rare that it is directed at humans unless the llama is improperly socialized. For example, in adult

males Berserk Male Syndrome may manifest as a consequence of the male being bottle fed and handled excessively by humans during early development. Spitting is more a means of determining male dominance and establishing social hierarchy and is used by the female to deter unwelcomed breeding partners.

Health Management

There are relatively few diseases characterized within the New World camelids. Llamas and alpacas are susceptible to the same internal and external parasites that affect other livestock species. Preventive measures, such as bi-annual deworming, proper nutrition, vaccinations, and proper sanitation practices should be routinely followed. If the llama or alpaca is kept on naturally soft terrain which does not wear down their hooves, hoof trimming may be necessitated to avoid lameness. In addition, males require the removal of their fighting, or canine, teeth around age two when the teeth erupt. The removal of these teeth reduces injuries within the herd and to the handler. Females also have fighting teeth; however, the teeth are smaller in size and are not considered a threat.

In moderate climates, llamas and alpacas require little housing because of their ability to both maintain and dissipate heat effectively. Some protection should be provided from inclement weather and strong winds. Often, a three sided structure is adequate and can be used as a source of shade in the warmer months. In general, llamas and alpacas are easy to care for because of their hardy nature. However, these animals are generally apathetic when an illness does present and therefore illness is often difficult to detect. For this reason, owners should carefully observe their stock in order to identify changes which might warrant intervention.

Capacity of Production

Unlike other mammalian livestock, llamas and alpacas do not have estrous cycles, but are induced ovulators. Induced ovulation allows the animals to be bred throughout the year so long as the females

■ **Fig. 14.8** Mating in both the llama and alpaca species takes place with the female in sternal recumbency. (© Nicolas Raymond, 2009. Under license from Shutterstock, Inc.)

■ **Fig. 14.9** Crias remain with the female until weaning at six to eight months of age. (© nouseforname, 2009. Under license from Shutterstock, Inc.)

are not gestating. The females will ovulate twenty-four to thirty-six hours after copulation. The actual act of mating can last anywhere from five to fifty-five minutes, averaging eighteen minutes for llamas and twenty minutes for alpacas. For this reason, the male will inspect the environment for any potential threats before mating. Once the male deems the environment safe, he begins to orgle in an effort to attract the female and the male continues this gurgling/snorting vocalization during copulation. Mating occurs with the female situated in sternal recumbency in which the female lies with her abdomen on the ground and her legs positioned beneath. The male rhythmically alternates and deposits semen in both horns of the uterus. Ovulation is stimulated by the seminal fluid of the male, which triggers the release of luteinizing hormone in the female.

The gestation period averages 350 days for llamas and 335 days for alpacas. Females are ideally rebred fourteen to twenty-one days after the birth of the cria. Longer rebreeding intervals reduce the chance of subsequent pregnancies. In her lifetime, a female may produce ten to twelve offspring. The females give birth standing and parturition typically occurs within the morning hours. In her native environment, birth during early daylight hours allows the cria to dry, stand, and nurse before the onset of the cooler night time temperatures of the higher altitudes, which contributes to a greater chance of survival for the offspring. Females also are reported to delay birth under unfavorable conditions. Singleton births are common, whereas twinning is rare, occurring only once in every two thousand live births.

Llamas as a Food Commodity

Llamas are an important source of meat in South America. In Bolivia, llama meat consumption increased 76% between 1985 and 2004. Llama meat is high protein, low in fat, and low in cholesterol compared to other red meats consumed, including beef and sheep. Furthermore, 42% of fatty acids are found as monounsaturated fatty acids and 7% as polyunsaturated fatty acids. Llama milk has reduced calories and fat compared to cow milk and increased concentrations of lactose. Although protein concentrations between cow and llama milk are similar, the latter does not contain detectable concentrations of β-lactoglobulin. The apparent lack of β-lactoglobulin makes llama milk an attractive alternative to cow milk for humans suffering from milk allergens as this protein is suggested to play a role in the allergen response. Llama milk is also greater in calcium with reduced concentrations of sodium when compared to cow milk. Despite the benefits of llama meat and milk, the current United States industry does not permit affordable production of either of these commodities, which is unlikely to change due to the high investment cost of purchasing animals.

■ **Fig. 14.10** A Bolivian couple prepares to slaughter a llama in order to sell the meat and skin. (© Carlos Cazalis/Corbis.)

Aquaculture

"Give a man a fish, you feed him for a day. Teach a man to fish, you feed him for a lifetime."

—*Chinese Proverb*

This chapter introduces aquaculture, the farming of organisms in controlled aquatic environments. The aquaculture industry in the United States is discussed along with the rudimentary production techniques for rearing aquatic organisms in fresh water and marine environments. Prominent species produced in the United States and their specific managerial requirements are highlighted. Lastly, the current challenges experienced by the industry and the nutritional benefits of fish are explored.

Aquaculture: A Historical Perspective

The cultivation of aquatic organisms under controlled or semi-controlled conditions has been practiced for centuries. Historians trace aquaculture to Egypt, with the culture of Tilapia, and China, with the culture of carp, occurring as early as 2000 B.C. in both regions. The Chinese are responsible for the first detailed writings of fish culture in a book titled *The Classic of Fish Culture* (500 B.C.). The text offered the first written description of pond structure and methods of propagation of common carp. Oyster farming by the Japanese and Romans trace to 100 B.C., and for many societies aquaculture was considered an extension of the agriculture practiced on land. Ancient forms of aquaculture involved harvesting of immature fish and shellfish from marine environments and continued rearing in confined systems. The first modern form of aquaculture was practiced in 1733 when a German farmer collected and fertilized fish eggs, and grew-out the hatched fish.

As early as the 1870s, some native aquatic animal populations were depleted or in decline in the United States. The United States Fish and Fisheries Commission was established in order to investigate and combat this issue by developing technology to mass produce, transport, and stock marine and freshwater fish and shellfish. However, early efforts in aquaculture within the United States were aimed primarily at the development of sport fishing. Interest and expansion of recreational fishing in the United States led to the creation of hatcheries that were responsible for stocking, or introducing, fish into lakes, rivers, or ponds to establish or enlarge fish populations. Post War World II, the commercial bait industry was established and by the 1950s the catfish industry began to take root. Important aspects relative to the aquaculture industry had not been defined at this point in time, including a fundamental understanding of feeds, diseases, and water quality requirements. It was not until the later part of the twentieth century that a scientific approach was brought to the field.

Early attempts of aquatic farming often failed as operators were inexperienced in fish culture, ponds were not properly built, low-value species were being raised, and there was limited technical support for the industry. This was to change in the 1970s, which was marked as an era of great advancement in aquaculture as government laboratories and universities became interested in and initiated efforts to understand management, nutrition, genetics, reproduction, and disease of aquatic species. This new knowledge would be applied to determining sustainable solutions to meeting the demand for seafood. By the 1980s, it was recognized that decades of unsustainable fishing and pollution had contributed to

70% of wild aquatic species being fully exploited, overexploited, depleted, or recovering from depletion. An increasing global demand for animal protein coupled with the static wild fish harvests that have persisted since the 1980s has increased the need for seafood through aquaculture. Today, aquaculture is considered one of the fastest growing sectors of the global food industries, increasing 10% per year and contributing to 51% of total seafood consumed globally and more than 30% of all fish consumed in the United States. In the last two decades, the value of the United States aqua-

culture has risen to nearly a one billion dollar industry. However, the United States is a minor player in the world aquaculture industry.

The Aquaculture Industry

Aquaculture is a means of farming aquatic organisms including fish, shellfish, reptiles, and aquatic plants. Successful cultivations are underway in fresh water, brackish water, and salt water. While the aquaculture industry in the United States was first created to support recreational fishing, in recent decades, it has evolved into an important food supplier to the nation. Growth of the industry has occurred in response to demand for fish and shellfish, concern over the shortage in trade of fishery products, and static commercial harvest of wild fish. Although per capita consumption of fish and shellfish in the United States has remained static at sixteen pounds per year, Americans consume nearly 4.9 billion pounds of fish and shellfish annually, with less than one-third of product consumed at home. This equates to $8.2 billion spent on fish and shellfish consumption. Approximately one-half of fish and shellfish consumed is imported. The United States ranks third as the largest consumer of fish and shellfish, following China and Japan, and is the second leading importer. The United States imported

■ **Fig. 15.1** Dam construction for a fish rearing pond, 1936, Michigan. (Library of Congress.)

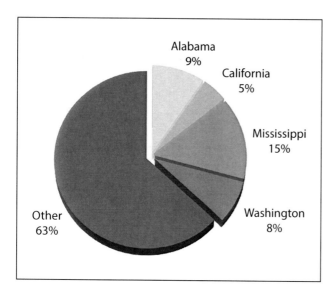

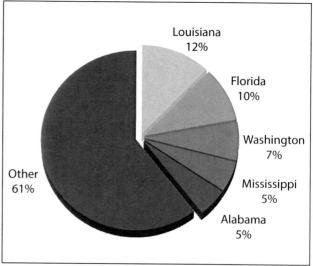

■ **Fig. 15.2** Food fish sales and distribution as percent of total dollars generated (left) and primary aquaculture farm locations (right) by state during 2012. Farm location reflects number of farms reporting sales across and includes all aquaculture farm types. (United States Department of Agriculture-National Agricultural Statistics Service.)

nearly $9.6 billion in fish and shellfish during 2013, this was a 75% increase in import value recognized a decade prior and equated to nearly 2.5 billion pounds. The top three imports are shrimp, Atlantic salmon, and tilapia representing 13, 7, and 5% of total imports. Although the United States does export food fish, the export market is small in light of domestic demand. By 2020 aquaculture is expected to provide the total supply of the top three fish and shellfish currently imported into the United States.

The United States Department of Agriculture Census of Agriculture in 2012 reported that the aquaculture industry in the United States contributed over 1.5 billion dollars to the economy with 55% attributed to the production of food fish. The top aquaculture states that reported over 100 million dollars in sales and distribution included Mississippi, California, Alabama, Louisiana, and Washington. The production of catfish, trout, salmon, tilapia, shrimp, mussels, clams, oysters, hybrid striped bass, crayfish, ornamental fish, and other fish species occured on 5530 aquaculture farms. Although the United States is considered the fifth leading nation in aquaculture contributions to the world market, the United States accounts for less than 1% of total world production. In recent years, the United States aquaculture industry has remained static despite national demand for fish and shellfish. China is the leading nation in aquaculture production holding

61% of production; other leading nations include India (7.8%) and Vietnam (4.5%).

Aquaculture Production Methods

Management techniques in aquaculture vary in labor intensity and those who participate in this industry range from the backyard hobbyists to the producer whose income rests with the success of the farm. When done properly, raising fish, shellfish, or other aquaculture species can be very efficient. Aquaculture species require less space in relation to land surface area because farms consist of three-dimensional rearing systems with the added dimension of depth in which to grow. In raising fish, fish are cold blooded and expend less energy on maintaining body temperature. Thus, fish are the most efficient at converting feed to flesh, even more so than poultry. This contributes to the high feed to gain ratio of fish. Even the feed that the fish eat lend to their efficiency as they are capable of utilizing wastes and waste by-products. In addition, fish have greater flesh to bone ratio than do land livestock. Accordingly, there are advantages of aquaculture production over land based agriculture; however, the industry lags behind land based agriculture technologically, which limits its expansion. The range and methods of aquaculture production greatly

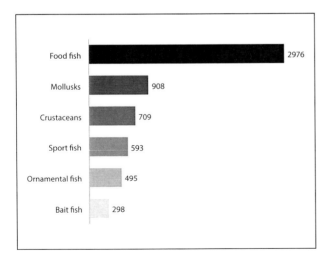

■ **Fig. 15.3** United States aquaculture farm type and number during 2012. (United States Department of Agriculture-National Agricultural Statistics Service).

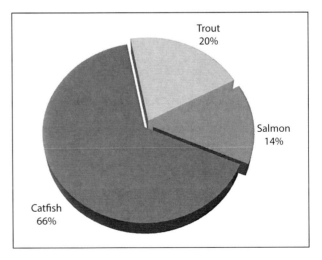

■ **Fig. 15.4** Top three food fish produced in the United States during 2013. (United States Department of Agriculture-Economic Research Service, National Oceanic and Atmospheric Association.)

varies according to environmental and social conditions and is greatly dependent upon the species raised. For these reasons, aquaculture has been molded into an exceptionally diverse enterprise and continues to be developed according to emerging discoveries. Although improvements in aquaculture are needed, there are some underlying commonalities in the production and management of aquaculture that has contributed to its current success. Aquatic animals and plants can be produced by utilizing naturally occurring water features or by artificially creating water reserves. These man-made structures can be earthen or concrete. In a natural freshwater or marine setting, fish pens, nets, or cages can be used as a means of raising and harvesting organisms. Depending upon the species and available resources, certain management systems are better suited to the environment.

Ponds

Ponds are the most prevalent method of production, particularly in the United States, and are the primary method of catfish production. Ponds may consist of spawning, fingerling, and finishing. Site location is important to ensure adequate fresh water, protection from flooding, and ease of drainage. Pond size of ten to twenty acres is common with a depth slope from three to six feet. Most ponds are rectangular in shape and smaller pond sizes are favored for ease of feeding.

Flow through Raceways

Raceways consist of a series of long, narrow channels through which there is continuous flow of fresh, oxygenated water. Trout are routinely produced using this system. Dimensions of raceways vary, but often retain a 30:3:1 ratio for length:width:depth. Raceways are constructed with an 8% to 10% slope to accommodate gravity driven water movement. The continuous movement of water provides aeration and removes waste. The facility costs of raceways are increased compared to other systems; however, these costs are reduced when water movement is supported by gravity and does not require manual pumping. Raceways may be constructed in conjunction with water recirculating systems in which water that has passed through the system is pumped to a processing facility that removes waste so that the water may be reused.

Recirculation Tanks

Recirculation tanks are an alternative system that supports greater control of water temperature and quality and do not require an existing water supply or land availability. The system has been widely used in research facilities, but use in aquaculture is limited. Primary use is restricted to broodstock production. The system consists of a filtration system for removal of solid and biological wastes and the addition of oxygen, a tank, and a pump system that connects the

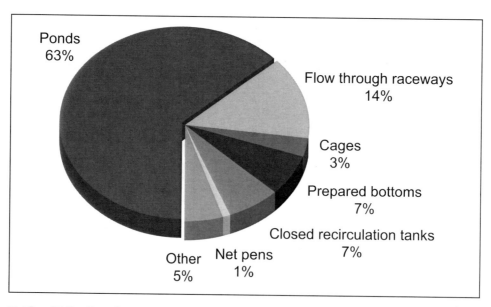

■ **Fig. 15.5** Leading methods of aquaculture production. (United States Department of Agriculture.)

tank and filtration system. Tanks range in size from a few hundred gallons to one hundred thousand gallons. The method recirculates 90% of the water contained within a closed loop system. Water that circulates through the system undergoes mechanical filtration of the solid waste and subsequent biological filtration for removal of metabolic wastes and gasses. With biological filtration ammonia and cabon dioxide are removed. Before water is reintroduced to the production tank, oxygen is added through aeration. Recirculation tanks are confined to indoor systems and as water temperatures can be monitored and adjusted to the needs of fish, the system is not influenced by fluctuating environmental temperatures. The system also offers improved disease and predator control; however, high-capital investment, increased maintenance costs, and increased management limits their commercial current use.

Net Pens

Net pens involve the suspension of nets, or cages, in existing water resources. The system is routinely used in marine culture of salmon and the freshwater culture of crustaceans. Water exchanges freely through the system; however, as stocking density is increased, the incidence of disease is increased. Furthermore, free exchange of water and waste from the system to the natural environment has raised concern over the ecological impacts of these commercial systems. In comparison to ponds and raceways, start-up costs are reduced as existing water resources are utilized.

Aquaculture Systems

Catfish

There are thirty-seven species of North American catfish; however, only seven species are suitable for

■ **Fig. 15.6** Ponds for catfish production in Mississippi. Ponds are above ground, formed from constructed levees. (© Jim Wark/AgStock / Corbis.)

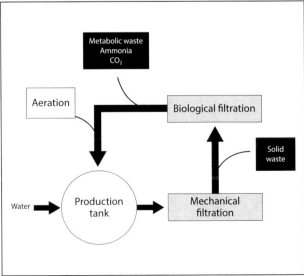

■ **Fig. 15.7** Recirculation tanks (left) use a closed loop system of filtration (right) for recirculation of existing water. (*Left:* © Jan Kaliciak, 2009. Under license from Shutterstock, Inc.)

■ **Fig. 15.8** Net pens are used in the culture of salmon and rely on existing water resources. Pens are commonly netted at the waters surface to deter predation, which contributes to major economic losses annually within the industry. (*Left:* © Konstantin Karchevskiy, 2009. Under license from Shutterstock, Inc. *Right:* © Kheng Guan Toh, 2009. Under license from Shutterstock, Inc.)

■ **Fig. 15.9** Brown Bullhead catfish. (© David Dohnal, 2009. Under license from Shutterstock, Inc.)

commercial production including blue, white, black bullhead, brown bullhead, yellow bullhead, flathead, and channel. Channel catfish are the most farmed and are the predominant aquaculture species farmed for the United States aquaculture industry. Efficient growth of catfish is dependent on water temperatures and optimal at temperatures that average 85° F. Metabolic rate, and therefore appetite and feed consumption, is reduced when water temperatures are decreased. At temperatures of 45° F an additional three to six months of growing time is required. For this reason, catfish production is mainly confined to Alabama, Arkansas, Mississippi, and Texas. Currently, Mississippi produces over half of all catfish grown in the United States and greatly contributes to the United States claim of the world's largest catfish producer.

Catfish production begins with domestic broodstock. Successful spawning requires fish of three years of age. Male and female catfish are stocked at ratios of 1:1 to 1:3 male to female in brood ponds in which spawning containers are placed. Females will spawn annually within the spring once temperatures reach a consistent 75°F. Fertilized eggs hatch within five to eight days. Hatching may occur within the brood ponds or fertilized eggs can be removed to hatchery facilities. After hatch the fry will rely on the nutrients of their yolk sac for the initial four to five days, after which the fry begin to surface and are fed. Fish that reach three inches in length are referred to as fingerlings and transferred to earthen ponds. Fish may be transferred one additional time before harvest at one to one and one half pounds in weight at a targeted age of eighteen months. Growth rates are increased in commercially raised fish in comparison to fish harvested from natural waters, which reach an average weight of one pound by twenty-four months of age.

The type of feed is especially important in catfish production and contributes to the increased growth rates of commercial fish. Many types of feed are available and classified as sinking, slow sinking, or floating. A primary factor in determining which feed is the most suitable is water temperature. Ideally, a floating feed is most desired because it allows the producer to observe feeding rates and patterns. However, floating feed is expensive and can only be fed in waters that are above 65° F. If the water is below this temperature, the catfish remain in the deeper depths of water and sinking or slow sinking feeds must be used. In order to reduce costs,

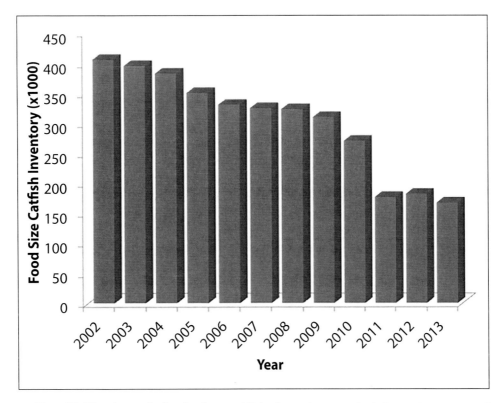

■ **Fig. 15.10** Annual food size catfish inventory (United States Department of Agriculture-Economic Research Service.)

most producers feed only once per day, usually in the morning. However, feeding schedules vary depending upon the season as some producers may choose to convert to every other day feedings in the fall and spring. A mechanical blower is commonly used to distribute the feed widely across the entire water surface to maximize the feeding opportunity.

Although catfish production is the largest segment of the United States aquaculture industry the industry has been in a state of decline since peak production in 2002. Food size fish inventory has decreased 2.6 fold and pounds of fish sold to processors has decreased 2.2 fold since peak production. Reduced profitability associated with high feed costs, competition with other food fish choices (catfish ranks ninth in the top ten food fish species consumed within the United States), infrastructure concerns, and lower price for imported product contribute to uncertainty in the future United States catfish industry. Canada is the leading importer of catfish from the United States, and efforts are underway to expand the current export market.

■ **Fig. 15.11** Rainbow trout. (© Evok20, 2009. Under license from Shutterstock, Inc.)

Trout

Raising trout is the oldest form of aquaculture in the United States and dates to the 1800s. Trout farming was originally developed as a way of supplementing the natural streams, lakes, and rivers with a greater population of recreational fish. Recreational fishing remains a market for farm raised trout today. Rainbow trout are the most popular raised species with Brook and Brown trout also being raised. Rainbow trout used as food fish are domestic strains developed for intensive farming and are raised in all female monoculture systems. Twenty states are significantly involved in trout production; however, Idaho is the most prominent trout producing state and accounts for over 50% of the total value of fish marketed. North Carolina ranks second in trout sales, followed closely by Pennsylvania. Prevalence of trout production in Idaho is attributed to the availability of natural springs. Considered greatest in overall water quality, natural springs contain reduced concentrations of dissolved oxygen and increased concentrations of dissolved gasses. Accordingly, water conditions must be closely monitored to ensure adequate oxygen concentrations and minimize disease, which is especially crucial to trout producers as disease is the leading cause of financial loss and accounts for 86% of total losses. The optimal temperatures for achieving the greatest growth rates are between 55 to 65° F. Temperatures below 38° F reduce growth as the fish fail to eat.

At temperatures above the optimal temperature range, reduced nutrient utilization by the digestive system occurs resulting in increased waste and nutrient loading of the water. The accompanying reduction in water quality compromises the health of the fish. Feeding rates must be monitored routinely and adjusted with changing water temperatures for optimal trout production. To this end, trout are routinely fed through demand feeders, which consist of hoppers located above the surface of the water that hold the feed pellets and a movable disc that extends into the water by attachment to a pendulum. Trout are trained to feed themselves by striking of the movable disc to releases feed into the water. The advantages of demand feeder, in comparison to hand feeding, include reduced labor and reduced fluctuations in dissolved oxygen saturation of the water that can occur with inefficient feed utilization associated with overfeeding. Disadvantages may include overfeeding and associated reduction in water quality if the feeders are not properly adjusted.

Tilapia

Worldwide, tilapia production is second in importance only to the production of carp and is the most influential of farmed species, farmed in more countries than any other aquaculture species. The affordable price and mild flavor contributes to the demand of tilapia, but domestic production fails to meet demand. Improved strain development and

technology are needed for domestic tilapia markets to be globally competitive, and tilapia is one of the leading aquaculture import products. As a member of the cichlid family, tilapia are native to Africa and intolerant to low temperatures. Tilapia stop feeding at temperatures below 65° F, and the fish are capable of withstanding temperatures of 50 to 52° F for only a few days without pursuant death. Optimal growth occurs at temperatures of 85 to 88° F and at reduced temperature of 72° F, growth is delayed three fold. Such a restriction only allows certain regions of the country to produce the fish outdoors. Furthermore, outdoor production is highly regulated to minimize the potential risk for contamination of native waters with this introduced species. Tilapia production in northern climates is confined indoors. Hawaii, and Florida dominate in total number of tilapia farms, whereas California ranks first in tilapia sales. When optimal temperatures are maintained, tilapia are easily managed and can withstand poor water quality with dissolved oxygen concentrations considerably below the tolerance limits for other fish. They are capable of digesting natural food organisms including plankton and decomposing organic matter, which may contribute to 50% of tilapia growth.

The tilapia industry typically conforms to two types of management systems, the mix sex culture and the monosex culture. In the mix sex culture, the males and females are reared together and harvested at the same time, once sexual maturity is reached. In the monosex culture, the males are separated from the females to allow them to attain a greater growth potential. All male culture is desired as male growth rates are two fold greater than females. In the monosex culture, males reach one pound in weight in five to six months

■ **Fig. 15.12**   Tilapia. (© Olga Lyubkina, 2009. Under license from Shutterstock, Inc.)

Salmon

There are six species of Pacific salmon (Sockeye, Pink, Chum, Coho, Chinook, and Masu) and one species of Atlantic salmon. In the United States, all farmed salmon is Atlantic salmon and over 90% of consumed Atlantic salmon is farmed. In contrast, over 80% of Pacific salmon consumed is wild caught, the remaining farmed species consisting primarily of Chinook and Coho salmon. Washington and Maine are current United States leaders in Atlantic salmon production, with Maine representing an area where Atlantic salmon occurs naturally. Salmon aquaculture is important in sustaining wild salmon populations, whose natural spawning is limited due to loss of habitat, and for supplying food fish. Salmon aquaculture, however, remains controversial over concerns that hatchery produced salmon are detrimental to wild populations. In Alaska, farming of salmon for food is banned as potential threats of disease, pollution, and interbreeding between farm raised and native, wild stock are considered to great of a risk.

Salmon are anadromous, maturing and spending the majority of their life in the ocean, but returning to fresh water to breed. Successful rearing relies on a two phase system of production: hatching and early growth in fresh water and the transfer of fish to marine net pens for continued growth until harvest. The fresh water phase may last between six months and two years. Milt and roe are harvested from commercially raised adult salmon. Once the eggs are fertilized they are transferred to hatcheries that are designed to mimic the natural spawning nests of salmon. Oxygenated water maintained at 39 to 47° F percolates through the fertilized eggs, supplying oxygen and removing waste. Once hatched, the alevins, which retain a yolk sac, will remain in the incubators. Once the yolk sac is depleted, the alevin will swim to the surface for feeding. At this stage they are referred to as parr and moved to fresh water tanks for feeding. The immature salmon will remain in fresh water tanks until the smolt stage is reached at approximately eighteen months of age, at this time the smolt are transferred to marine net pens that average nearly one hundred feet in diameter with a depth of sixty-five feet and are anchored to the ocean floor. After twenty-four months in the marine net pens, the salmon are harvest at within a weight range of four to ten pounds.

Shellfish

Crawfish is the most established of the shellfish industries in the United States, dating to the late 1800s. Louisiana accounts for 95% of domestic crawfish produced and is the world leader in crawfish production. Of the thirty-two species of crawfish identified in Louisiana, there are currently only two types farmed: red swamp and white river crawfish. Crawfish are farmed in woodland or rice-field ponds, with rice pond production dominating the industry. Ponds range in size from ten to twenty acres and may be permanent, in which only crawfish are produced, or rotational, in which crawfish production is commonly rotated with rice production. Permanent ponds rely on hold-over crawfish from the prior year as a source of broodstock, whereas continual restocking of broodstock is necessary in rotational pond systems. Ponds must be managed to mimic the wet and dry periods experienced in natural habitats. Thus ponds are flooded from autumn to spring and drained in the summer. Crawfish are capable of breeding year round, however, peak breeding season occurs during the spring wet-cycle. Sperm is transferred to the female, after which the female will burrow and within the burrow the eggs are fertilized and maintained. Crawfish over-summer within the burrows. It is not until autumn flooding that the crawfish will emerge. Summer drainage further promotes the establishment of vegetation that will provide a food source during the wet cycles. Unlike many other farmed species, crawfish do not receive supplemental feed. Market size is attained after three to fourth months of growth. Crawfish are harvested using baited traps and live sales dominate the market. Although crawfish are the dominant shellfish produced within the United States, national per capita consumption is less than one-quarter of a pound per person. However, local consumption exceeds ten pounds per person and attributes to 70% of crawfish produced within the state being consumed within the state. In 2012,

■ **Fig. 15.13** Atlantic salmon. (© Tischenko Irina, 2009. Under license from Shutterstock, Inc.)

110.9 million pounds of Louisiana crawfish were marketed at a value of $168.5 million.

Mollusks

Mollusks, including clams, mussels, and oysters are the largest segment of marine aquaculture. There are over four hundred species of oyster throughout the world, with only seventy species edible. The native Eastern oyster and non-native Pacific oyster are the two of commercial importance in the United States. Both Chesapeake Bay oysters of Maryland and Virginia and Gulf Coast oysters of Louisiana are of the Eastern variety, whereas oysters of Washington are Pacific oysters. Approximately 60% of total oyster harvest is attributed to cultivation of oysters using aquaculture practices. Oyster cultivation uses hatchery produced or wild-caught broodstock that are used for the production of immature or seed oysters that are subsequently transferred to estuarian or tidal water habitats.

The most simple of oyster cultivation involves providing a suitable substrate for cementation of larvae where oysters naturally spawn. Previously harvested oyster shells are the most common substrate; however, clam shells may be used as well. More intensive oyster production involves spawning oysters and hatching larvae within tanks. Both hatchery produced and wild-caught broodstock are used for the production of immature spat or seed oysters. Oysters spawn between temperatures of 60 to 68° F. Sperm and eggs are released and fertil-

ization occurs within the water. Fertilized eggs develop rapidly and the resulting larvae will feed on algae. As the larvae feed, they become increasingly heavy and can no longer be supported by the water. Consequently, the immature oyster will settle and find a suitable substrate in which to attach and become cemented and grow to maturity.When ready to set, the larvae are transferred to tanks containing substrate for cementation. Set oysters, known as cultched, must be transferred to natural waters for the grow-out phase, which are often estuarine or tidal waters, though oysters may be maintained in managed tanks until maturity. Transferred oysters are cultured using either bottom or off-bottom practices. Bottom culture relies on transfer of cultched oysters directly to firm estuarine bottoms. Off-bottom refers to suspended techniques where racks, trays, or stakes are used for grow-out of cultched oysters. Off-bottom culture may also use floating docks for oyster suspension. Off-bottom culture is the preferred culture of single oysters that are used in the freshly shucked market. Oysters reach harvest size between twelve and thirty-six months, the range being dependent on natural water conditions. and breeding. Triploid oysters, which contain an extra set of chromosomes as a result of breeding between tetraploid and diploid oysters, reach market size sooner. The triploid oyster is considered a summer oyster. As it is sterile and incapable of spawning, the oyster remains highly edible throught the year, eliminating the addage that oysters should

■ **Fig. 15.14** Off-bottom oyster beds provide the substrate for the culture of oysters in their traditional aquatic environment. (*Left:* © Christian Musat, 2009. Under license from Shutterstock, Inc. *Right:* © Arwyn, 2009. Under license from Shutterstock, Inc.)

only be consumed during months that contain the letter R.

Health Management

Many aspects contribute to a healthy environment which allows aquatic species to flourish. No organism is more dependent on their environment than those living in water. Water quality, oxygen content of the water, and pH levels are all crucial factors that attribute to a healthy environment. Correct balances and attentiveness to these factors help to reduce the stress placed on the aquatic organisms in culture. When an animal is stressed, it is significantly more susceptible to disease. Stress can be brought on by rapid changes in water temperature, exposure to low oxygen levels, a fluctuation in pH, poor nutrition, deficient water quality, handling, overstocking, and pollutants. Signs that a fish is stressed include listlessness, decreased feeding, loss of the fish's internal equilibrium and erratic swimming, excessive bottom resting, and rubbing against objects.

A disease is most likely to manifest itself in the immediate days after the stress stimulus has occurred. Decreasing stress greatly correlates to a decrease in disease among the stock. Disease control is especially relevant to the aquaculture industry because of the concentrated nature in which fish are kept. Disease can be transmitted rapidly from organism to organism. Illness may originate form bacteria, viruses, fungi, and parasites. A change in fish behavior, lesions, or increased mortality rates are indications of a potential disease problem. The earlier the disease is identified, the greater the likelihood that the disease can be controlled. Chemicals, antibiotics, and vaccines help to control and prevent most outbreaks. Medications can even be distributed through the feed. However, all drugs administered to the fish are closely regulated by the United States Department of Agriculture and must be in compliance with established standards.

Proper sanitation practices can greatly reduce the risk of a disease epidemic. Supplies that are used in one tank should not be used in another unless sanitized between usages. People also can introduce diseases and some facilities may require special clothing to be worn. Iodine foot baths are common to disinfect the bottoms of shoes before

■ **Fig. 15.15**   Members of the cichlid family protect the offspring through mouth brooding. Many of the tilapia species are maternal mouth brooders, in which the female gathers and incubates the eggs in her mouth either before or after fertilization. During mouth brooding the female will seldom eat and weight loss during the incubation period is common. Tilapia species commonly used as food fish extend mouth brooding to the fry stage and must be forced to release the developing fry for production rearing. (© Mitch Aunger, 2009. Under license from Shutterstock, Inc.)

entering a culture chamber room. The strictest facilities are regimented around a color system of sanitized, coverall clothing. In an effort to avoid cross contamination, no person is allowed to enter a room unless wearing the corresponding color.

Capacity of Production

Reproduction in aquatic animals varies by species. All species produce sexually. Most aquatic species comply with a male/female classification. Some however, are hermaphrodites and possess both male and female sex organs. If a fish has both functioning gonads at the same time, it is known as a *synchronous hermaphrodite*. If one gonad functions at the beginning of life and then alternates to the other, opposite gonad, the fish is known as a *sequential hermaphrodite*. In addition, the sex of aquatic species is not determined at birth and can be influenced by environ-mental factors, including temperature, in the early life stages.

The actual reproduction process also varies greatly in aquatic species. Most fish lay thousands

Fig. 15.16 Consumption of fish has remained static at sixteen pounds per person annually in recent years despite the recognition of the health benefits afforded by adding fish to the diet. (© Dusan Zidar, 2009. Under license from Shutterstock, Inc.)

of eggs, less than a millimeter in diameter. For example, the adult catfish lays approximately three thousand eggs per pound of body weight or eighteen thousand eggs for a mature female. Many fish deposit their eggs into the water current and do not provide parental care to the fertilized eggs or resulting larvae. Although, some fish are known to prepare a spawning site and guard the fertilized eggs until hatch, as is the case with male catfish. Some species of female tilapia brood the eggs in their mouth. In addition, fish may spawn multiple times per year, once per year, or in the case of the Pacific salmon, once in their lifetime.

Challenges to Industry Expansion

There is certainly room for expansion within the United States aquaculture industry as the country is a leading global importer of edible fish and shellfish. However, expansion poses challenges to the industry, especially in marine segments. Many environmental groups are closely guarding native waters and are fearful of ecological changes that might ensue if the marine aquaculture industry expands. These concerns include worry over increased diseases that could devastate natural fish populations, genetic alteration to existing natural species as farmed species escape commercial enclosures and interbreed, pollution from concentrated rearing systems, and decreasing aesthetics. Commercial fishermen are wary of expansion as some perceive aquaculture as competition. Concerns of the effects of aquaculture sites on the navigation of the seaways also have been voiced. Aquaculture is regulated at the federal and state level. The primary federal agencies include: the Food and Drug Administration, Department of Health and Human Services, Department of Agriculture, and the Environmental Protection Agency. Additional federal agencies provide further input governing the regulation, research, and expansion of the industry. Until a consensus is reached on policy and regulation, growth in the industry will be limited.

Fish as a Food Commodity

Like other animal meats, fish is a nutritious food that contains vitamins, minerals, and high value protein. However, fish contains fewer calories, less fat, and less cholesterol than do other meats. It is a general concept that fish is a healthy choice and can be a beneficial addition to a balanced diet. The exact nutritional makeup of fish varies by the species consumed; however, many fish contain little, if any, saturated fats. Some fish species have a high oil content that can be beneficial to the consumer. Contained within this oil are omega three fatty acids, which can help to reduce blood choles-terol and improve heart health. Some fish are also a plentiful source of natural iodine and fluorine in addition to magnesium, phosphorus, iron, and copper. Because of these components, in addition to its high quality protein and its low levels of fat and cholesterol, some nutritionists suggest replacing a portion of traditional meats with fish.

Welfare

"The greatness of a nation and its moral progress can be judged by the way its animals are treated."
—Mahatma Ghandi (1869–1948)

Animal Welfare: A Historical Perspective

There is no universally accepted definition for animal welfare. In general, animal welfare represents the view that animals under human care should be maintained under conditions that do not pose unnecessary harm. Concern over animal welfare dates to 2000 B.C. and the Indus Valley Civilization. There was a religious belief that ancestors return to earth in animal form, and therefore animals must be treated with respect by humans. Throughout the world, many cultures have developed codes, laws, and regulations that protect animals. In 1876, the Cruelty to Animals Act was established and created a central governing body that reviewed and approved all animal use in research. In 1966, the United States Animal Welfare Act was enacted and was the first federal law protecting the welfare of laboratory animals. Subsequent amendments to the Act have expanded coverage to animals in commerce, exhibition, teaching, and testing. In part, the Animal Welfare Act was initiated following public outcry after the release of stories of dogs used in biomedical research originating from theft, and the unsuitable living conditions of dog facilities that supplied animals for research. The initial passing of the Animal Welfare Act set minimum standards for the handling, sale, and transport of cats, dogs, rabbits, nonhuman primates, guinea pigs, and hamsters. The amendment of 1976 ensured proper handling during transportation and established standards for feed, water, and temperature during transport. In 2002, the definition of animal recognized by the Act was expanded to include mice, rats, and birds, except those bred for research. The 2002 amendment also introduced language that began restriction on animal fighting ventures. In

■ **Fig. 16.1** Concern over the living conditions and sources of animals housed for research prompted the enactment of the Animal Welfare Act in 1966. (© Jose AS Reyes, 2009. Under license from Shutterstock, Inc.)

2007, the buying, selling, or transportation of instruments used in animal fighting was prohibited through the Animal Welfare Act, and in 2008, the possession, training, or advertising of animals for fighting was prohibited.

Ethics

Ethics are a system of moral values that provide the basis for the determination of what a person believes is right or wrong. Thereby, this system provides a set of principles that outline acceptable conduct and ultimately underlies an individual's decision to do what is acceptable with respect to animal welfare. An ongoing moral dilemma that has been under consideration for hundreds of years is whether humans should raise and kill animals for food; and if so, how should animals be raised and killed. While many people, especially those of agrarian societies see the raising and harvesting of animals for meat as a perfectly acceptable practice; it is still important to consider the moral and ethical implications of raising and killing animals for human consumption.

Ethics are intertwined with animal welfare and animal rights. As rational beings, with the ability to reason and form a system of moral beliefs, the question arises: Do humans have the right to use other organisms for their own benefit?

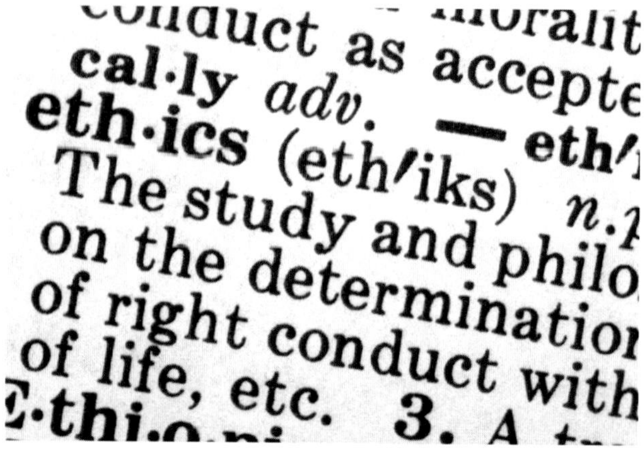

Animal Welfare Versus Animal Rights

Animal welfare is the viewpoint that animals should not suffer unnecessarily and, under human care, should be provided with an adequate environment and adequate provisions to meet their physiological and behavioral needs. Those that support animal welfare believe that it is acceptable to use animals for food, research, recreation, and to keep as companions assuming that the animals are maintained humanely. As defined by the American Veterinary Medical Association (AVMA), animal welfare is,

> A humane responsibility that encompasses all aspects of animal well-being, including proper housing, management, nutrition, disease prevention and treatment, responsible care, humane handling, and when necessary humane euthanasia.

■ **Fig. 16.2** Ethics, an individual's values of right and wrong, underpin the decisions to accept or reject the practices of animal use. (© robyn Mackenzie, 2009. Under license from Shutterstock, Inc.)

On the other hand, animal rights is the belief that animals have intrinsic rights to life and liberty, just as humans. Animal rights theorists state that nonhuman animals should be allowed to live according to their own nature, free from harm, abuse, and exploitation by man. Support for animals rights leads to the exclusion of animals for farming, research, and entertainment. As there is no uniformly accepted definition of animal welfare, there is no universally accepted definition of animal rights. Amongst supporters of animal rights, the defining line that distinguishes between which animals deserve these rights and which animals do not is debated. Within the animal rights movement advocates may be categorized, either as reformists or abolitionists. Animal rights abolitionists are more extreme in their viewpoints and advocate the total abandonment of any animal use. Reformists are generally less extreme in their viewpoints and are usually willing to work within the system to change the methods of animal use.

Approaches to Animal Welfare

In order to understand and be informed of the impact of animal care practices on the physiological and behavioral needs of animals, there must be a framework to assess, study, and define animal welfare. Currently, there are five primary approaches to the study of animal welfare: feelings based, animal-choices, nature of the species, the five freedoms, and biological-functioning based.

Feelings

Feelings based approach defines animal welfare in terms of emotions. Most agree that animals are capable of basic emotions: anger, fear, joy and happiness. Feelings evolved as a means of protecting the basic needs of animals and are suggested to play a role in survival. For example, fear evolved as a flexible means of avoiding danger. The feelings based approach emphasizes a reduction in negative emotions such as pain and fear and an increase in positive emotions such as comfort and pleasure. Although the mental well-being of an animal is considered an essential component of animal welfare, the problem arises with defining emotions. Feelings are subjective and only available to the animal experiencing the emotion. As there is no common language between humans and other animals, animals can not relate their emotions and only by analogy can humans assume the emotions experienced by animals. Emotions are often tied to a bodily response; however, most animals react physiologically in the same manner to positive and negative experiences. Heart rate may increase in response to the presence of a mate or the presence of a predator.

Animal Choices

In the animal-choices approach, the animal is allowed to choose between certain aspects of its environment and the assumption is made that the animal will choose according to what is in its best interests. Choices made by an animal, however, may be influenced by prior experiences and not represent the best interest of the animal. Furthermore, animal choices are difficult to interpret and the choices made may have adverse outcomes. For the wild animal, many choices are based on survival. Without natural selection, how are preferences driven and does this reflect welfare? For example, feast or famine is a survival mechanism of the lion. When prey is secured, the lion feasts and will gorge as food is plentiful and the certainty of the next meal is not guaranteed. Gluttony is important to allow the animal to survive periods of food uncertainty. For the domesticated cat, the instinct to feast may still reside. If the cat is fed *ad-libitum* it may feast; however,

■ Fig. 16.3 Allowing an animal to make its own choice may not be in the animals best interest. (© Sue Smith, 2009. Under license from Shutterstock, Inc.)

without the period of famine to follow, the animal will put on excess weight and obesity may occur. The animal's choice, therefore, is not in its best welfare interests.

Nature of the Species

The nature of the species approach requires the animal to be raised in an environment that allows it to express its complete range of behaviors. In general, this approach states that animals should be raised in their natural environment in natural ways. The term natural is not clearly defined, but is often equated with the wild environment and behaviors displayed by non-domesticated animals. Although a common approach early in the study of animal welfare, it does not consider the modification of behaviors as a consequence of domestication and some behaviors that occur in the wild could be detrimental. For example, pre-weaning pig mortality exceeds 20% in wild pig populations; however, mortality rates are less than 10% in indoor farrowed pigs. Is it acceptable to permit increased mortality simply because the animals are considered free to express what is interpreted as normal behaviors?

Five Freedoms

The five freedoms approach to animal welfare outlines the elements necessary for ideal welfare and defines the husbandry and resources required to promote this ideal welfare state. The five freedoms and provisions are as follows:

1. Freedom from thirst, hunger, and malnutrition by ready access to fresh water and a diet to maintain full health and vigor.
2. Freedom from discomfort by providing a suitable environment including shelter and a comfortable resting area.
3. Freedom from pain, injury, and disease by prevention or rapid diagnosis and treatment.
4. Freedom from fear and distress by ensuring conditions that avoid mental suffering.
5. Freedom to express normal behavior by providing sufficient space, proper facilities, and company of the animal's own kind.

The five freedoms approach is adopted widely and has had a significant impact on the minimum standards required for animal welfare. This approach, however, does not consider social behaviors and there is not a universally accepted definition of what constitutes normal behaviors.

■ **Fig. 16.4** Defining what constitutes normal behavior is a primary limitation to the five freedoms approach to animal welfare. (© Jana Shea, 2009. Under license from Shutterstock, Inc.)

Biological Functioning

Another approach to animal welfare is the biological-functioning based approach. Because of difficulties associated with some of the aforementioned practices in defining welfare, many scientists prefer a more traditional measure based on body function. Animals will adapt to a new environment or situation and a failure to adapt will lead to a loss in biological functioning. This approach aims to define the biological costs associated with the environment. For example, does the environment result in decreased growth, reproductive failure, injury, disease, or death? It is suggested under this approach that welfare reflects the status of the animal considering it attempts to contend with its environment. Welfare is then appraised by how much has to be done to cope (the biological responses elicited) and is coping successful (measured by a lack of biological costs, ie deterioration of growth). Although a preferred method of welfare assessment by scientists, it only considers the physical and not the emotional needs of the animals.

Animal Welfare Issues

As the stance to the use of animals differs, approaches to animal welfare differ, emphasizing different aspects of an animal's well being. With the advent of industrialized farming techniques, the growth in the number of companion animals, and the use of animals for research; animal welfare will remain a major social and political issue. The key to improving animal welfare resides in the attitudes and behaviors of humans, since humans have dominion. Although science attempts to express animal welfare as a scientific concept, quality of life is difficult to measure. Between the demands of an ever increasing population and the welfare of the sentient species entrusted in human care, striking the correct balance may never be universally attained.

References

Adams, G.P., M.H. Ratto, W. Huanca, and J. Singh. 2005. Ovulation-inducing factor in the seminal plasma of alpacas and llamas. Biol. Reprod. 73:452-457.

Agriculture Marketing Resource Center. Commodities and Products. 2014. Retrieved from agmrc.org /commodities_products/.

Alaska Department of Fish and Game. Gaudet, D. 2002. Atlantic salmon: a white paper.

Albarella, U., K. Dobney, A. Ervynck, and P. Rowley-Conway. 2008. Pigs and humans: 10,000 years of interaction. Oxford; Oxford University Press.

Alberts, B., D. Bray, J. Lewis, M. Raff, K. Roberts, J.D. Watson. 1994. Molecular biology of the cell. 3rd ed. New York: Garland Publishing.

Alpaca registry statistics. From Alpaca Registry, Inc. 2014. Retrieved from http://www.alpacaregistry.com/.

American Horse Council. 2005. The economic impact of the horse industry on the United States. Washington, D.C.: Deloitte Consulting, LLP.

American Sheep Industry Association. Fast Facts. 2014. Retrieved from http://www.sheepusa.org/Research Education_FastFacts.

American Veterinary Medical Association. American College of Animal Welfare Organizing Committee. Animal Welfare. Retrieved from http://www.avma.org/issues/animal_welfare/default.asp.

Animal and Plant Health Inspection Services. 2014. Feral Swine. Retrieved from: http://www.aphis.usda. gov/wps/portal/banner/help?1dmy&urile=wcm%3Apath%3A/APHIS_Content_Library/SA_Our_Focus/ SA_Wildlife_Damage/SA_Operational_Activities/SA_Feral_Swine.

Andersen, I.L., S. Berg, and K.E. Boe. 2005. Crushing of piglets by the mother sow (*Sus scrofa*)-purely accidental or a poor mother? Appl. Anim. Behav. Sci. 93:229-243.

Bahr, J. 2008. The Chicken as a Model Organism. From the Sourcebook of Models for Biomedical Research. Totowa: Humana Press Inc.

Baldwin, J.M. 1896. Heredity and instinct (I). Science.

Barnett, J.L., P.H. Hemsworth, G.M. Cronin, E.C. Jongman, and G.D. Hutson. 2001. A review of the welfare issues for sows and piglets in relation to housing. Aust. J. Agric. Res. 52:1-28.

Baumans, V. 2004. Use of animals in experimental research: an ethical dilemma? Gene Therapy. 11:S64-S66.

Bauman, D.E., I.H. Mather, R.J. Wall, and A.L. Lock. 2006. Major advances associated with the biosynthesis of milk. J. Dairy Sci. 89:1235-1243.

Bearden, H.J., J.W. Fuquay, and S.T. Willard. 2004. Applied animal reproduction. 6th ed. Upper Saddle River: Pearson Prentice Hall.

Beaumont, W. 1833. Experiments and observations on the gastric juice and the physiology of digestion. Pittsburgh: Allen.

Behringer, R.B. G.S. Eaking, and M.B. Renfree. 2006. Mammalian diversity: gametes, embryos, and reproduction. Reprod. Fertil. Dev. 18:99-107.

Bell, F.R. 1972. Sleep in the larger domesticated animals. Proc. Roy. Soc. Med. 65:176-177.

Belloc, H. 1957. Stories, essays, and poems. San Diego, CA: Dent.

Berger, Y.M. and D.L. Thomas. 1997. Early experimental results for growth of East Friesian crossbred lambs and reproduction and milk production of East Friesian crossbred ewes. Proc. 3rd Great Lakes.

Dairy Sheep Symp., Wisconsin Sheep Breeders Cooperative, Madison, Wisconsin:12-21.

Board on Agriculture and Natural Resources. 2008. Changes in the sheep industry in the United States. Making the transition from tradition. Washington, D.C: The National Academies Press.

Bradford, G.E. 1999. Contributions of animal agriculture to meeting global food demand. Livest. Prod. Sci. 59:95-112.

Brady, C. 2008. An illustrated guide to animal science terminology. Clifton Park: Thomson Delmar Learning.

Brillat-Savarin, J-A. 1985. The philosopher in the kitchen. Middlesex: Penguin Books.

Broom, D.M. 1986. Indicators of poor welfare. Br. Vet. J. 142:524-526.

Broom, D.M. 1988. The scientific assessment of animal welfare. Appl. Anim. Behav. Sci. 20:5-19.

Broom, D.M. 1991. Animal welfare: concepts and measurement. J. Anim. Sci. 69;4167-4175.

Bruford, M., D. Bradley, G. Luikart. 2003. DNA markers reveal the complexity of livestock domestication. Nature. 4:900-910.

Cain, K. and D. Garling. 1993. Trout culture in the north central region. North Central Regional Aquaculture Center. Fact Sheet 108.

Campbell, K.L. and J.R. Campbell. 2009. Companion animals: their biology, care, health, and management. 2nd ed. Upper Saddle River: Pearson Prentice Hall.

Campbell, J.R. and J.F. Lasley. 1969. The science of animals that serve mankind. New York: McGraw-Hill.

Capuco, A.V. and R.M. Akers. 2009. The origin and evolution of lactation. J. Biol. 8:37-40.

Carpenter, K.J. 2003. A short history of nutritional science: Part 1 (1785-1885). J. Nutr. 133:638-645.

Carpenter, K.J. 2003. A short history of nutritional science: Part 2 (1885-1912). J. Nutr. 133:975-984.

Carpenter, K.J. 2003. A short history of nutritional science: Part 3 (1912-1944). J. Nutr. 133:3023-3032.

Carpenter, K.J. 2003. A short history of nutritional science: Part 4 (1945-1985). J. Nutr. 133:3331-3342.

Cheeke, P.R. 2004. Contemporary issues in animal agriculture. 3rd ed. Upper Saddle River: Pearson Prentice Hall.

Chen, C. J.E. Sanders, and N.M. Dale. 2003. The effect of dietary lysine deficiency on the immune response to Newcastle disease vaccination in chickens. Avian Dis. 47:1346-1351.

Chessa, B. et. al. 2009. Revealing the history of sheep domestication using retrovirus integrations. Science. 324:532-536.

Clarke, A.S. 1996. Maternal gestational stress alters adaptive and social behavior in adolescent rhesus monkey offspring. Infant Behav. Dev. 19:451-461.

Clauss, M., A. Schwarm, S. Ortmann, D. Alber, E.J. Flach, R. Kühne, J. Hummel, W.J. Streich, and H. Hofer. 2004. Intake, ingesta retention, particle size distribtution and digestibility in the hippopotamidae. Comp. Biochem. Physiol. A. 139:449-459.

Combs, G.F. Jr. 1998. The Vitamins: Fundamental Aspects in Nutrition and Health. 2nd ed. San Diego: Academic Press.

Correa, J.E. 2008. Nutritive value of goat meat. Alabama Cooperative Extension System. UNP-0061.

Costa, D.A. and D.J. Reinemann. 2004. The purpose of the milking routine and comparative physiology of milk removal. Paper presented at the National Mastitis Council Meeting.

Crawfish production manual. LSU Ag. Center. Publication 2637.

Crawford, R.D. 1990. Origin and history of poultry species. In R.D. Crawford (Ed.) Poultry breeding and genetics. Amsterdam: Elsevier.

Cunningham, M., M.A. Latour, and D. Acker. 2005. Animal Science and Industry. 7th ed. Upper Saddle River: Pearson Education, Inc.

Damron, W.S. 2006. Introduction to Animal Science: Global, Biological, Social, and Industry Perspectives. 3rd ed. Upper Saddle River: Pearson Education Inc.

Daniel, C.R., A.J. Cross, C. Koebnick, and R. Sinha. 2011. Trends in meat consumption in the United States. Public Health Nutr. 14:575-583.

Darmon, N. and A. Drewnowski. 2008. Does social class predict diet quality. Am. J. Clin. Nutr. 87:1107-1117.

Darwin, C. 1872. The expression of the emotions in man and animals. London: John Murray.

Dehoux, J-P. and P. Gianello. 2007. The importance of large animal models in transplantation. Frontiers in Bioscience. 12:4864-4880.

Dekkers, J.C.M. 2004. Commercial application of marker- and gene-selected selection in livestock: strategies and lessons. J. Anim. Sci. 82:E313-E328.

DeVries, A., M. Overton, J. Fetrow, K. Leslie, S. Eicker, and G. Rogers. 2007. Exploring the impact of sexed semen on the structure of the dairy industry. J. Dairy Sci. 91:847-856.

Diamond, J. 1999. Guns, germs, and steel: The fates of human societies. W.W. Norton & Co.

Drake, A., D. Fraser, D.M. Weary. 2008. Parent-offspring resource allocation in domestic pigs. Behav. Eco. Sociobiol. 62:309-319.

Dryden, G. McL. 2008. Animal Nutrition Science. Wallingford:CABI Publishing.

Duncan, I.J.H. 2005. Science-based assessment of animal welfare: farm animals. Rev. Sci. Tech. Off. Int. Epiz. 24:483-492.

Engle, C., G. Greaser, and J. Harper. 2000. Agriculture alternatives: Meat goat production. Pennsylvania State University Extension. ps37409.

Ensminger, M.E. and R.C. Perry. 1997. Beef cattle science. 7th ed. Danville: Interstate Publishers Inc.

Evans, J.P., A. Borton, H.F. Hintz, and L.D. van Vleck (Ed.). 1990. The Horse. 2nd ed. New York: W.H. Freeman and Co.

Evans, P. 2005. Equine vision and its effects on behavior. Utah State University Cooperative Extension Service. AG/Equine/2005-03.

Evershed, R.P., S. Payne, A.G. Sherratt, M.S. Copley, J. Coolidge, D. Urem-Kotsu, et. al. 2008. Earliest date for milk use in the Near East and southeastern Europe linked to cattle herding. Nature. 455:528-531.

Fernandez, S.R., S. Aoyagi, Y. Han, C.M. Parsons, and D.H. Baker. 1994. Limiting order of amino acids in corn and soybean meal for growth of the chick. Poultry Sci. 73:1887-1896.

Field, T.G. and R.E. Taylor. 2008. Scientific farm animal production: an introduction to animal science. 9th ed. Upper Saddle River: Pearson Prentice Hall.

Food and Agriculture Organization of the United Nations, Fisheries and Aquaculture Department. Cultured Aquatic Species Fact Sheets. Retrieved from http://www.fao.org/fishery/culturedspecies/search/en.

Food and Agricultural Organization of the United Nations, FAO Corporate Document Repository. Petrie, O.J. Harvesting of textile animal fibres. Bulletin No.122.

Food and Agricultural Organization of the United Nations, FAOSTAT. Retrieved from: http://faostat3.fao.org/faostat-gateway/go/to/home/E.

Food and Agriculture Organization of the United Nations, FAO Fisheries Department. 2006. State of the world aquaculture. Tech.500.

Food and Agriculture Organization of the United Nations, International Committee on Animal recording. Cardellino, R., A. Rosati, and C. Mosconi. 2004. Current status of genetic resources, recordings, and production systems in African, Asian, and American camelids. TS.no.2.

Food and Agriculture Organization of the United Nations, United Nations Development Program, Regional Small Scale -Coastal Fisheries Development Project. Rabanal, H. 1988. History of Aquaculture. ASEAN/SF/88/Tech.7.

Foote, R.H. 1987. In vitro fertilization and embryo transfer in domestic animals: Applications in animals and implications for humans. J. Assisted Repro. and Genetics. 4:73-88.

Foote, R.H. 2002. The history of artificial insemination: selected notes and notables. J. Anim. Sci. 80:1-10.

Frandson, R.D., W.L. Wilke, and A.D. Fails. 2003. Anatomy and physiology of farm animals. 6th ed. Baltimore: Lippincott Williams and Wilkins.

Fricker, J. 2001. The pig: a new model of atherosclerosis. DDT. 6;921-922.

Fumihito, A., T. Miyake, S. Sumi, M. Takada, S. Ohno, and K. Kondo. 1994. One subspecies of the red junglefowl (Gallus gallus gallus) suffices as the matriarchic ancestor of all domestic breeds. PNAS. 91:12505-12509.

Gatlin L.A., M. T. See, D.K. Larick, X. Lin, and L. Odle. 2002. Conjugated linoleic acid in combination with supplemental dietary fat alters pork fat quality. J. Nutr. 132:3105-3112.

Geraert, P.A. and Y. Mercier. 2010. Amino acids: beyond the building blocks. Antony ADISSEO France SAS.

Guenther, P. H. Jensen, S.P. Batres-Marquez, and C-F. Chen. 2005. Sociodemographic, knowledge, and attitudinal factors related to meat consumption in the United States. J. Am. Diet. Associ. 105:1266-1274.

Gillespie, J.R. 1998. Animal Science. Albany: Delmar Publishers.

Grant, R. 2007. Taking Advantage of Natural Behavior Improves Dairy Cow Performance. Presented at: Western Dairy Management Conference Reno, NV. March 7-9. Retrieved from http://www.extension.org/pages/Taking_Advantage_of_Natural_Behavior_Improves_Dairy_Cow_Performance.

Grémillet, D., A. Prudor, Y. Maho, and H. Weimerskirch. 2012. Vultures of the Seas: hyperacidic stomachs in wandering albatrosses as an adaptation to dispersed food resources, including fishery wastes. PLoS One 7:e37834.

Griffiths, J.T. 2008. Equine Science: basic knowledge for horse people of all ages. Gaithersburg: Equine Network Publishers.

Grützner, B., Nixon, and R.C. Jones. 2008. Reproductive bioloy in egg-laying mammals. Sex. Dev. 2:115-127.

Hänninen, L. 2007. Sleep and rest in calves-relationship to welfare, housing and hormonal activity. University of Helsinki: Academic Dissertation.

Hendricks, B. 1996. International encyclopedia of the horse. Norman: University of Oklahoma Press.

Hewson, C. 2003. What is animal welfare? Common definitions and their practical consequences. Can. Vet. J. 44:496-499.

Hinshaw, J. 1990. Trout production: feeds and feeding methods. Southern Regional Aquaculture Center. SRAC223.

Hodges, J. 1999. Animals and value in society. Livest. Prod. Sci. 58:159-194.

Hogan, J.S. and K.L. Smith. 1987. A Practical Look at Environmental Mastitis. Compendium on Continuing Education for the Practicing Veterinarian. 9: F342.

Holden, P.J. and M.E. Ensminger. 2006. Swine science. 7th ed. Upper Saddle River: Pearson Prentice Hall.

Hooke, R. 1667. Micrographia: or some physiological descriptions of minute bodies made by magnifying glasses: with observations and inquiries thereupon. London: John Martyn.

Horton, H.R., L.A. Moran, K.G. Scrimgeour, M.D. Perry, and J.D. Rawn. 2006. Principles of Biochemistry. 4th ed. Upper Saddle River: Pearson Education Inc.

Hovey, R.C., J.F. Trott, and B.K. Vonderhaar. 2002. Establishing a framework for the functional mammary gland: from endocrinology to morphology. J. Mammary Gland Biol. 7:17-38.

Jansen, T., P. Forster, M.A. Levine, H. Oelke, M. Hurles, C. Renfrew, J. Weber, and K. Olek. 2002. Mitochondrial DNA and the origins of the domestic horse. PNAS. 99:10905-10910.

Jiang, Z. and M.F. Rothschild. 2007. Swine genomics comes of age. Int. J. Biol. Sci. 3:129-131.

Johannsen, W. 1909. Elemente der exakten Erblichkeitslehre. Gustav Fischer, Jena.

Johnson, D.E. 2007. Contributions of animal nutrition research to nutritional principles: energetic. J. Nutr. 137:698-701.

Johnson, L.W. 1994. Llama nutrition. Veterinary Clinics of North America: Food Animal Practice. 10:187-201.

Joyce, C. 2010. Meat, fire, and the evolution of man. NPR. Retrieved from: http://www.npr.org/blogs/health/2010/07/30/128877628/meat-fire-and-the-evolution-of-man.

Josephson, M. 2005. Stendhal or the pursuit of happiness. New York: Jorge Pinto Books.

Jurgens, M.H. 1993. Animal Feeding and Nutrition., 7th ed. Dubuque: Kendall Hunt Publishing Company.

Kadwell, M., M. Fernandez H.F. Stanley, R. Baldi, J.C. Wheeler, R. Rosadio, and M.W. Bruford. 2001. Genetic analysis reveals the wild ancestors of the llama and the alpaca. Proc. R. Soc. Land. B. 268:2575-2584.

Kaushik, S.J. 1999. Animals for work, recreation, and sports. Livest. Prod. Sci. 59:145-154.

Keeling, L.J. and H.W. Gonyou. 2001. Social behavior in farm animals. Wallingford:CABI Publishing.

Kessel, A.L. and L. Brent. 1998. Cage toys reduce abnormal behavior in individually housed pigtail macques. J. Applied Anim. Welfare Sci. 1:227-234.

Kjaer, J.B., P. Sorensen, G. Su. 2001. Divergent selection on feather pecking behavior in laying hens (*Gallus gallus domesticus*). App. Anim. Behaviour Sci. 71:229-239.

Kues, W.A. and H. Niemann. 2004. The contribution of farm animals to human health. Trends in Biotech. 22:286-294.

Kunz, T.H. and D.J. Hosken. 2008. Male lactation: why, why not and is it care? Trends Ecol. Evol. 24:81-85.

Lacy, M. and M. Czarick. 1998. Mechanical harvesting of broilers. Poultry Sci. 77:1794-1797.

Lai L., J.X. Kang, R. Li, L. Wang, W.T. Witt, H.Y. Yong, et. al. 2006. Generation of cloned transgenic pigs rich in omega-3 fatty acids. Nat. Biotechnol. 24:435-6.

Lacy, R.C. 1997. Importance of genetic variation to the viability of animal population. J. Mammal. 320-335.

Land, M.R. and J.W. Wells. 1987 A review of egg shell pigmentation. World's Poultry Sci J. 43:238-246.

Lavoisier, A.L. and P.S. Laplace. 1780. Mémoire sur la chaleur. Mém. Acad. Sci. Paris. 355-408.

Lecce, J. 1979. Intestinal barriers to water-soluble macromolecules. Environ. Health Perspec. 33:57-60.

Lechner-Doll, M., W. von Engelhardt, A.M. Abbas, H.M. Mousa, L. Luciano, and E. Reale. 1995. Particularities in forestomach anatomy, physiology and biochemistry of camelids compared to ruminants. In Tisserand J.-L. (ed.). *Elevage et alimentation du dromadaire = Camel production and nutrition* (19-32). Zaragoza:CIHEAM-IAMZ.

Lewandowski, R. 2003. Goat: The other red meat. Buckeye Meat Goat Newsletter, The Ohio State University Extension. 1:7-8.

Lindsay, S.R. and G.E. Burrows. 2000. Handbook of applied dog behavior and training, volume 1: adaptation and learning. Ames: Iowa State Press.

Leeuwenhoek, A. 1678. Des natis é semine genital animalculis. R. Soc. (Lond.) Philos. Trans. 12:1040-1043.

Logothetis, N.K. 2004. Francis Crick 1916-2004. Nature Neuroscience. 7:1027-1028.

Ludwi, A., M. Pruvost, M. Reissmann, N. Benecke, G.A. Brockmann, P. Castaños, M. Cieslak, S. Lippold, L. Llorente, A-S. Malaspinas, M. Slatkin, and M. Hofreiter. 2009. Coat color variation at the beginning of horse domestication. Science. 324:485.

Luginbuhl, J-M. 1998. Breeds and production traits of meat goats. North Carolina Cooperative Extension Service. ANS00-603MG.

Luginbuhl, J-M, J.T. Green, J.P. Mueller, and M.H. Poore. 1996. Meat goats in land and forage management. Southeast Regional Meat Goat Production Symposium. Florida A&M University, Tallahassee.

Lunney, J.K. 2007. Advances in swine biomedical model genomics. Int. J. Biol. Sci. 3:179-184.

MacHugh, D.E. and D.G. Bradley. 2001. Livestock genetic origins: goats buck the trend. PNAS. 98:5382-5384.

Marchant-Forde, J.N. 2009. The welfare of pigs. Netherlands: Springer.

Masataka, N., H. Koda, N. Urasopon, and K. Watanbe. 2009. Free-ranging macaque mothers exaggerate tool-using behavior when observed by offspring. PloS One. 4:e4768.

Mason, G.J. and N.R. Latham. 2004. Can't stop, won't stop: is stereotypy a reliable animal welfare indicator? Animal Welfare. 13:S57-S69.

Matthew, W.D. 1926: The evolution of the horse: a record and its interpretation. Quart. Rev. Biol. 1:139-185.

Matilla-Sandholm, T. and M. Saarela (Ed.) 2003. Functional Foods. Cambridge: Woodhead Publishing Limited.

Maynard, L.A. 1951. Animal Nutrition. New York: McGraw-Hill Inc.

McLain, W.R. 2012. Crawfish production: pond construction and water requirements. SRAC Publication No: 240.

McNeilly, A.S. 2001. Lactational control of reproduction. Reprod. Fertile. Dev. 13:583-590.

McPherron, A.C. and S-J. Lee. 1997. Double muscling in cattle due to mutations in the myostatin gene. Proc. Natl. Acad. Sci. 94:12457-12461.

Mepham, T.B. (Ed.) 1983. Biochemistry of lactation. Amesterdam: Elsevier.

Meyers, R.A. and B. Worm. 2003. Rapid worldwide depletion of predatory fish communities. Nature. 423:280-283.

Min, B.R., S.P. Hart, T. Sahlu, and L.D. Satter. 2005. The effect of diets on milk production and composition, and on lactation curves in pastured dairy goats. J. Dairy Sci. 88:2604-2615.

Morin, D.E., L.L. Rowan, W.L. Hurley, and W.E. Braselton. 2005. Composition of milk from llamas in the United States. J. Dairy Sci. 78:1713-1720.

Mosher, D., P. Quignon, C.D. Bistamante, N.B. Sutter, C.S. Mellersh, H.G. Parker, and E.A. Ostrander. 2007. A mutation in the myostatin gene increases muscle mass and enhances racing performance in heterzygote dogs. PLoS Genetics. 3:e79.

Mozdziak, P.E. and J.N. Petitte. 2004. Status of transgenic chicken models for developmental biology. Dev. Dynamics. 229:414-421.

Nash, C.E. (Ed.). 2001. The net-pen salmon farming industry in the Pacific Northwest. U.S. Dept. Commer., NOAA Tech. Memo. NMFS-NWFSC-49, 125.

National Oceanic and Atmospheric Administration. Aquaculture in the United States. Retrieved from: http://www.nmfs.noaa.gov/aquaculture/aquaculture_in_us.html.

National Oceanic and Atmospheric Administration. Status of fishery resources off the Northeastern U.S. Retrieved from: http://www.nefsc.noaa.gov/sos/spsyn/af/salmon/.

National Sustainable Agriculture Information Service. Appropriate Technology Transfer in Rural Areas. Beetz, A. Grass-based and seasonal dairying. 1998: ATTRA Publication #CT079. Retrieved from http://attra.ncat.org/attra-pub/gbdairy.html.

Neil, J. 2002. Farming triploid oysters. Aquaculture. 10:69-88.

Olson, T.A. 1999. Genetics of colour variation. In Genetics of Cattle. CAB International.

Orwell, G. 1996. Animal Farm: a fairy story, 50th Anniversary ed. New York: Penguin Books Ltd.

Osborne, L. 2002. Got silk? The New Work Times. 6:49.

Pacheco, A.A., A. García-Amado , C. Bosque, and M. Domínguez-Bello. 2004. Bacteria in the crop of the seed-eating green-rumped parrotlet. The Condor. 106:139-143.

Pavlov, I. 1967. In Nobel Lectures, Physiology or Medicine 1901-1921. Amsterdam: Elsevier Publishing Company.

Parker, R. 2007. Equine Science. 3rd ed. Clifton Park: Thompson Delmar Learning.

Pauly, D., V. Christensen, S. Guénette, T. Pitcher, U.R. Samaila, C. Walters, R. Watson, and D. Zeller. 2002. Towards sustainability in world fisheries. Nature. 418:689-695.

Peaker, A. 2002. The mammary gland in mammalian evolution: a brief commentary on some of the consepts. J. Mammary Gland Biol. 7:347-353.

Pennington, J. and M.McCarter. 2007. Marketing of meat goats. University of Arkansas Cooperative Extension Services. FSA3094.

Perkins, B.E. 1995. Aquacultured oyster products: inspection, quality, handling, storage, safety. Southern Regional Aquaculture Center. SRAC434.

Perry, T.W., A.E. Cullison, and R.S. Lowrey. 2003. Feeds and Feeding, 6th ed. Upper Saddle River: Pearson Education, Inc.

Pigs domesticated 'many times'. 2005. BBC News. Retrieved from http://news.bbc.co.uk/2/hi/science/nature/4337435.stm.

Pigs: Webster's quotations, facts, and phrases. 2008. San Diego: Icon Group International Inc.

Polidori, P., C. Renieri, M. Antonini, P. Passamonti, and F. Pucciarelli. 2007. Meat fatty acid composition of llama (*Lama glama*) reared in the Andean highlands. 75:356-358.

Pond, C.M. 1977. The significance of lactation in the evolution of mammals. Evolution. 31:177-199.

Pond, W.G., D.C. Church, K.R. Pond, and P.A. Schoknecht. 2005. Basic Animal Nutrition and Feeding. 5th ed. Hoboken: John Wiley and Sons, Inc.

Porteus, J. 2002. Slick birds are wearing wool. Knitted pullovers protect penguins from oil discharge. Nature News. Retrieved from http://www.nature.com/news/2002/020122/full/news020121-3.html.

Pray, L.A. 2008. Eukaryotic genome complexity. Nature Educ. 1:96.

Price, E.O. 1984. Behavioural aspects of animal domestication. Rev. Biol. 59:1-32.

Price, E.O. 1999. Behavioral development in animals undergoing domestication. Appl. Anim. Behav. Sci. 65:245-271.

Prout, W. 1824. On the nature of the acid and saline matters usually existing in the stomach of animals. Philos. Trans. R. Soc. Lond. 114:45-49.

Purser, J. and N. Forteath. 2003. Salmonids. In J.S. Lucas & P.C. Southgate (Eds.), Aquaculture: Farming Aquatic Animals and Plants. Oxford: Blackwell Publishing, 295-320.

Regional multi-state interpretation of small farm financial data from the third year report on 2002 great lakes grazng network grazing dairy data. 2004. Great Lakes Grazing Network. Fact Sheet 5: Grazing versus Confinement Farms-Year 3.

Reimer, J.J. 2006. Vertical integration in the pork industry. Amer. J. Agr. Econ. 88:234-248.

Reiner, R. and F. Bryant. 1983. A different sort of sheep. Rangelands. 5:106-108.

Robinson, R. 2008. For mammals, loss of yolk and gain of milk went hand in hand. PLoS Biology. 6:e77.

Rooney, M.B. and J. Sleeman. 1998. Effects of selected behavioral enrichment devices on behavior of western lowland gorillas. J. Applied Anim. Welfare Sci. 1:339-351.

Scanes, C.G. 2003. Biology of growth of domestic animals. Ames: Iowa State Press.

Scanes, C.G., G. Brant, and M.E. Ensminger. 2004. Poultry Science. 4th ed. Upper Saddle River: Prentice Education, Inc.

Schilo, K.K. 2009. Reproductive physiology of mammals: from farm to field and beyond. Clifton Park: Delmar Cengage Learning.

Schook, L., C. Beattie, J. Beaver, S. Donovan, R. Jamison, F. Zuckerman, et. al. 2005. Swine in biomedical research: creating the building blocks of animal models. Animal Biotechnology. 16:183-190.

Schwann, T. 1836. Ueber das wesen des verdauungsprocesses. Müllers Arch. Anat. Physiol. 90-138.

Schwann, T. and M.J. Schleiden. 1847. Microscopical Researches Into the Accordance in the Structure and Growth of Animals and Plants. London: The Sydenham Society.

Seaman, J. and T.J. Fangman. 2001. Biosecurity for today's swine operation. University of Missouri Cooperative Extension Service. G2340.

Sell, R. 1993. Llama. North Dakota State University Extension Service. Alternative Agriculture Series, Number 12.

Stahler, C. 2012. How often do Americans eat vegetarian meals? And how many adults in the U.S. are vegetarian. Vegetarian Resource Group. Retrieved from http://www.vrg.org/blog/2012/05/18/how-often-do-americans-eat-vegetarian-meals-and-how-many-adults-in-the-u-s-are-vegetarian/.

Starkey, P. Moving forward with animal power for transport: how people, governments, and welfare organisations can make an impact: examples from Africa and Madagascar. World Assoc. Transport Anim. Welfare Studies.

Stevens, C.E. and I.D. Hume. 2004. Comparative physiology of the vertebrate digestive system. Cambridge University Press.

Stickney, R.R. 2005. Aquaculture: an introductory text. Wallingford:CABI Publishing.

Stoka, G., J.F. Smith, J.R. Dunham, T. Van Anne. 1997. Lameness in dairy cattle. Kansas State University Agricultural Experiment Station and Cooperative Extension Service. MF-2070.

Swenson, M.J. (Ed.). 1977. Duke's physiology of domestic animals. Ithaca: Cornell University Press.

Teweldmehidin, M. and A.B. Conroy. 2010. The economic importance of draught oxen on small farms in Namibia's Eastern Caprivi region. African J. of Ag. Research. 5:928-934.

Thiruvenkadan, A.K., N. Kandasamy, and S. Panneerselvam. 2008. Coat color inheritance in horses. Livest. Sci. 117:109-129.

Thomas, D. 1996. Dairy Sheep Basics for Beginners. In: Proceedings of the Great Lakes Dairy Sheep Symposium. Cornell University, Ithaca, NY. 0-77. Retrived from www.ansci.wisc.edu/Extension-New%20copy/sheep/Publications_and_Proceedings/Pdf/Dairy/Management/Daily%20sheep%20basics%20for%20beginners.pdf

Trut, L., I. Oskina, and A. Kharlamova. 2009. Animal evolution during domestication: the domesticated fox as a model. BioEssays. 31:349-360.

Tucker, C.C. and E.H. 1991. Channel catfish farming handbook. Springer.

Tyler, H.D. and M.E. Ensminger. 2004. Dairy science. 4th ed. Upper Saddle River: Pearson Prentice Hall.

Tyndale-Biscoe, C.H. 2001. Australasian marsupials - to cherish and to hold. Reprod. Fertil. Dev. 13:477-485.

United States Department of Agriculture, Agriculture Research Service. 2006. Livestock Behavior Research Unit. Mission statement. Retrieved from http://www.ars.usda.gov/main/site_main.htm?modecode=36-02-20-00.

United States Department of Agriculture, Animal Plant Health and Inspection Services. 2009. Dairy 2007. Part IV: Reference of dairy cattle health and management practices in the United States. N494.0209.

United States Department of Agriculture, Animal Plant Health and Inspection Services. 2006. The goat industry: structure, concentration, demand and growth. Retrieved from http://www.aphis.usda.gov/vs/ceah/cei/bi/emergingmarketcondition_files/goatreport090805.pdf.

United States Department of Agriculture, Census of Agriculture. Census of Aquaculture. 2012. Retrieved from http://www.agcensus.usda.gov/Publications/2012/.

United States Department of Agriculture, Economic Research Service. Data sets; Aquaculture data. 2014. Retrieved from http://www.ers.usda.gov/data-products/aquaculture-data.aspx#.U6Oww0DDUeE.

United States Department of Agriculture, Economic Research Service. Food availability (per capita) data system. 2014. Retrieved from http://www.ers.usda.gov/data-products/food-availability-%28per-capita%29-data-system/.aspx#.U6OxO0DDUeE.

United States Department of Agriculture, Economic Research Service. Data sets; Livestock and meat domestic data. 2014. Retrieved from http://www.ers.usda.gov/data-products/livestock-meat-domestic-data.aspx#.U6OxmEDDUeE.

United States Department of Agriculture, Economic Research Service. Data sets; Livestock and meat international trade data. 2014. retrieved from: http://www.ers.usda.gov/data-products/livestock-meat-international-trade-data.aspx#.U6OxzEDDUeE.

United States Department of Agriculture, Economic Research Service. Blayney, D.P. 2002. The changing landscape of US Milk production. SBN978.

United States Department of Agriculture, Economic Research Service. 2014. McBride, W. and N. Key. Productivity growth slows for specialized finishing operations. Retrieved from: http://www.ers.usda.gov/amber-waves/2014-januaryfebruary/productivity-growth-slows-for-specialized-hog-finishing-operations.aspx#.U6OsSkDDUeE.

United States Department of Agriculture, Economic Research Service. Dimitir, C. and C. Greene. 2002. Recent growth patterns in the US organic foods market. AIB777.

United States Department of Agriculture, Economic Research Service. Matthews, F. 2014. Livestock, Dairy, and Poultry Outlook. LDPM-237.

United States Department of Agriculture, Economic Research Service. Miller, J.J. and D.P. Blayney. 2006. Dairy Backgrounder. LDP-M-145-01.

United States Department of Agriculture, Economic Research Service. Stewart, H., D. Dong, and A. Carlson. 2013. Why are Americans consuming less fluid milk? A look at generational differences in intake. ERR-149.

United States Department of Agriculture, Forest Service. Bisson, P.A. 2006. Assessment of the risk of invasion of national forest streams in the Pacific Northwest by farmed Atlantic salmon. PNW-GTR-697.

United States Department of Agriculture, National Agricultural Library. Adams, B. and J. Larson. 2007. Legislative history of the animal welfare act. AWIC Resource Series No. 41.

United States Department of Agriculture, National Agricultural Statistics Service. Quick Stats: Agricultural Statistics Database. 2014. Retrieved from http://quickstats.nass.usda.gov/.

United States Department of Agriculture, Natural Resources Conservation Service. Aschmann S. and J. Cropper. 2007. Profitable grazing-based dairy systems. Range and Pasture Tech. Note. 1.

United States Pork Center of Excellence. 2013. The pork industry at a glance. Retrieved from: http://www.pork.org/Resources/95/QuickFacts.aspx#.U6OucEDDUeE.

Vilá, C., J.A. Leonard, A. Götherström, S. Marklund, K. Sandberg, K. Lidén, R. Wayned, and H. Ellegren. 2001. Widespread origins of domestic horse lineages. Science. 291:474-477.

Vodičřka, P., K. Smetana, B. Dvo_ánková, T. Emerick, Y.Z. Xu, J. Ourednik, V. Ourednik, and J. Motlík. 2005. The miniature pig as an animal model in biomedical research. Ann. N.Y. Acad. Sci. 1049:161-171.

Wallace, R.K. 2001. Cultivating the eastern oyster *Crassostrea virginica*. Southern Regional Aquaculture Center. SRAC432.

Ward, C. 1997. Vertical integration comparison: beef, pork, and poultry. Oklahoma Cooperative Extension Service. WF552.

Watson, J.D., M. Gilman, J. Witkowski, and M. Zoller. 1992. Recombinant DNA. 2nd ed. New York: W.H. Freeman and Company.

Wayne, R.K. and E.A. Ostrander. 2007. Lessons learned from the dog genome. Trends in Genetics. 23:557-567.

Webster, J. 2005. Animal welfare: limping towards eden. Oxford: Blackwell Publishing.

White, K., B. O'Neill, and Z. Tzankova. 2004. At a crossroads: will aquaculture fulfill the promise of the blue revolution? Report for the SeaWeb Aquaculture Clearinghouse.

Whittier, J.C. 2011. Urea and NPN for cattle and sheep. Colorado State University Extension. Retrieved from: http://www.ext.colostate.edu/pubs/livestk/01608.html.

Wilmut, I., L. Young, P. DeSousa, and T. King. 2000. New opportunities in animal breeding and production: an introductory remark. An. Repro. Sci. 60-61:5-14.

Wilson, E.B. 1925. The cell. New York: Macmillan.

Wilson, E.O. 1999. The Diversity of Life. New York: Norton, W. W. & Company, Inc.

Wilstach, F.J. 1990. A dictionary of similes. 2nd ed. Detroit: Omingraphics, Inc.

Wolff, J.A. and J. Lederberg. 1994. An early history of gene therapy and transfer. Human Gene Therapy. 5:469-480.

Wood, G. 1937. Wood, hard-bitten. The Art Digest. 18.

World Health Organization. Epidemic and Pandemic Alert and Response. Avian influenza. Retrieved from http://www.who.int/csr/disease/avian_influenza/en/.

Wright, S. 1978. The relation of livestock breeding to theories of evolution. J. Anim. Sci. 46:1192-1200.

Zawistowski, S. 2008. Companion animals in society. Clifton Park: Thomson Delmar Learning.

Glossary

abomasum   The glandular stomach compartment of ruminant animals located prior to the small inestine.

acid-base balance   The body's balance between acidity and alkalinity. It is precisely controlled by biological buffering systems, excretion of compounds through the kidneys, and respiration processes of the lungs.

ad libitum   Having feed available at all times.

alevin   The larval stage of salmon development that follows hatching and precedes the absorption of the yolk sac. Alleles.

allele   One of two or more alternative forms of a gene occupying corresponding sites (loci) on homologous (similar) chromosomes.

allometric   Unequal growth rates of part of an organism relative to the whole organism.

altricial   Hatched or born developmentally immature. Altricial young are generally hairless or featherless with eyes closed.

amylase   Enzyme that catalyzes the break-down of starch.

anadromous   Migrating from sea to fresh water to spawn. Characteristic of the reproductive behavior of salmon.

anabolism   Metabolic pathways that construct large complex molecules from simple units

anaerobic   Conditions that lack oxygen.

anestrus   Period of time when a female is not displaying regular estrous cycles.

aneuploidy   Abnormal chromosome number, a chromosome number that is not an exact multiple of the monoploid (single set—haploid).

anion   An ionic species that carries a negative charge.

anthropomorphism   Attributing human thoughts, emotions, and characteristics to animals.

antioxidant   Substances that protect cells against cellular damage caused by free radicals, which are produced during metabolism and may be increased by environmental factors. Free radicals are unstable compounds that lack complete paired electrons and will attempt to stabilize by removing electrons from other molecules, including those within the cell membrane, and will decrease the stability of the molecule from which the electron was obtained.

aquaculture   The farming of aquatic orginisms including fish, mollusks, crustaceans, and plants.

artificial insemination   Means of placing semen in the female reproductive tracts by means other than natural mating.

ataxia   Lack of muscle coordination and movement.

atresia   Degeneration of follicles that do not make it to the mature, or Graafian stage.

autosomes   All chromosomes other than sex chromosomes.

bantam   A small sized variety of poultry.

barr body   The inactivated X chromosome in normal females (or Z in male poultry). Inactivation is a random event that occurs early in embryonic development as a result of folding into an inactive chromatin structure.

barrow   A castrated male hog.

binocular vision   The use of both eyes together to maintain visual focus on an object.

biosecurity   Prodcedures designed to minimize disease transmission from outside and inside a production unit.

blastocyst   The structure of early embryogenesis prior to implantation. It consists of an inner and outer cell mass that gives rise to the embryo and placenta, respectively.

blastomere   The cells that result from the cleavage of a fertilized ovum, preceding the formation of the blastocyst.

bloat   Abnormal quantities of gas collecting in the fermentive portion of the digestive tract.

boar   An intact male pig.

bolus   A rounded mass formed during mastication that initiates the swallowing reflex.

bovine somatotropin (BST)   A naturally produced growth hormone in cattle. Artificially produced bovine somatotropin, otherwise known as recombinant bovine somatotropin (rBST) is manufactured from bacteria and approved for use in dairy cattle for increased milk yield.

breeding value   The worth of an individual as a parent.

broiler   A chicken of either sex used for meat purposes.

calorie   The measue of food energy. The amount of energy needed to raise one gram of water by one degree Celsius. One thousand calories are needed for one kilocalorie or one food calorie (Calorie).

candling   Inspection of the inside of an intact egg with a light to detect defects.

carbohydrate   Any of a group of organic compounds that includes sugars, starches, celluloses, and gums and serves as a major energy source in the diet of animals. These compounds are produced by photosynthetic plants and contain only carbon, hydrogen, and oxygen, usually in the ratio 1:2:1.

carnivore   Animals that subsist on flesh/meat.

caruncle   Outgrowth of the endometrium that serves as the maternal contribution to the placenta in bovines.

catabolism   Metabolic pathways that breakdown large complex molecules into simple units

catalysis   The process by which enzymes increase the rate of a reaction without being consumed in the process.

cation   An ionic species that carries a positive charge.

caudal   At or near the posterior end or tail of the body.

cellulose   Carbohydrate composed of glucose molecules that forms the support structure of plants.

chromatin   The folded complex of nucleic acids and primarily histone proteins within cells.

chromosome   Deoxyribonucleic acid (DNA) containing structures within cells.

chyme   Mixture of food, saliva, and gastric secretions as it is ready to leave the stomach.

circadian rhythm   The daily rhythmic cycle of activity that occurs within a 24 hour window.

cis   On the same side.

cloning   The process of producing a genetic copy of a gene, DNA segment, embryo, or animal.

coagulation   To thicken or form a clot as in the process of blood coagulation.

collagen   Fibrous extracellular proteins that form the connective tissues of the skin, bone, cartilage, tendons, and teeth.

colostrum   Specialized milk produced in the initial days after parturition. Colostrum is higher in vitamins, mineral, and protein compared with normal milk and generally carries antibodies depending on the species.

conception   The point of time the ovum is fertilized by sperm.

coprophagy   The act of eating feces.

copulation   The act in which the male reproductive organ enters the females reproductive organ.

corpus luteum   Progesterone secreting structure of the ovary developed after the release of the follicle during ovulation.

cotyledon   The fetal contribution to the placenta of bovines.

cranial   Toward the head.

cria   The newborn offspring of llamas and alpacas.

cribbing   A stereotypical behavior observed in stabled horses that involves grasping an object with the incisors, pulling against the object, and sucking in air.

cross-breeding   Mating of genetically diverse breeds within a species.

cryptorchid   Failure of one or both testes to descend into the scrotum.

cull   Removal of inferior or undesirable animals from the group.

deglutination   The act of swallowing.

denature   To disrupt the structure of a native protein causing it to lose its ability to perform its required function.

dentition   Teeth, including the type, number, and positioning.

deoxyribonucleic acid (DNA)   The primary component of chromosomes that represents the blue-print of life and is the material of genetic inheritance.

dermatitis   Inflammation of the skin.

digestion   The physical, chemical, and enzymatic means the body uses to prepare a feedstuff for absorption.

diploid   Having two sets of chromosomes.

discoidal   Having a flat, disk-shape. In primates, refers to the type of placentation whereby maternal and fetal attachments are restricted to a circular plate.

diurnal   Active by day.

dominant   When one member of an allele pair is expressed to the exclusion of the other.

dorsal   Toward or near the back or upper surface of an animal.

double helix   The spiral arrangement of complementary strands of DNA

dry matter   Everything in a feed excluding water.

dystocia   Abnormal or difficult labor

ectoderm The outer of the three primary germ layers of the embryo that will develop into skin, nervous tissue, and sensory tissues.

embryo transfer Collecting embryos from a female and transferring to a surrogate for gestation.

emulsify To form an emulsion or suspension of lipid droplets in a aqueous phase.

endocrine Cellular signaling by which an extracellular signaling molecule is released and distributed through the body by blood to act on a remote target.

endoderm The inner of the three primary germ layers of the embryo that will develop into gastrointestinal and reproductive tracts, and endocrine organs.

endogenous Pertaining to inside or within an organism or system.

enzymes Proteins capable of catalyzing reactions.

epididymis Duct connecting testis with the vas deferens. Responsible for sperm storage, transport, and maturation.

epistasis Interaction of genes at different loci. Expression of genes at one loci depends on alleles present at another.

epithelium Tissue that covers the external and lines the internal surface of the body and its organs.

eructation The act of belching that is common amongst ruminant animals.

estrogen Any of a group of steroid hormones that regulate the growth, development, and function of the female reproductive tract. The major active estrogen compound in females is estradiol.

estrous cycle Time from one period of sexual receptivity in the female to the next.

estrus Period when a female is receptive to mating.

ethogram A descriptive catalog of the behaviors displayed by a species.

ethology Study of animals in their natural or most common environment.

etiology Study of factors that cause disease.

exocrine Pertaining to the release of a secretion into a duct.

exogenous Originating from outside of the organism or system.

expression Manifestation of characteristics that is specified by a gene.

farrow Act of giving birth in swine.

feed Food given to animals, which contain nutrients and meet the demands of living organisms.

feedback The process whereby a product effects the input. Feedback maybe negative or positive. In negative feedback, the product brings about a reduction or inhibition of a process. In positive feedback, the product brings about a continuation or magnification of a process.

felt (felting) Textile produced from the matting and pressing of fibers. The process of producing such textiles is known as felting. Felting may be applied to natural fibers, including wool, or synthetic fibers.

feral An animal in a wild state whose lineage is from domesticated stock.

filly A young female horse.

finishing The final feeding stage when animals are readied for market.

flight zone The distance which an animal is caused to flee from an intruder.

fluid mosaic model   In cell biology, the dynamic structure of the cell membrane that occurs with the interactions of hydrophobic and hydrophilic macromolecules including lipids, proteins, and carbohydrates.

follicle-stimulating hormone   FSH. Hormone responsible for growth, development, and maintenance of follicles in females.

food   Substances that are consumed, which contain nutrients and meet the demands of living organisms.

forestomachs   The digestive compartments of the ruminant preceeding the glandular stomach.

founder   The initial transgenic animal used to establish a transgenic line.

freemartin   In cattle, condition where the female born twin to a bull is sterile due to masculinization of the female by testosterone of the male.

gametes   Mature sperm in the male and ova in the female.

gametogenesis   Formation of gametes.

gander   Male goose.

gastrulation   Restructuring of the embryonic cells that leads to the formation and positioning of the germ layers that include the endoderm, mesoderm, and ectoderm.

gelding   A castrated, male horse.

gene   A short segment of chromosome that is expressed.

gene map   Location of specific genes along a chromosme that are marked with a probe.

genetic drift   A change in gne frequency due to chance.

genetic markers   Biochemical labels used to identify specific alleles on a chromosome.

genome   The complete genetic material of an organism.

gentoype   The genetic make-up of an organism.

gestation   Period of time when a female is pregnant.

glycogen   Storage form of glucose in liver and muscle tissues. Glycogen storage in liver serves to replenish and maintain circulating glucose concentrations.

gonadotropin   Protein hormones secreted by the pituitary including luteinizing hormone and follicle stimulating hormone.

gonads   Sex organs, testes in male and ovaries in female.

grass tetany   Metabolic disorder resulting from inadequate circulating magnesium and characterized by a staggering, uncontrolled gait. May lead to coma and death and is frequently observed in cattle nursing calves on Spring pasture.

gregarious   Inclination to associate with others in a group.

haploid   A cell with $\frac{1}{2}$ the usual number of chromosomes. Characteristic of the gametes.

hemorrhage   Rapid loss of blood from the circulatory system.

hemoglobin   Oxygen carrying pigment found in erythrocytes (red blood cells).

hen   A mature, female chicken.

herbivore   An animal that eats a diet of only plant material.

heritable   Capable of being passed from one generation to the next.

heterosis   Superiority of an outbred individual relative to the average performance of the parent populations. Increased performance associated with cross-breeding. Also known as hybrid vigor.

heterzygous Contrasting alleles at a given locus.

homologous chromosomes Similar chromosomes that pair during meiosis.

homozygous Identical alleles at a given locus.

hoof-and-mouth Also known as foot-and-mouth disease, is a highly contagious viral disease of hooved animals that result in blistering of the mouth and feet and may lead to death in few infected animals.

hydrolysis A chemical reaction that involves water molecules.

hydroxylation Introduction of a hydroxyl (-OH) group during a chemical reaction.

hyperkeratosis Thickening of the outerlayer of the skin due to increased accumulation of keratin.

hyperplasia An increase in cell number.

hypertrophy An increase in cell size.

hypothalamus A region of the brain responsible for many homestatic funcitons.

ileum The last, short portion of the small intestine.

Implantation Attachment of the embryo to the lining of the uterus.

imprinting A rapid learning process by which a newborn or very young animal establishes a behavior pattern of recognition and attraction to another animal of its own kind or to a substitute or an object identified as the parent.

in vitro Outside the body.

inbreeding Mating system in which mated individuals have one common ancestor appearing several time at least three to four generations back in the pedigree.

incomplete dominance When neither allel is dominant to the other.

inheritance The transfer of gene containing chromosomes from parent to offspring.

isomers Compounds with the same molecular formula, but different structural or spatial configurations.

isometric Growth that occurs at the same rate for all parts of an organism so that the shape and relative size remains consistent throughout development.

jejunum The second and longest portion of the small intestine where the majority of absorption occurs.

keratin Fibrous, structural proteins that form hard surfaces including the dental pad of ruminants, horns, hooves, claws, hair, wool, and nails. It also lines the teat sphincters and protects against infectious pathogens and accompanying mastitis in dairy cattle.

kidding Parturition in goats.

kosher Meat that is ritually fit for use as sanctioned by Jewish religious law.

lactation The process of milk production from the mammary glands of mammals.

lactose Milk sugar comprised of one glucose and one galactose unit.

lambing Parturition in sheep.

leucopenia A decrease in the number of white blood cells in circulation, which is associated with increased risk of infection.

leutinize Formation of the corpus luteum.

linebreeding The mating of related individuals. Considered a mild form of inbreeding practiced for the retention of desired traits.

lipase Enzymes that catalyze the degradation of triglycerides through release of fatty acids from the glycerol backbone.

locus The specific location of a gene on a chromsome.

luteinizing hormone LH. Hormone primarily responsible for providing the signal to disrupt the mature follicle in females (ovulation) and production of testosterone in males.

luteolysis Degeneration of the corpus luteum.

marker assisted selection Selection for certain alleles using markers such as linked DNA sequences.

mastication The process of chewing.

mastitis Inflammation of the mammary gland, most often the result of bacterial infection.

meiosis The formation of haploid cells, sex cells.

megoblastic anemia The presence of large, immature, dysfunctional red blood cells. May be caused by a deficiency in vitamin B_{12}.

mesoderm The middle of the three primary germ layers of the embryo that will develop into bones, cartilage, connective tissue, and muscles.

metabolic water The water generated inside an organism as a consequence of metabolic processes.

metabolism The chemical reactions of a living organism that are required to sustain life. Involved both anabolic (constructive) and catabolic (destructive) reactions.

milk fever Otherwise known as parturient paresis, a disease of dairy cattle associated with reduced postpartum blood calcium.

milt Seminal fluid of fish

mitochondrial DNA mtDNA. Maternally inherited DNA associated with the mitochondria.

mitosis The process of somatic (body) cell division

molting The replacement of worn feathers by birds. It is a naturally occurring process as feathers are incapable of repair (similar to nails) and must be replaced.

monocular vision Vision in which each eye is used separately and there is an accompanying increase in the visual field.

monosomy Absence of one chromosome from an otherwise diploid cell.

morula Early stage of embryonic development that follows rapid cleavage of the zygote.

muscular dystrophy A generalized term that refers to a group of genetic diseases that contribute to muscle weakness.

myelin Electrical insulating material that surrounds axons, the neuronal extensions or nerve fibers. Essential in muscle and nerve cell transmissions.

natural selection Selection based on factors that favor individuals better suited to living and reproducing in a given environment.

necrosis Non-programmed, premature, cellular death. Cause by external factors including infection or trauma.

nucleotides The building blocks of nucleic acids. They include a nitrogenous base (adenine, guanine, cytosine, thymine or uracil), a five carbon sugar backbone (ribose), and a phosphate group.

nutraceuticals Products perceived to have both nutrient and pharmaceutical properties.

nutrient A chemical substance (including water, carbohydrates, lipids, proteins, vitamins, and minerals) that provides nourishment to the body. May be classified as essential or nonessential according the requirements by the body for normal function and the body's ability to synthesize in adequate quantities to meet these requirements.

omnivore   Animals that eat both plant and animal based foods.

opisthotonus   Hyperextension of the head, neck and vertebrate that leads to an arching position. Causes include thiamine deficiency.

osmotic pressure   Pressure produced by the differences in concentration of solutes across a membrane. In biological systems it contributes to the movement of water across a cell membrane and is important in maintaining cellular volume and processes of cell shrinkage and expansion.

osteomalacia   A softening of bone due to defective bone mineralization. In developing animals it is associated with the onset of rickets.

osteoporosis   A disease of bone characterized by reduced bone mineral density and associated increased susceptibility to fracture.

outbreeding   Mating of less closely related individuals when compared to the average of the population.

ovulation   Release of the ovum from the ovary.

ovum   Egg. pl. ova

oxidation   A chemical reaction that involves the loss or transfer of electrons, usually accompanied by the loss of H^+ ions.

pair-bonding   The affinity between a male and female that leads to mating in some species. Pair-bonding relationships also may form within the same sex as seen in horses and reduces social tension within a group.

parturition   Process of giving birth.

pepsin   Gastric digestive enzyme that degrades proteins into the smaller units of peptides.

peristalsis   Progressive, wave like muscular contractions of the digestive tract that directs movement of ingested material.

perosis   Deformity of the leg bones of growing chicks as a consequence of manganese deficiency.

pharming   A term used to coin the production of pharmaceuticals from livestock or crops.

phenotype   The observable physical or biochemical characteristics of an organism, as determined by both genetic makeup and environmental influences.

photoperiod   The duration of daily exposure to light and its impacts on growth, development, and physiology.

pica   A behavior of neurological and nutritional origins that involves eating of nonnutritive and unnatural feedstuffs.

pinocytosis   The process of cellular internalization by pinching off of vesicles from the cell membrane into the intracellular space.

placentome   The structures of connection between maternal (caruncle) and fetal (cotyledon) tissues.

placenta   The temporary organ that envelops the fetus and provides sites of attachment to the uterine lining for the exchange of nutrients and waste.

pluripotent   Having more than one potential outcome. In cell biology it refers to the potential of cells to differentiate into more than one cell type depending on external influences and environment.

polyneuritis   Inflammation of multiple peripheral nerves.

postpartum   Following parturition.

precocious   Born or hatched mature in development. Young are born feathered or with hair, capable of standing, and in some cases, running soon after birth or hatch, with eyes open.

prehension   The act of seizing and grasping food.

progesterone   Female sex steroid produced by corpus luteum or the placenta.

prostaglandin   Hormone which breaks down the corpus luteum and allows the return of estrus.

prothombin   A blood clotting protein that is essential to the initial formation of blood clots.

proventriculus   The glandular stomach in fowl.

puberty   Transition from immature reproductive and hormonal state the a mature state.

purine   A six-membered and five-membered nitrogen containing ring that is the base of adenine and guanine formation.

pyrimidine   A six-membered nitrogen containing ring that is the base of thymine, cytosine, and uracil formation.

qualitative traits   Traits of quality, such as coat color, by which phenotypes can be classified into groups.

quantitative traits   Numerically measurable traits.

ration   Specific feed allotment given to an animal in a twenty-four hour period.

recessive   The member of an allele pair that is expressed only when the dominant allele is absent.

recombinant DNA   DNA molecules that have had new genetic material inserted into them.

rennin   A protein digesting enzyme secreted by the abomasums of cattle that is responsible for curdling of milk. In its commercial form it is referred to as rennet.

ribonucleic acid   RNA. Single stranded molecules that are essential in decoding DNA. Include messenger or mRNA, which carries the instructions for the synthesis of proteins; ribosomal or rRNA, which is the catalytic unit of mRNA translation into protein; and transfer or tRNA, which transfers the appropriate amino acids encoded in the DNA.

roe   The mature egg masses of fish.

rooster   A mature male chicken.

rumen   The largest of the ruminants forestomachs, which contains the microorganisms that degrade complex carbohydrates.

rumination   The process in ruminants where a cud or bolus of rumen contents is regurgitated, remasticated, and reswallowed for further digestion.

saturated fats   Fatty acids which contain no double bonds.

selection differential   The phenotypic advantage of those chosen to be parents.

semen   Fluid from the male that contains sperm and secretions.

somatic cells   A cell in the body other than gametes.

sperm   Male gamete.

stereotypic behavior   Repetitive, intentional behaviors.

stigma   The site of follicular rupture of the hen ovary.

sustainable agriculture   Agriculture that meets the needs of the present generation without jeopardizing the ability of further generations to meet their needs.

tame   Display of docility and submissiveness.

testosterone   A steroid hormone that is the principle male sex hormone that contributes to masculine characteristics. It also increases protein synthesis as an anabolic steroid.

thermoregulation   The maintenance of a narrow body temperature range under variable external conditions.

trans   On the opposite side.

transcription   Assembling RNA that is complementary to DNA.

transgenic   An animal or plant that has had DNA from an external source inserted into its genetic code.

translation   Assembling an amino acid sequence according to the code specified by mRNA.

unsaturated fats   Fatty acids with at least one double bond.

vasoconstriction   The narrowing of the blood vessels.

vasodilation   The widening of the blood vessels.

vertical integration   A process by which several steps in the production or distribution of a product are controlled by one or few entities. It is a common practice of the poultry industry and is increasing within the swine industry.

wean   Cessation of nursing.

white muscle disease   A degenerative muscle disease caused by a deficiency in selenium and/or vitamin E.

withers   The highest point on the back of an upright animal located between the shoulders.

xanthophyll   Yellow pigment of the carotenoids (vitamin A precursors) that contribute to the yellow coloration of egg yolks. Animals cannot produce xanthophylls and must obtain from their diet.

xenotransplantation   Transplantation of living cells, tissues, or organs from one species to another.

zoonosis   Transmission of an infectious disease from animals to humans or vice versa.

zygote   Organism produced from the fertilization of the ovum by the sperm.

Index